Multi-mechanism Modeling of Inelastic Material Behavior

Multi-mechanism Modeling of Inelastic Material Behavior

Georges Cailletaud
Kacem Saï
Lakhdar Taleb

iSTE

WILEY

First published 2018 in Great Britain and the United States by ISTE Ltd and John Wiley & Sons, Inc.

ISTE Ltd
27-37 St George's Road
London SW19 4EU
UK

www.iste.co.uk

John Wiley & Sons, Inc.
111 River Street
Hoboken, NJ 07030
USA

www.wiley.com

Library of Congress Control Number: 2017955720

British Library Cataloguing-in-Publication Data
A CIP record for this book is available from the British Library
ISBN 978-1-84821-580-1

Contents

Preface

It is common to say that each industrial sector relies on the performance of its materials. Examples of this type are multiple, think of the railways (development of rail steels at the end of the 19th Century), civil construction (development of advanced concrete formulations), the space shuttle (composites, carbon–carbon tiles) or aeronautics, where the performance of reactors depends on the maximum temperature supported by the materials in the hottest areas. But in fact, it would be more accurate to say that the performance obtained also depends on knowledge of the material used. An improved knowledge of the material paves the way for a structure design that, in addition to its elegance, has two important aspects: (1) safety improvement, insofar as having a good knowledge of the physical phenomena is better than applying large safety parameters, which often sound like coefficients to hide ignorance; (2) better ecological performance, since weight reduction decreases the fuel consumption of cars or aircrafts.

This modern approach has been greatly facilitated by the tremendous increase in the power of computers and the robustness of the numerical algorithms. Materials are too often forgotten in this process, and engineers' fear is then to apply the "garbage in, garbage out" proverb when there is such a weak point in the calculation chain. Yet researchers have made considerable progress in the field of material modeling. The time has now come for efforts to popularize the models obtained and to encourage their use by providing examples on materials of current use.

About 50 years after Mandel's paper on the "Généralisation de la théorie de la plasticité de W. T. Koiter" in Int. J. Solids Struct. (1965), the authors of the present book decided that it was time to gather the most recent results on the field of the so-called multi-mechanism models (MM). After Zarka and his co-workers, they have been active developers of this model class, which was reformulated in a thermodynamic framework, introduced in finite element codes, and adapted for a large number of materials and loading conditions. As a result, they have reached a good level of maturity. They offer a versatile toolbox to develop new constitutive equations for metals, polymers and geomaterials under monotonic or cyclic loading paths.

The implementation of the ideas was rather fast, and it was then time to start writing. This was spread over several years. It gave rise to several discoveries and developments of the models that had remained unexploited in the original versions. It has been enriched by ongoing research, which ensures that the document is fully up to date. Being able to resume the work begun every time a niche was released in the schedule was only possible, thanks to the mutual encouragements that the authors gave each other and to the sincere friendship that now crowns their efforts.

The hope is that the text now meets its readers, that it allows the sharing of results and a certain know-how, and that it is useful for students, researchers and engineers who will have it in hand.

The authors extend their sincere thanks to Vladislav A. Yastrebov, who designed the cover illustration from the map of a plastic deformation field in a cobalt matrix tungsten carbide composite. They would also like to thank Odile Adam for the careful attention she payed to the review of the bibliographies.

Georges CAILLETAUD
Kacem SAÏ
Lakhdar TALEB
October 2017

Introduction

"Plasticity" is a word with various meanings that change from brain reconstruction to metal deformation processes depending on the context and speaker. As far as materials are concerned, the aim of researchers is to develop not only models with a physical basis, but also models that are able to deliver results in minutes rather than hours. This is why the literature contains two types of approaches, namely (poly-)crystal plasticity and macroscopic models. The former are inherited from the early work of Taylor (1938), then Koiter (1953). It is clearly acknowledged in this case that plasticity has a lot of sources at the microstructure scale, and that plastic flow has to be built by collecting the contribution of each slip system. After von Mises (1911), the latter have been clearly formulated by Hill (1950), then Rice (1972) and Mandel (1972), with the introduction of internal variables to describe hardening, and Germain (1973) who elaborated the thermodynamic formulation. In the 1950s, a few authors attempted to bridge the gap between the two model classes. What remains from crystal plasticity when embedded in a macroscopic approach is the concept of "mechanism" and the fact that plastic flow may have various origins, possibly at different scales in the material. Accordingly, there is a loss of information, as the morphology is no longer explicit, and the variables represent averages more than the precise physical mechanisms. Extensions of the corresponding models can be made by considering that a mechanism does not directly refer to a slip system, but is the result of a more complex, combined plastic flow.

The purpose of the book is to provide readers with a rigorous framework showing the limits of the validity of the multi-mechanism (MM) models, and also to exhibit a series of "success stories" where the models correctly represent complex material behavior. Depending on the context, the mechanisms can be associated with different phases or follow various deformation regimes. Each mechanism will be represented by a (pseudo-) potential, from which plastic or viscoplastic flow can be derived. In all chapters, precise explanations are given, such that the reader can reproduce the calculations and apply the equations to his own experimental data base.

After a short state of the art (Chapter 1), the model formulation shown in Chapter 2 first mentions the thermodynamic framework, then derives two sets of models, namely (i) where each mechanism has its own plasticity criterion or (ii) all the mechanisms contribute to an unique plasticity criterion. In both cases, isotropic and kinematic hardening can be introduced. Chapter 3 contains a series of typical mechanical responses. It describes how time-dependent and time-independent flow can be combined to produce relevant histories in plasticity and creep. It also discusses the behavior of the various versions for the case of specific phenomena such as ratcheting or static recovery, and in the presence of non-proportional loading paths. Practical examples are provided in Chapter 4. A number of applications are devoted to metallic alloys (Ni-based, steels, aluminum alloys, zirconium alloys), but polymers (semi-crystalline, glassy) are also considered. The identification process is detailed, and the material parameters are available. Chapter 5 is devoted to damage. A few applications demonstrated that MM models may appear as good companions for expressions that introduce damage in metallic materials or mortar–rubber mixtures, under monotonic or cyclic loadings. The goal of Chapter 6 is to demonstrate that the models can be easily implemented in a finite element code in order to be used in structural calculations. After showing the framework for explicit and implicit integration, a few applications are shown, together with the full open source code.

1

State of the Art

The construction of models in the multi-mechanism (MM) family uses a series of generic tools ("building bricks") that are also introduced in the classical plastic or viscoplastic formulations. They are recalled here, which allows us to define the main notations in the book. Since various MM models can be either seen as purely phenomenological or present a physical background dealing with multiphase models, scale transition rules are also recalled. The models have been applied to a large set of materials, some of them being submitted to large deformation. A simple framework has been used in this case, which is also presented here. The chapter concludes with a brief description of the main steps in the development of MM models. All these elements relate to the mechanical aspects of the models. Nevertheless, some of them are developed with a view to providing a specific material microstructure. This is the reason why the first section of this chapter is devoted to the description of deformation mechanisms at the microscale.

1.1. Motivation from the microstructure

Shear deformation and volume change observed at the macroscale have various origins at the microscale, depending on material microstructure. Metallic alloys usually deform by dislocation slip, climbing and grain boundary sliding, for instance. Each of these deformation sources leads to a type of nonlinearity and a specific shape for the constitutive equations. In the following, the basic mechanisms of plastic deformation are enumerated for (1) single crystals, (2) polycrystals, (3) amorphous polymers and (4) semi-crystalline polymers:

1) *Single crystals*: the plastic deformation of metallic single crystals results from the evolution of the crystal network.

- Deformation by slip: this mechanism of deformation occurs when dislocations move on certain crystallographic (dense) planes and directions (slip system). According to Schmid's law, a slip system becomes active provided the resolved shear stress reaches a threshold value called the critical resolved shear stress. Slip produces slip lines, so that the classical models taking into account crystal plasticity represent the average of the deformation produced in a small material element. Dislocation may interact with a number of defects that are present in the material.

- Mechanical twinning: twinning is a deformation mechanism mainly observed at low temperature. Crystal twinning occurs when a crystal can jump from its initial configuration to a new one, involving the original grain and a twinned part, with an interface compatible with the two crystal networks. The deformation process is time independent, and produces locally both instantaneous shear deformation and eventually volumetric change. Twinning is one of the most active deformation mechanisms in crystals involving a reduced number of slip systems. It is pronounced in HCP crystals, and also observed at low temperature in higher symmetry BCC and FCC crystals.

2) *Polycrystals*: the grains of polycrystalline aggregates do not deform in the same way as single crystals, due to the local plastic flow that generates multiaxial stress states. The behavior is affected by grain boundaries, and by the presence of subgrain boundaries within the grains. Particles may be placed on purpose in the material, in order to produce hardening. The related mechanisms are solid solution strengthening or dislocation–precipitate interaction (Figure 1.1). Hard second-phase particles and ductile phases will change the behavior according to their size, shape, number and spatial distribution. In the classical literature, local fields in the phases are estimated by means of two opposite assumptions, uniform stress or uniform strain. This is a key issue of the MM models and it will be discussed in detail in the next chapters. In addition to the previous effects, grain boundary sliding may become a significant mechanism in specific cases, like the presence of nanograins, at high temperature that promotes diffusion at grain boundaries.

3) *Amorphous polymers*: they are polymers whose molecular structure lacks a definite repeating form, shape or structure. In some glassy polymers, the tensile stress–strain curve can be separated into three regimes: (i) initially,

the deformation is linear until the yield stress is reached, (ii) the stress decreases due to softening associated with the formation of a neck and (iii) molecular orientation provides a mechanism for hardening that predominates at large strains (Figure 1.2).

4) *Semi-crystalline polymers*: semi-crystalline polymers (SCP) belong to a family of materials that combine a crystalline molecular structure and amorphous regions. Their microstructure can then be seen as a two-phase composite material consisting of crystalline and amorphous phases. Spherulites are spherical regions inside non-branched linear polymers. They are composed of highly ordered lamellae. The macroscopically homogeneous deformation results from various deformation mechanisms in the heterogeneous microstructure (Figure 1.3). The main deformation mechanisms responsible for the plastic deformation of SCP are crystallographic in nature. For small levels of deformation, the crystals are distorted but the crystalline lamellae are not damaged, whereas, under large plastic deformation, the distortion in the spherulitic region creates a strong preferential orientation. The elastic and viscoplastic behavior of SCP is mainly affected by the degree of crystallinity, the initial microstructure and the evolution in this microstructure during the deformation process.

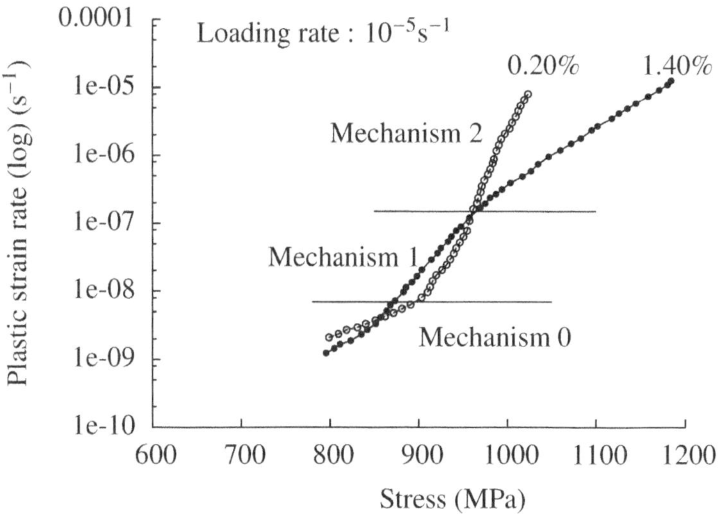

Figure 1.1. *N18 alloy: stress relaxation at 650°C for two levels of initial plastic strain [SAÏ 04]*

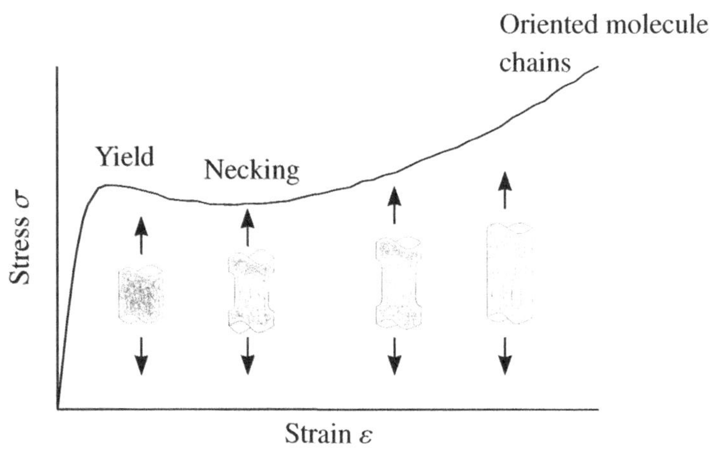

Figure 1.2. *Typical stress–strain curve for glassy polymers*

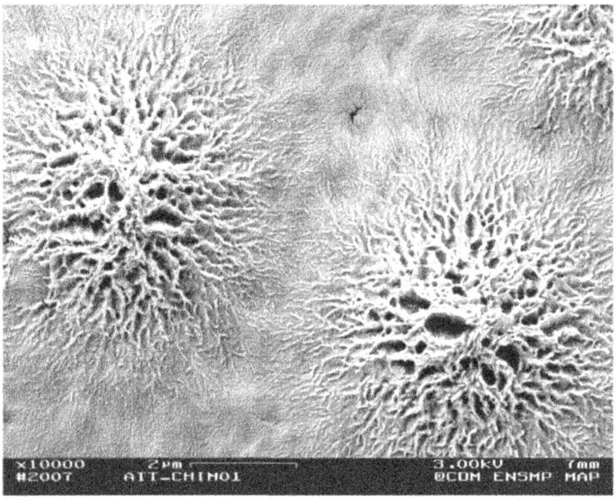

Figure 1.3. *SEM observation of the spherulitic microstructure of a semi-crystalline polymer, polyamide 6 (PA6) [REG 09b]*

Depending on the application, the MM models presented later in this book can be seen as *multi-mechanism*, *multi-phase* or *multi-potential*. The corresponding terminology can be characterized as follows:

– "Multi-phase" is used when different phases can be distinguished in a material such as the SCP.

– "Multi-mechanism" is used when different regimes can be observed in a material. These behaviors may be linked to different strain or stress ranges or different temperatures. The term "multi-mechanism" can also be used if nonlinearities on the macroscale come from different sources of deformation at the crystal level.

– "Multi-potential" refers to potentials in the thermodynamic formalism.

Beside the mechanisms related to metallic materials and polymers enumerated above, phase transformation is a source of deformation that deserves to be included in the MM modeling. "Phase transformation" refers to solid–solid metallurgical transformations that change the volume fraction of crystallographic phases. According to experimental observations, the transformation can be classified into (i) diffusional transformations, (ii) displacive transformations and (iii) hybrid transformations:

– In *diffusional transformations*, the diffusion of one or several species allows the creation of a new balance between several phases (like austenite–ferrite in steels) or to change the morphology of precipitates (like gamma prime in Ni-base alloys). The resulting microstructure will minimize the energy of the full system.

– The *martensitic transformation* is a time-independent phase transformation that occurs in some metals, resulting in the formation of martensite. The martensite is generally formed as thin plates with a predefined orientation with respect to the parent phase.

– *Hybrid transformations* incorporate characteristics related to both diffusional and displacive transformations. The ferrous bainitic transformation is a typical example because of its morphology, kinetics and mechanism of growth.

All these transformations are regulated by thermodynamic considerations. The driving forces for the transformation depend on temperature and deformation. Twinning-induced plasticity (TWIP) is an example where the phase change is produced by deformation (Figure 1.4). The well-known TWIP steels are increasingly used due to their high strength and good ductility.

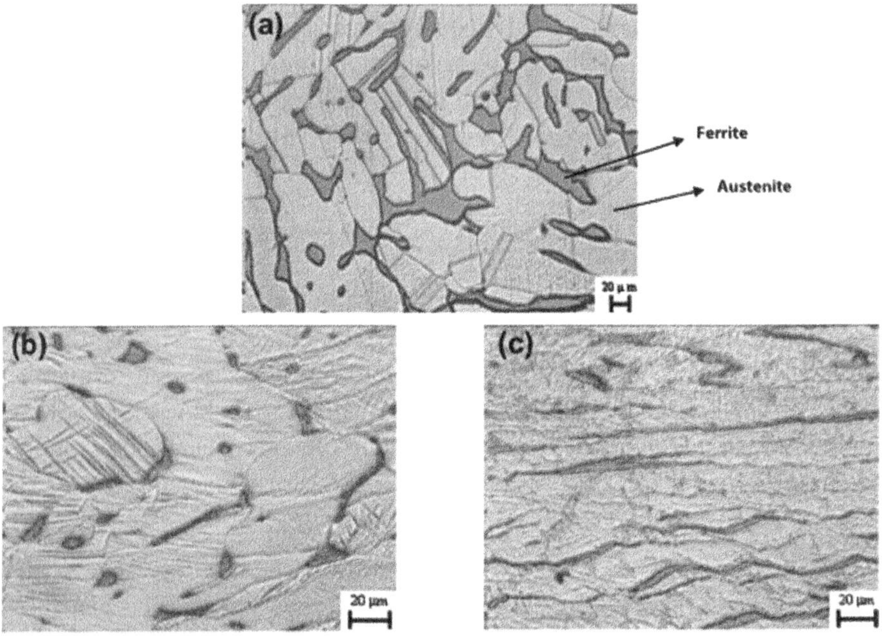

Figure 1.4. *The optical microstructures of an experimental alloy: (a) as-received dual-phase TWIP steel, (b) 40% cold-rolled steel and (c) 70% cold-rolled steel [TOR 11]. For a color version of the figure, see www.iste.co.uk/cailletaud/multi-mechanism.zip*

1.2. Building bricks

This section provides an analytical presentation of the various numerical tools that will be used in any model development. Having these elements in hand, the plastic and viscoplastic models can be constructed by assembling various blocks. What is needed is the definition of:

– the initial yield surface, defined by a plasticity criterion;

– the way in which this surface will change during inelastic flow, defined by hardening rules. Classical models introduce isotropic and/or kinematic hardening. Specific models are able to reproduce surface distortions;

– the flow direction and intensity. They come directly from the two previous items in the case of time-independent plasticity. The intensity is defined by a viscosity rule for viscoplastic models.

For the sake of brevity, most of the blocks are presented without long comments, which allows us to mention rather a large number of possibilities.

1.2.1. *Criteria*

A yield criterion is the function f characterizing the surface that delimits the elasticity domain under multiaxial stress states. For the case of materials with no hardening, it is defined in the space formed by the six components of the stress tensor, or by the eigen stresses or stress invariants. Hardening will deform the surface. It means that, when hardening is present, the function f depends not only on stress, but also on hardening variables. Only initial yield is addressed for the moment. After defining the notations, two tables will first introduce classical choices for isotropic materials and then for anisotropic materials.

– The most general form of the criterion is as follows:

$$f(\sigma_{11}, \sigma_{22}, \sigma_{33}, \sigma_{12}, \sigma_{23}, \sigma_{31}) = 0$$

– The principal stresses σ_1, σ_2 and σ_3 can also be used in some cases:

$$f = (\sigma_1, \sigma_2, \sigma_3) = 0.$$

– The three invariants of the stress tensor or the two invariants of its deviator can also be of interest, specifically for isotropic materials:

$$f(I_1, I_2, I_3) = 0 \quad \text{or} \quad f(J_2, J_3) = 0$$

where I_1 is the first principal invariant of the stress tensor and I_2, I_3 (resp. J_2, J_3) are the second and third invariants of the stress tensor (resp. its deviatoric part).

Depending on whether the plastic flow depends on the hydrostatic pressure or not, different classes of models can be considered:

– For a large class of materials, specifically those that deform by crystallographic slip, the hydrostatic pressure does not influence plastic behavior. The models can then be expressed in terms of deviatoric stress $\underset{\sim}{s}$

and its invariants instead of the stress tensor $\underset{\sim}{\sigma}$:

$$\underset{\sim}{\mathbf{s}} = \underset{\sim}{\sigma} - (I_1/3)\underset{\sim}{\mathbf{I}} \qquad [1.1]$$

$$J_1 = Tr\left(\underset{\sim}{\mathbf{s}}\right) = 0 \qquad [1.2]$$

$$J_2 = (1/2)Tr\left(\underset{\sim}{\mathbf{s}}^2\right) \qquad [1.3]$$

$$J_3 = (1/3)Tr\left(\underset{\sim}{\mathbf{s}}^3\right) \qquad [1.4]$$

– On the other hand, polymeric materials, metallic alloys experiencing damage, or powders, show a significant sensitivity to hydrostatic pressure. f is then function of the stress invariants:

$$I_1 = Tr\left(\underset{\sim}{\sigma}\right) \qquad [1.5]$$

$$I_2 = (1/2)Tr\left(\underset{\sim}{\sigma}^2\right) \qquad [1.6]$$

$$I_3 = (1/3)Tr\left(\underset{\sim}{\sigma}^3\right) \qquad [1.7]$$

The most widely used criterion is the von Mises criterion [MIS 13], even though it was formulated after Maxwell (1865) and Huber (1904). It can be applied to isotropic materials that are insensitive to hydrostatic pressure. For this case, the simplest formulation based on invariants introduces the second invariant of the deviatoric stress tensor. In the following, the critical variable retained to express the criterion will be J instead of J_2:

$$f(\underset{\sim}{\sigma}) = J - \sigma_y \quad \text{with} \quad J = \left((3/2)\,\underset{\sim}{\mathbf{s}} : \underset{\sim}{\mathbf{s}}\right)^{1/2} \qquad [1.8]$$

where σ_y is the initial yield limit. The interest of using J is that the 3D expression (used for structural computations) is the same as the 1D expression (used for the material parameter identification), provided the absolute value of the stress (1D) is replaced by J (3D).

Another very simple criterion is due to Tresca [TRE 64]. It can be expressed as a function of the eigenstresses as follows:

$$f(\underset{\sim}{\sigma}) = Max_{i,j}|\sigma_i - \sigma_j| - \sigma_y \qquad [1.9]$$

or as a combination of the second and third invariants:

$$f(\underset{\sim}{\sigma}) = J \sin\left(\theta + \frac{\pi}{3}\right) - \frac{\sqrt{3}}{2}\sigma_y \qquad [1.10]$$

where θ is the so-called Lode angle:

$$\cos(3\theta) = \frac{3\sqrt{3}}{2}\frac{J_3}{J_2^{3/2}} = \left(\frac{K}{J}\right)^3 \quad \text{with} \quad K^3 = \frac{27}{2}J_3 \qquad [1.11]$$

Table 1.1 shows a list of classical criteria for isotropic materials. A non-exhaustive list of anisotropic yield criteria is given in Table 1.2.

Isotropic criteria without hydrostatic pressure
von Mises [MIS 13]
$f(\underset{\sim}{\sigma}) = \left(\frac{1}{2}(\lvert\sigma_1 - \sigma_2\rvert^2 + \lvert\sigma_2 - \sigma_3\rvert^2 + \lvert\sigma_3 - \sigma_1\rvert^2)\right)^{1/2} - \sigma_y$
Hosford [HOS 72]
$f(\underset{\sim}{\sigma}) = \left(\frac{1}{2}(\lvert\sigma_1 - \sigma_2\rvert^m + \lvert\sigma_2 - \sigma_3\rvert^m + \lvert\sigma_3 - \sigma_1\rvert^m)\right)^{1/m} - \sigma_y$
Cazacu04 [CAZ 04]
$f(\underset{\sim}{\sigma}) = J_2^{3/2} - cJ_3 - (\tau_y)^3$
τ_y is the yield stress in pure shear. c is a material parameter related to the uniaxial yield stresses in tension and compression.
Cazacu06 [CAZ 06]
$f(\underset{\sim}{\sigma}) = (\lvert s_1\rvert - ks_1)^a + (\lvert s_2\rvert - ks_2)^a + (\lvert s_3\rvert - ks_3)^a - F$
a and k are material parameters. F gives the size of the elastic domain.
Drucker
$f(\underset{\sim}{\sigma}) = t(\underset{\sim}{\sigma}) - \sigma_y$
$t(\underset{\sim}{\sigma}) = \frac{J}{2}\left((1 + K) + (1 - K)\left(\frac{J_3}{J}\right)^3\right)$
K is a material parameter $0.778 < K \leqslant 1$
Criteria taking into account the hydrostatic pressure
Mohr–Coulomb [MOH 00]
$f(\underset{\sim}{\sigma}) = \sigma_1 - \sigma_3 + (\sigma_1 + \sigma_3)\sin\Phi - 2C\cos\Phi$
The material parameter C is called cohesion.
The angle Φ denotes the internal friction in the material.
Drucker–Prager
$f(\underset{\sim}{\sigma}) = (1 - \alpha)\, t(\underset{\sim}{\sigma}) + \alpha I_1(\underset{\sim}{\sigma}) - \sigma_y$
t is already defined in the table above. α is a material parameter.

Table 1.1. *Examples of plastic isotropic criteria*

Nouailhas [NOU 92]

$$f(\underset{\sim}{\sigma}) = \left(\left(\left[\frac{3}{2}(I_1 + 2a_4 I_4)\right]^2 + 3a_8 I_8\right)^3 - (a_6 I_6)^4\right)^{1/12} - \sigma_y$$

$$I_1 = s_{11}^2 + s_{22}^2 + s_{33}^2, I_6 = s_{11}s_{22} + s_{22}s_{33} + s_{33}s_{11}$$

$$I_4 = s_{12}^2 + s_{23}^2 + s_{31}^2, I_8 = s_{12}^4 + s_{23}^4 + s_{31}^4$$

Hill [HIL 76]

$$f(\underset{\sim}{\sigma}) = [F(\sigma_{11} - \sigma_{22})^2 + G(\sigma_{22} - \sigma_{33})^2 + H(\sigma_{33} - \sigma_{11})^2 + 2L\sigma_{12}^2 + 2M\sigma_{23}^2 + 2N\sigma_{13}^2]^{1/2} - \sigma_y$$

Hu [HU 05]

$$f(\underset{\sim}{\sigma}) = \left(J'^{(1)}_2\right)^2 + \left(J'^{(2)}_2\right)^2 + \left(J'^{(3)}_2\right)^2 - \frac{1}{C^2}$$

$$\begin{pmatrix} \left(J'^{(1)}_2\right)^2 \\ \left(J'^{(2)}_2\right)^2 \\ \left(J'^{(3)}_2\right)^2 \end{pmatrix} = \frac{1}{6} \begin{bmatrix} 1 & 1 & 1 & 1 & 1 & 1 \\ a_{11}^2 & a_{22}^2 & a_{33}^2 & a_{12}^2 & a_{23}^2 & a_{31}^2 \\ b_{11}^2 & b_{22}^2 & b_{33}^2 & b_{12}^2 & b_{23}^2 & b_{31}^2 \end{bmatrix} \begin{pmatrix} (\sigma_{11} - \sigma_{22})^2 \\ (\sigma_{22} - \sigma_{33})^2 \\ (\sigma_{33} - \sigma_{11})^2 \\ 6\sigma_{12}^2 \\ 6\sigma_{23}^2 \\ 6\sigma_{31}^2 \end{pmatrix}$$

Gotoh [GOT 77]

$$f(\underset{\sim}{\sigma}) = A_0(\sigma_{11} + \sigma_{22})^2 + \Sigma$$

$$\Sigma = [A_1\sigma_{11}^4 + A_2\sigma_{11}^3\sigma_{22}^2 + A_3\sigma_{11}^2\sigma_{22}^2 + A_4\sigma_{11}\sigma_{22}^3 + A_5\sigma_{22}^4 + (A_6\sigma_{11}^2 + A_7\sigma_{11}\sigma_{22} + A_8\sigma_{22}^2)\sigma_{12}^2 + A_9\sigma_{12}^4]$$

K–B [KAR 93]

$$\Phi = (1 - c)\Phi_1 + c\Phi_2 = 2Y^m$$

$$\Phi_1 = (|S_1 - S_2|^m + |S_2 - S_3|^m + |S_3 - S_1|^m)$$

$$\Phi_2 = \frac{3^m}{2^{m-1}+1}(|S_1|^m + |S_2|^m + |S_3|^m)$$

Y is the average yield stress in uniaxial tension

$\underset{\approx}{\mathbf{L}}$ is a fourth-order tensorial operator that introduces material anisotropy

$\underset{\sim}{\mathbf{S}} = \underset{\approx}{\mathbf{L}}(\underset{\sim}{\sigma} - \underset{\sim}{\mathbf{B}})$

Banabic [BAN 04]

$$f(\underset{\sim}{\sigma}) = \left[a(b\Gamma + c\Psi)^{2k} + a(b\Gamma - c\Psi)^{2k} + (1 - a)(2c\Psi)^{2k}\right]^{1/2k}$$

$$\Gamma = M\sigma_{11} + N\sigma_{22}, \Psi = \sqrt{(P\sigma_{11} + Q\sigma_{22})^2 + R\sigma_{12}^2}$$

Tsaï $f(\underset{\sim}{\sigma}) = f_H(\underset{\sim}{\sigma}) + Q(\sigma_{22} - \sigma_{33}) + P(\sigma_{11} - \sigma_{33})$

Bron [BRO 04]

$$\overline{\sigma} = \left(\alpha\left(\overline{\sigma}^1\right)^a + (1-\alpha)\left(\overline{\sigma}^2\right)^a\right)^{(1/a)}$$
$$\overline{\sigma}^k = \left(\psi^k\right)^{1/b^k}$$
$$\psi^1 = \tfrac{1}{2}\left(|S^1_2 - S^1_3|^{b^1} + |S^1_3 - S^1_1|^{b^1} + |S^1_1 - S^1_2|^{b^1}\right)$$
$$\psi^2 = \frac{3^{b^2}}{2^{b^2}+2}\left(|S^2_1|^{b^2} + |S^2_2|^{b^2} + |S^2_3|^{b^2}\right)$$

$S^k_{i=1,2,3}$ are the principal values of a modified stress deviator $\underset{\sim}{\mathbf{s}}^k$ defined by $\underset{\sim}{\mathbf{s}}^k = \underset{\approx}{\mathbf{L}}^k : \underset{\sim}{\sigma}$

$\underset{\approx}{\mathbf{L}}^k$ are fourth-order tensorial operators that introduce material anisotropy

Yld2004-18p [BAR 05]

$$\Phi(\widetilde{\mathbf{S}}', \widetilde{\mathbf{S}}'') = |\widetilde{S}'_1 - \widetilde{S}''_1|^a + |\widetilde{S}'_1 - \widetilde{S}''_2|^a + |\widetilde{S}'_1 - \widetilde{S}''_3|^a + |\widetilde{S}'_2 - \widetilde{S}''_1|^a + |\widetilde{S}'_2 - \widetilde{S}''_2|^a + |\widetilde{S}'_2 - \widetilde{S}''_3|^a$$
$$+ |\widetilde{S}'_3 - \widetilde{S}''_1|^a + |\widetilde{S}'_3 - \widetilde{S}''_2|^a + |\widetilde{S}'_3 - \widetilde{S}''_3|^a = 4\overline{\sigma}^a$$
$$\underset{\sim}{\widetilde{\mathbf{S}}'} = \underset{\approx}{\mathbf{C}'}\underset{\sim}{\mathbf{S}} = \underset{\approx}{\mathbf{C}'}\underset{\approx}{\mathbf{T}}\underset{\sim}{\sigma} = \underset{\approx}{\mathbf{L}'}\underset{\sim}{\sigma}$$
$$\underset{\sim}{\widetilde{\mathbf{S}}''} = \underset{\approx}{\mathbf{C}''}\underset{\sim}{\mathbf{S}} = \underset{\approx}{\mathbf{C}''}\underset{\approx}{\mathbf{T}}\underset{\sim}{\sigma} = \underset{\approx}{\mathbf{L}''}\underset{\sim}{\sigma}$$
$$\underset{\approx}{\mathbf{C}'} = \begin{pmatrix} 0 & -\alpha_1 & -\alpha_2 & 0 & 0 & 0 \\ -\alpha_3 & 0 & -\alpha_4 & 0 & 0 & 0 \\ -\alpha_5 & -\alpha_6 & 0 & 0 & 0 & 0 \\ 0 & 0 & 0 & -\alpha_9 & 0 & 0 \\ 0 & 0 & 0 & 0 & -\alpha_7 & 0 \\ 0 & 0 & 0 & 0 & 0 & -\alpha_8 \end{pmatrix}$$
$$\underset{\approx}{\mathbf{C}''} = \begin{pmatrix} 0 & -\alpha_{10} & -\alpha_{11} & 0 & 0 & 0 \\ -\alpha_{12} & 0 & -\alpha_{13} & 0 & 0 & 0 \\ -\alpha_{14} & -\alpha_{15} & 0 & 0 & 0 & 0 \\ 0 & 0 & 0 & -\alpha_{18} & 0 & 0 \\ 0 & 0 & 0 & 0 & -\alpha_{16} & 0 \\ 0 & 0 & 0 & 0 & 0 & -\alpha_{17} \end{pmatrix}$$
$$\underset{\approx}{\mathbf{T}} = \frac{1}{3}\begin{pmatrix} 2 & -1 & -1 & 0 & 0 & 0 \\ -1 & 2 & 2 & 0 & 0 & 0 \\ -1 & -1 & 2 & 0 & 0 & 0 \\ 0 & 0 & 0 & 3 & 0 & 0 \\ 0 & 0 & 0 & 0 & 3 & 0 \\ 0 & 0 & 0 & 0 & 0 & 3 \end{pmatrix}$$

Table 1.2. *Examples of plastic anisotropic criteria*

Figures 1.5, 1.6 and 1.7 show classical yield surfaces plotted according to some common criteria. In these examples, the yield stress is 100 MPa. The Drucker criterion is first compared with the von Mises criterion for two values of the parameter K. There is a limit value for K, since, for the smaller values, the surface is no longer convex, as shown by Figure 1.5. The shape of the yield surface predicted by the Hosford criterion is illustrated in Figure 1.6. It can be seen that, when $m = 1$ (or m goes to infinity), the Hosford criterion reduces to the Tresca yield criterion. When $m = 2$, the Hosford criterion reduces to the von Mises yield criterion. The so-called "Yld2004-18p" anisotropic model, known to capture the anisotropic behavior of a material sheet to a reasonable degree of accuracy is compared with Hill's anisotropic criterion in Figure 1.7. In this example, the r-values are $r_0 = r_{90} = 0.6$. The Hill parameters $F = r_0/(r_0(1 + r_{90}))$, $G = 1/(1 + r_0)$ and $H = r_0/(1 + r_0)$ can then be deduced. The "Yld2004-18p" constants correspond to a fictitious material investigated by [YOO 06]. ($a = 8, \alpha 1 = 0.7297, \alpha_2 = 0.8777, \alpha_3 = 0.4252, \alpha_4 = 0.7268, \alpha_5 = 1.1386, \alpha_6 = 1, \alpha_{10} = 1.0513, \alpha_{11} = 1.0389, \alpha_{12} = 1.3289, \alpha_{13} = 1.1775, \alpha_{14} = 0.7651, \alpha_{15} = 0.9169$).

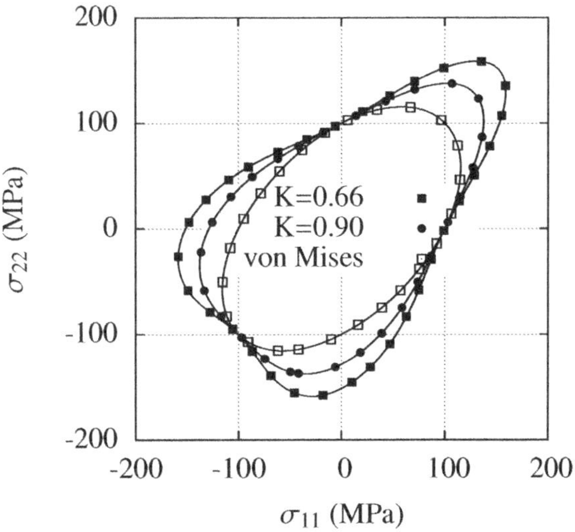

Figure 1.5. *Yield surface: comparison of von Mises and Drucker criteria*

1.2.2. *Isotropic hardening rules*

In this section, the word *isotropic* goes with the scalar variable R that defines the size of the elastic domain. Consequently, the rules providing isotropic hardening evolution also exist for anisotropic materials. First, we restrict ourselves to the case of isotropic materials. The relevant state variable that is associated with R is also a scalar. According to the way the theory is built, this variable may be the plastic work, or it may depend on the accumulated plastic strain, or be the accumulated plastic strain itself. For the sake of simplicity, we adopt the latter assumption here. The variable will be denoted by p, with $\dot{p} = \left((2/3)\dot{\underset{\sim}{\varepsilon}}^p : \dot{\underset{\sim}{\varepsilon}}^p\right)^{1/2}$. Note that p is not exactly the accumulated inelastic strain for the MM models as it will be shown later.

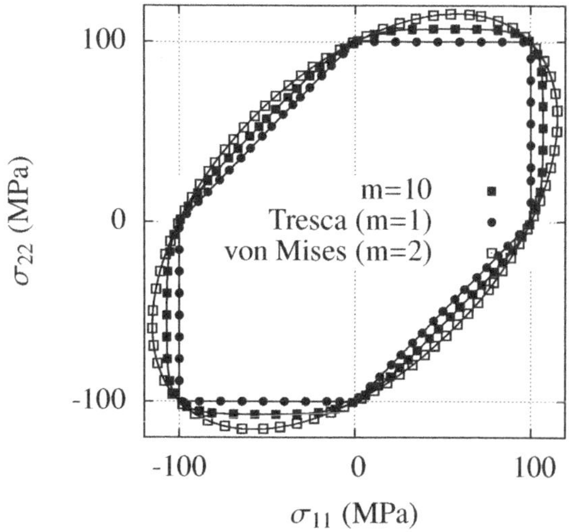

Figure 1.6. *Yield surface: comparison of von Mises, Tresca and Hosford criteria*

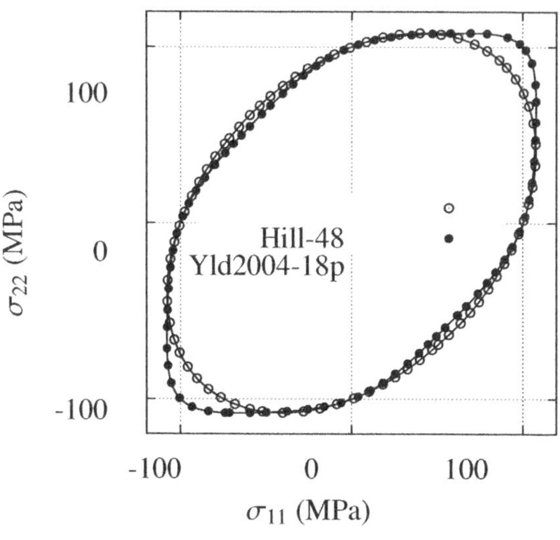

Figure 1.7. *Yield surface: comparison of the anisotropic criteria Hill-48 and Yld2004-18p*

In the framework of time-independent plasticity, the simplest model uses the following von Mises yield function:

$$f(\underset{\sim}{\sigma}, R) = J(\underset{\sim}{\sigma}) - R - R_0 \qquad [1.12]$$

where R_0 is a function of the initial size of the yield surface. Besides the simplest rules (constant size, $R = 0$, or linear evolution of the yield surface, $R = Hp$), classical rules take into account softening or hardening of materials:

$$R = Q(1 - e^{-bp}) \qquad [1.13]$$

where the material parameter Q is positive for the hardening effect and negative for the softening effect, and b controls the rate of saturation of the hardening with respect to plastic flow. To allow fine tuning of isotropic hardening, the following two-term evolution rule introduces short- and long-term mechanisms, with possible positive or negative values for Q_1 and Q_2:

$$R = Q_1(1 - e^{-b_1 p}) + Q_2(1 - e^{-b_2 p}) \qquad [1.14]$$

Isotropic hardening has been used in the literature to account for specific effects due to loading path. We have retained the most popular effects, namely non-proportional effect and strain memory effect. The description of the mechanical hardening has also been associated with microstructural changes, so that isotropic hardening is classically linked to dislocation densities, as shown at the end of this section.

1.2.2.1. *Non-proportional effect*

Non-proportional loading paths have been found to produce additional hardening, if compared with reference one-dimensional cyclic loadings. The landmark paper in this field [LAM 78] used combined axial–torsional fatigue tests and specifically "out-of-phase" tests, where the sinusoidal signals for the axial strain and the shear strain present a 90° phase lag. The hardening was compared with "in-phase" loadings, with a zero phase lag. "Additional hardening" is present if the equivalent stress range obtained in "out-of-phase" tests is larger than that for "in-phase" tests with the same equivalent strain range. After initial studies on copper and 304 stainless steel, dozens of studies have shown that the effect is present each time the stacking fault energy (SFE) is large: in these conditions, the dislocation pair must remain close to each other, so that they can easily change their slip planes. This mechanism is reinforced by non-proportionality, and leads to additional hardening. Consequently, the largest effects are observed on FCC materials with high SFE.

A specific variable is created to generate a ranking of the various loading path histories. It must be determined by an ordinary differential equation (ODE), and take a fixed reference value for in-phase loadings. A measurement of the non-proportionality comes from the relative direction of the actual state and the actual rate, in terms of stress, or plastic strain rate, for instance. Benallal and Marquis [BEN 87] have tested many different angles, before proposing a model based on the angle α between the actual plastic strain rate and the deviatoric stress rate. The model has a good quality/price ratio, even if it is not able to describe cross-effect for certain specific loading histories. This angle is equal to zero for any proportional loading.

The isotropic strain hardening variable R then takes the following form:

$$R = Q_1(1 - e^{-b_1 p}) + Q_2(1 - e^{-b_2 p}) \quad [1.15]$$

Q_1, b_1 and b_2 are material parameters. Q_2 is used to express the non-proportionality through the angle α:

$$\dot{Q}_2 = [(d - f)A + f](Q_s - Q_2)\dot{p} \quad [1.16]$$

$$Q_s = \frac{gAQ_m + (1 - A)Q_0}{gA + (1 - A)} \quad [1.17]$$

$$\cos\alpha = \frac{\dot{\underset{\sim}{s}} : \dot{\underset{\sim}{\varepsilon}}^p}{\sqrt{\dot{\underset{\sim}{s}} : \dot{\underset{\sim}{s}}}\,\sqrt{\dot{\underset{\sim}{\varepsilon}}^p : \dot{\underset{\sim}{\varepsilon}}^p}} \quad [1.18]$$

$$A = 1 - \cos^2\alpha \quad [1.19]$$

d, f, g, Q_0 and Q_m are material parameters.

It is worth noting that the evaluation of the non-proportionality by means of the angle between plastic strain rate and the deviatoric stress rate has two main drawbacks related to certain specific situations. This angle will be left undefined each time the deviatoric stress rate is zero; this case corresponds, for instance, to classical creep tests. The second drawback can be seen in situations where non-proportional effects may exist. Such cases may be encountered in multistep tests such as cyclic tension compression (sequence 1) followed by cyclic torsion (sequence 2). In such a configuration, if, after the first sequence, an adequate path is applied in order to come back to zero axial stress and strain [JIA 08, MUR 82], the application of the second sequence may be accompanied by extra hardening despite the fact that $\alpha=0$. When the MM model is used for simulating creep tests, we let $A=0$, which corresponds to proportional loading conditions.

1.2.2.2. *Strain memory effect*

After non-proportionality of strain path, we consider here the effect of the strain range history. For the previous case, it has been demonstrated that the variation in the strain state is able to activate mechanisms that are not present for in-phase loadings. In the present case, the application of a large prestrain, either monotonic or cyclic, produces an additional hardening in subsequent loading cases of smaller strain ranges.

The experimental demonstration of this effect on 316 stainless steel is shown in [CHA 79]. It is shown that the value of the stress range for a given cyclic strain range is augmented if a larger cyclic strain range has been applied first. It means that the value of the parameter Q that controls the maximum reached by the isotropic hardening variable R of equation [1.13] should in fact depend on strain history. An additional variable, q, is then introduced [CHA 79, OHN 82] in order to keep the memory of the maximum inelastic strain range. A plasticity-type formalism is used in the plastic strain space, with a non-hardening surface F defined by

$$F(\underset{\sim}{\varepsilon}^p, \underset{\sim}{\xi}, q) = \frac{2}{3} J\left(\underset{\sim}{\varepsilon}^p - \underset{\sim}{\xi}\right) - q \qquad [1.20]$$

The memory surface is activated each time plastic strain overpasses its previous maximum. Its transformation is controlled by two variables, the tensor $\underset{\sim}{\xi}$ that defines its center, and the scalar q, figuring its radius. The respective evolution rules can be written as follows:

$$\dot{q} = \eta H(F)\langle \underset{\sim}{\mathbf{n}} : \underset{\sim}{\mathbf{n}}^* \rangle \dot{p} \qquad \dot{\underset{\sim}{\xi}} = \frac{3}{2}(1-\eta) H(F) \langle \underset{\sim}{\mathbf{n}} : \underset{\sim}{\mathbf{n}}^* \rangle \underset{\sim}{\mathbf{n}}^* \dot{p} \qquad [1.21]$$

$$\underset{\sim}{\mathbf{n}}^* = \left(\partial F / \partial \underset{\sim}{\varepsilon}^p\right) / \|\partial F / \partial \underset{\sim}{\varepsilon}^p\| \qquad \underset{\sim}{\mathbf{n}} = (\partial f / \partial \underset{\sim}{\sigma}) / \|\partial f / \partial \underset{\sim}{\sigma}\| \qquad [1.22]$$

where η is a material parameter ($0 < \eta < 1$) and where $\underset{\sim}{\mathbf{n}}$ and $\underset{\sim}{\mathbf{n}}^*$ are, respectively, the unit normal to the classical yield surface and to the non-hardening region. $H(.)$ stands for the Heaviside function. A value of 1 for η would mean that the material has a "total memory" of the strain path, so that the value of the strain memory reaches the maximum plastic strain value. Lower values for η mean that the increase in the plastic range is partially accommodated by a translation of the memory surface, so that only a fraction of it is memorized in the variable. This "partial memory" case corresponds to the experimental observations.

The variable q is finally used to update the value of Q, which has an impact on the size of the elastic domain:

$$\dot{R} = b(Q - R)\dot{p} \quad \text{and} \quad \dot{Q} = 2\mu(Q_M - Q)\dot{q} \qquad [1.23]$$

1.2.2.3. *Dislocation density-based hardening rules*

When plastic flow is governed by dislocation motion, a series of mechanistic models are available in the literature. Almost all of them are based on the assumption that flow stress is proportional to the square root of the dislocation density, at a temperature higher than $0.2T_m$. As a result, an approximate form of the hardening curve is

$$\sigma = \sigma_y + M\alpha\mu b\sqrt{\rho} \qquad [1.24]$$

where M is the Taylor factor, μ the shear modulus, b the modulus of the Burgers vector, α a material constant and ρ the dislocation density [MEC 81].

1.2.3. *Kinematic hardening rules (KHR)*

The concept of internal stress was developed independently by researchers in metallurgy and mechanics in the 1960s. The idea supporting the concept is that, due to material heterogeneities, the effective stress that is applied to deformation mechanisms differs from the macroscopic stress. On a microstructural level, several kinematic hardening types should be considered, since the kinematic effect is present on a given level each time heterogeneities are present, namely kinematic hardening is created at the intergranular level, due to residual stress redistribution between the various grains. Kinematic hardening is also present inside grains, due to dislocation–precipitate or dislocation–dislocation interactions. Adopting the mechanical point of view, the kinematic hardening rule specifies the translation of the yield surface, then the shape of the hysteresis loop for cyclic loadings, and also regulates the ratcheting effect.

Kinematic hardening rules (KHR) have been considered as the main features introduced in the MM models, and their coupling is a critical aspect of the model behavior. It is then interesting to analyze the individual response of the various formulations with respect to uniaxial and biaxial loading (i.e. tension–torsion loading).

Following the basic nonlinear hardening rule of Armstrong–Frederick (AF) [ARM 66], a major contribution to the model evolution has been made by Chaboche [CHA 08] who proposed among other ideas the decomposition of the total backstress into a sum of several variables. A generalized form of the backstress increment was proposed by [JIA 96].

$\dot{\underset{\sim}{X}} = \frac{2}{3} C \dot{\underset{\sim}{\varepsilon}}^p$	[PRA 56]
$\dot{\underset{\sim}{X}} = (\underset{\sim}{\sigma} - \underset{\sim}{\mathbf{X}})\dot{p}$	[ZIE 59]
$\dot{\underset{\sim}{X}} = \frac{2}{3} C(p) \dot{\underset{\sim}{\varepsilon}}^p - D(p)\underset{\sim}{\mathbf{X}}\dot{p}$	[MRO 67]
$\dot{\underset{\sim}{X}} = \frac{2}{3} C \dot{\underset{\sim}{\varepsilon}}^p - D\underset{\sim}{\mathbf{X}}\dot{p}$	[ARM 66, CHA 77]
$\dot{\underset{\sim}{\mathbf{X}}} = \frac{2}{3} C \dot{\underset{\sim}{\varepsilon}}^p - D\phi(p)\underset{\sim}{\mathbf{X}}\dot{p} - D^* \left[J(\underset{\sim}{\mathbf{X}})\right]^{m-1} \underset{\sim}{\mathbf{X}}$	[MAL 72, CHA 75]
$\dot{\underset{\sim}{\mathbf{X}}} = \frac{2}{3} C \dot{\underset{\sim}{\varepsilon}}^p - D(\underset{\sim}{\mathbf{X}} - \underset{\sim}{\mathbf{Y}})\dot{p}$ $\dot{\underset{\sim}{\mathbf{Y}}} = \frac{1}{K}(\underset{\sim}{\mathbf{X}} - \underset{\sim}{\mathbf{Y}})(\frac{3}{2}\dot{\underset{\sim}{\mathbf{X}}} : \dot{\underset{\sim}{\mathbf{X}}})^{1/2}$	[ROU 85a]
$\dot{\underset{\sim}{\mathbf{X}}} = \frac{2}{3} C \dot{\underset{\sim}{\varepsilon}}^p - \frac{D^2}{C}\left[J(\underset{\sim}{\mathbf{X}})\right]^{m-1} \underset{\sim}{\mathbf{X}}\dot{p}$	[HEN 87]
$\dot{\underset{\sim}{\mathbf{X}}} = \frac{2}{3} C \dot{\underset{\sim}{\varepsilon}}^p - D\phi(p)\underset{\sim}{\mathbf{X}}\dot{p}$	[MAR 89]
$\dot{\underset{\sim}{\mathbf{X}}} = \frac{2}{3} C \dot{\underset{\sim}{\varepsilon}}^p - D\left\langle 1 - \frac{X_l}{J(\underset{\sim}{\mathbf{X}})}\right\rangle \underset{\sim}{\mathbf{X}}\dot{p}$	[CHA 91]
$\dot{\underset{\sim}{\mathbf{X}}} = \frac{2}{3} C \dot{\underset{\sim}{\varepsilon}}^p - D\left(\frac{J(\underset{\sim}{\mathbf{X}})}{C/D}\right)^m \left\langle \dot{\underset{\sim}{\varepsilon}}^p : \frac{\underset{\sim}{\mathbf{X}}}{J(\underset{\sim}{\mathbf{X}})}\right\rangle \underset{\sim}{\mathbf{X}}$	[OHN 93]
$\dot{\underset{\sim}{\mathbf{X}}} = \frac{2}{3} C \dot{\underset{\sim}{\varepsilon}}^p - D\left((1-\delta)(\underset{\sim}{\mathbf{X}} : \underset{\sim}{\mathbf{n}})\underset{\sim}{\mathbf{n}} + \delta\underset{\sim}{\mathbf{X}}\right)\dot{p}$	[BUR 87, DEL 95]
$\langle\rangle$ denotes the MacCauley brackets $< u >= u$ if $u > 0$ and $< u >= 0$ if $u < 0$	
$\phi(p) = \phi_\infty + (1 - \phi_\infty)e^{-\omega p}$	$J(\underset{\sim}{\mathbf{x}}) = \sqrt{\frac{3}{2}\underset{\sim}{\mathbf{x}} : \underset{\sim}{\mathbf{x}}}$

Table 1.3. *Historical review of KHR for unified models*

As far as ratcheting is concerned, the basic AF nonlinear kinematic hardening rule is known to overpredict the amount of accumulated strain. This is why many authors have proposed modifications of the NLK rule (Table 1.3). Ohno and Wang [OHN 93] assumed that the dynamic recovery term in backstress evolution vanishes for low plastic strain ranges. Chaboche [CHA 91] introduces a nonlinearity in the recall term via a threshold, that will affect the limit state producing plastic shakedown. Burlet and Cailletaud [BUR 87] and Delobelle *et al.* [DEL 95] used a combined model in which the nonlinear recall term of the AF model is superimposed to a radial fading memory term. This compromise was also used by Chen and Jiao [CHE 04] who proposed a modified kinematic hardening rule connecting the dynamic recovery term of [OHN 93] and the evanescence term of the model of [BUR 87].

1.2.4. *Plastic modulus*

In the following, the constitutive equations of four classical models [MRO 67, DAF 76, CHA 86, LUB 91] are briefly recalled. The resulting plastic modulus, that is, $H = d\sigma/d\varepsilon^p = \dot{\sigma}/\dot{\varepsilon}^p$, the slope of the curve $\sigma = h(\varepsilon^p)$, for one-dimensional loading, is expressed for each case.

1.2.4.1. *Mroz's model [MRO 67]*

Mroz's model involves multiple yield surfaces, with linear hardening for each of them. The predicted stress–strain curve is then piecewise linear. The resulting curve tends to be a smooth nonlinear curve when the number of yield surfaces increases. The model is based on several surfaces $f_0, f_1, \ldots f_a$, with f_0 being the initial yield surface:

$$f_i = J(\underset{\sim}{\sigma} - \underset{\sim}{\mathbf{X}}_i) - R \qquad [1.25]$$

In the following, $f_a = 0$ denotes the active surface (loading surface) and the inner surfaces $f_0, f_1, \ldots f_{a-1}$ are tangent at the loading point:

$$f_0 = f_1 = \ldots = f_a = 0 \qquad [1.26]$$

The plastic strain rate is given by

$$\underset{\sim}{\dot{\varepsilon}}^p = \underset{\sim}{\mathbf{n}}^a \dot{p}^a \quad \text{with} \quad \underset{\sim}{\mathbf{n}}^a = \frac{\partial f_a}{\partial \underset{\sim}{\sigma}} \quad \text{and} \quad \dot{p}^a = \frac{\left\langle \underset{\sim}{\mathbf{n}}^a : \underset{\sim}{\dot{\sigma}} \right\rangle}{C^a \underset{\sim}{\mathbf{n}}^a : \underset{\sim}{\mathbf{n}}^a} \qquad [1.27]$$

The chosen rule imposes the same translation for the inner surfaces:

$$\dot{\underset{\sim}{X}}^1 = \dot{\underset{\sim}{X}}^2 = \ldots = \dot{\underset{\sim}{X}}^{a-1} = \dot{\underset{\sim}{X}}^a \quad [1.28]$$

$$\dot{\underset{\sim}{X}}^a = \underset{\sim}{\Delta}_a^{a+1} \dot{\mu} \quad [1.29]$$

$$\underset{\sim}{\Delta}_a^{a+1} = \frac{\left(R^{a+1} - R^a\right)\underset{\sim}{\sigma} - \left(R^{a+1}\underset{\sim}{\mathbf{X}}^a - R^a\underset{\sim}{\mathbf{X}}^{a+1}\right)}{R^a} \quad [1.30]$$

$$\dot{\mu} = \frac{\langle \underset{\sim}{\mathbf{n}}^a : \dot{\underset{\sim}{\sigma}} \rangle}{\underset{\sim}{\mathbf{n}}^a : \underset{\sim}{\Delta}_a^{a+1}} \quad [1.31]$$

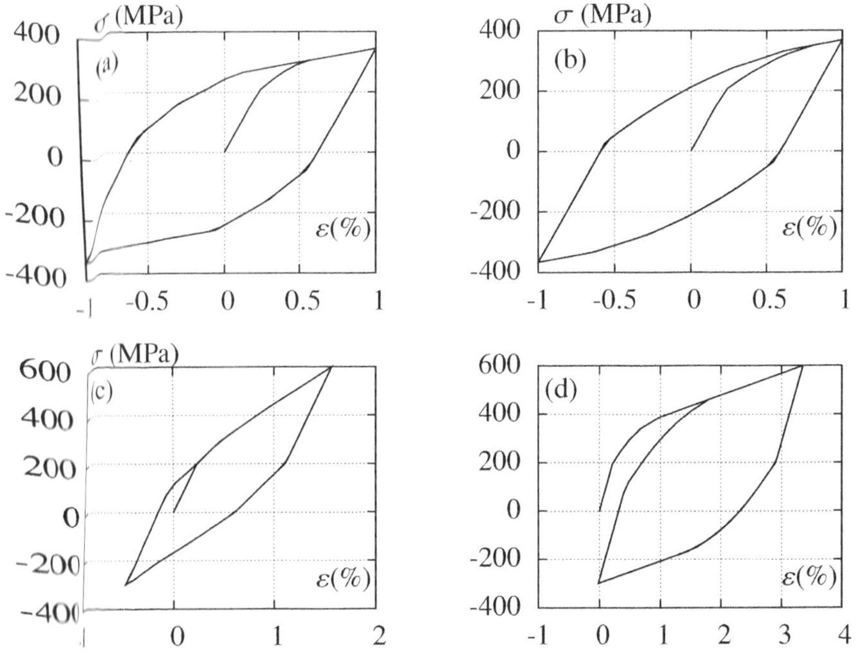

Figure 1.8. *Response of Mroz's model. Top: symmetric loading under prescribed strain; bottom: unsymmetric loading under prescribed stress: (a) 4 surfaces, $R_{\varepsilon-1}$; (b) 8 surfaces, $R_{\varepsilon-1}$; (c) 4 surfaces $-300 < \sigma < 600$; (d) 8 surfaces $-300 < \sigma < 600$*

oz's model is a generalization of the linear kinematic hardening rule. It it possible to correctly describe the nonlinear stress–strain behavior and uschinger effect. However, under asymmetric stress loading, shakedown place after the first cycle. Figure 1.8 shows the response of the model for a strain controlled test and a ratcheting test, using four surfaces and eight surfaces, respectively. The material parameters used for these simulations are given in Table 1.4.

E	C_1	C_2	C_3	C_4	R_1	R_2	R_3	R_4
90,000	70,000	40,000	10,000	5,000	10	20	30	∞
E	C_1	C_2	C_3	C_4	C_5	C_6	C_7	C_8
90,000	60,000	50,000	40,000	30,000	20,000	10,000	5,000	4,000
R_1	R_2	R_3	R_4	R_5	R_6	R_7	R_8	
200	230	260	290	320	350	380	∞	

Table 1.4. *Material parameters (MPa) used for the simulation of Mroz's model in Figure 1.8*

1.2.4.2. *Dafalias and Popov's model [DAF 76]*

The model proposed by Dafalias and Popov can be seen as a Mroz model with two surfaces, but with a variable plastic modulus $K = K(\delta, \delta_{in})$ instead of the constant moduli C_i. The parameter δ is the distance (in terms of von Mises equivalent) between the actual stress and its "image", which is the point on the bounding surface with the same normal. δ_{in} is the value of δ at the initial stage of the loading process. The plastic modulus varies between K_0 and ∞ using, for example,

$$K(\delta, \delta_{in}) = K_0 + h\frac{\delta}{\delta_{in} - \delta} \qquad [1.32]$$

The plastic flow is then classically derived from the normality rule

$$\dot{\underset{\sim}{\varepsilon}}^p = \frac{1}{K}\langle \dot{\underset{\sim}{\sigma}} : \underset{\sim}{\mathbf{n}} \rangle \underset{\sim}{\mathbf{n}} \qquad [1.33]$$

1.2.4.3. *Chaboche's model [CHA 86]*

The yield criterion is

$$f = J(\underset{\sim}{\sigma} - \underset{\sim}{\mathbf{X}}) - R - R_0 = \left[\frac{3}{2}(\underset{\sim}{\mathbf{s}} - \underset{\sim}{\mathbf{X}}) : (\underset{\sim}{\mathbf{s}} - \underset{\sim}{\mathbf{X}})\right]^{(1/2)} - R - R_0 \qquad [1.34]$$

where $\underset{\sim}{\mathbf{s}}$ is the deviatoric stress tensor and $\underset{\sim}{\mathbf{X}}$ is the current yield surface center in the deviatoric space. R_0 and R are respectively the initial size and the evolution of the yield surface. The plastic flow follows the normality rule

$$\dot{\underset{\sim}{\varepsilon}}^p = \dot{p}\frac{\partial f}{\partial \underset{\sim}{\sigma}} = \dot{p}\underset{\sim}{\mathbf{n}} = \frac{3}{2}\dot{p}\frac{\underset{\sim}{\mathbf{s}} - \underset{\sim}{\mathbf{X}}}{J(\underset{\sim}{\sigma} - \underset{\sim}{\mathbf{X}})} \qquad [1.35]$$

In the case of a time-independent framework, the plastic multiplier is given by the consistency condition $f = df = 0$ when plastic flow occurs ($\dot{p}$=0 when $f < 0$ or $df < 0$):

$$\frac{\partial f}{\partial \underset{\sim}{\sigma}} : d\underset{\sim}{\sigma} + \frac{\partial f}{\partial \underset{\sim}{\mathbf{X}}} : d\underset{\sim}{\mathbf{X}} + \frac{\partial f}{\partial R} : dR = 0 \qquad [1.36]$$

The basic nonlinear kinematic hardening rule of Armstrong–Frederick can be written as

$$\dot{\underset{\sim}{X}} = \frac{2}{3} C \dot{\underset{\sim}{\varepsilon}}^p - D\underset{\sim}{\mathbf{X}}\dot{p} \qquad [1.37]$$

The isotropic hardening variable R that characterizes the size change of the elastic domain can be written in integrated form:

$$R = Q(1 - e^{-bp}) \qquad [1.38]$$

The expression of the plastic multiplier $\dot{p}$ comes from the consistency condition $\dot{f} = 0$:

$$\dot{p} = \frac{\underset{\sim}{\mathbf{n}} : \dot{\underset{\sim}{\sigma}}}{h} \quad \text{with} \quad h = Qbe^{-bp} - D\,\underset{\sim}{\mathbf{n}} : \underset{\sim}{\mathbf{X}} \qquad [1.39]$$

The plastic strain rate is then given by

$$\dot{\underset{\sim}{\varepsilon}}^p = \frac{< \underset{\sim}{\mathbf{n}} : \dot{\underset{\sim}{\sigma}} >: \underset{\sim}{\mathbf{n}}}{h} \qquad [1.40]$$

For uniaxial case, assuming that only σ_{11} is non-null, and $x = (2/3)X_{11}$:

$$\dot{\varepsilon}^p = \frac{\dot{\sigma}}{C + Qbe^{-bp} - Dx} \qquad [1.41]$$

A close form solution can be obtained at each half cycle, beginning at ε_0^p, x_0:

$$\sigma = \nu\frac{C}{D} + \left(x_0 - \nu\frac{C}{D}\right)e^{-D\nu(\varepsilon^p - \varepsilon_0^p)} + Qbe^{-bp} + R_0 \qquad [1.42]$$

where $\nu = \pm 1$ is positive in the tension going branch and negative in the compression going branch.

1.2.4.4. *Lubliner model [LUB 91]*

This approach is motivated by the behavior of some solids which, after a tensile plastic flow and a partial elastic unloading, present a new plastic flow during reloading, for a stress value lower than the level previously reached in tension (Figure 1.9). In this so-called "generalized plasticity" model, the relation between the plastic strain ε^p and the stress σ on initial loading under uniaxial stress results in the integration of the differential equation:

$$\frac{d\varepsilon^p}{d\sigma} = \frac{1}{H\beta}\langle \sigma - \sigma_y - H\varepsilon^p \rangle \qquad [1.43]$$

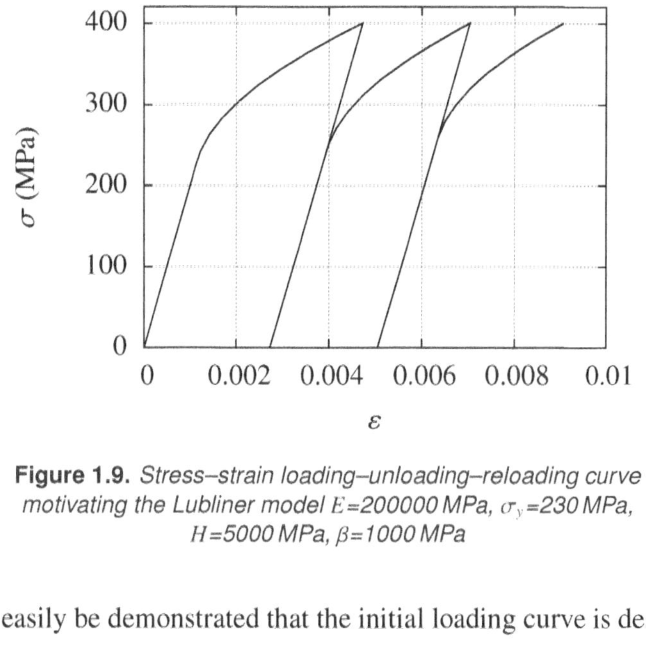

Figure 1.9. *Stress–strain loading–unloading–reloading curve motivating the Lubliner model E=200000 MPa, σ_y=230 MPa, H=5000 MPa, β=1000 MPa*

It can easily be demonstrated that the initial loading curve is described by

$$\varepsilon^p = \frac{\beta}{H} e^{-(\sigma-\sigma_y)/\beta} + \frac{\sigma - \sigma_y - \beta}{H} \qquad [1.44]$$

For a test under prescribed stress between 0 and σ_M, the plastic flow occurs for the stress level

$$\sigma = \sigma_M + \beta\left(-1 + e^{-(\sigma-\sigma_y)/\beta}\right) \qquad [1.45]$$

Lubliner *et al.* [LUB 93] have generalized the previous model to account for asymptotically perfectly plastic or strain softening by postulating the following uniaxial model:

$$\frac{d\varepsilon^p}{d\sigma} = \frac{\langle \sigma - \sigma_y - H\varepsilon^p \rangle}{H\beta + R[\beta - (\sigma - \sigma_y - H\varepsilon^p)]} \quad [1.46]$$

1.2.5. *Viscosity*

Viscoplastic constitutive equations are mainly characterized by a flow function to describe the relation between the viscous stress and the norm of the viscoplastic strain rate. Besides the intensively used power law (known as Norton's law), several approaches may be used to determine the flow function. Table 1.5 gives classical flow rules taken from the literature.

1.2.5.1. *Portevin–Le Chatelier effect (PLC)*

PLC describes a serrated stress–strain curve "jerky flow", which some materials show as they undergo plastic deformation, specifically inhomogeneous deformation. The PLC effect is due to dynamic strain aging and can be associated with a negative strain rate sensitivity for some temperature and strain rate ranges. Two classes of models can be found in the literature:

1) According to [PEN 72], the PLC effect can be mathematically expressed by the empirical law:

$$\sigma = h\varepsilon + F(\dot{\varepsilon}) \quad [1.47]$$

h: work hardening rate considered constant and positive

$F(\dot{\varepsilon})$: function that gives rise to a strain rate dependency of σ.

$\dot{\varepsilon}$ is summed up over all cross-sections along the specimen length. For deformation with imposed strain rate $\dot{\varepsilon}_a$

$$\dot{\varepsilon}_a = \frac{\dot{\sigma}}{M} + \frac{1}{L}\int_0^L \dot{\varepsilon}(x)dx \quad [1.48]$$

L: length of the specimen and M is the combined elastic modulus of the specimen and the testing machine.

Kubin and Estrin [KUB 85] have investigated the PLC effect in the case of loading with constant stress rate and creep loading.

2) On the other hand, the PLC effect attributed to dynamic strain aging is described in [KUB 85] by including the time dependency of the solute concentration C_s at temporarily arrested dislocations:

$$\sigma = \sigma_f + M_i\left(ln\frac{\dot{\varepsilon}}{\dot{\varepsilon}_0} + H'_c C_0 (KDt_a)^{2/3}\right) \qquad [1.49]$$

σ_f: internal stress usually taken as a function of the forest dislocation density
M_i: instantaneous strain rate sensitivity
C_0: alloy solute composition
D: solute diffusion coefficient
t_a: effective aging time of arrested mobile dislocations
K: constant that includes dislocation binding energy
$H'_c = \partial H/\partial C_s$
H: activation enthalpy
C_s: dislocation motion

This model was extended to the 3D case by [ZHA 01] and used by [MAZ 10]:

$$f = J(\underset{\sim}{\sigma}) - R(p) - P_1 C_s(p, t_a) \qquad [1.50]$$

$$C_s = C_m\left(1 - e^{-P_2 p^\alpha t_a^n}\right) \qquad [1.51]$$

$$R(p) = R_0 + Q(1 - e^{-bp}) \qquad [1.52]$$

$$\dot{t}_a = 1 - \frac{t_a}{w}\dot{p} \qquad [1.53]$$

$$\dot{p} = \mu(f) = \dot{\gamma}_0 \sinh\left(\frac{< f >}{K}\right) \qquad [1.54]$$

A closed-form solution giving t_a and the stress value has been reported by [MAZ 06] for a strain controlled test:

$$t_a = \frac{w}{\dot{\varepsilon}}\left(1 - e^{-\frac{p}{w}}\right) + \frac{R_0}{E\dot{\varepsilon}}e^{-\frac{p}{w}} \qquad [1.55]$$

$$\sigma = Kasinh\left(\frac{\dot{p}}{\dot{\gamma}_0}\right) + R_0 + Q\left(1 - e^{-bp}\right) + P_1 C_m\left(1 - e^{-P_2 p^\alpha t_a^n}\right) \qquad [1.56]$$

	Time-independent plasticity theory	
Plasticity	$f = 0 \; \dot{f} = 0$	
	Norton rule	
Power law	$\dot{p} = \langle f/K \rangle^n$	[NOR 29]
Two-term law	$\dot{p} = \langle f/K_1 \rangle^{n_1} + \langle f/K_2 \rangle^{n_2}$	[CHA 89]
	VBO	
Krempl	$\dot{p} = \frac{\sigma}{EK_1}\left(1 + \frac{\sigma}{K_2}\right)^{K_3}$	[KRE 86]
	Strain hardening	
Strain hardening	$\dot{p} = \langle f/K \rangle^n (p + p_0)^m$	
	Summation of inverse Norton rates	
Flow sum inv	$1/\dot{p} = \sum_i 1/\dot{p}_i$	
	Summation of Norton terms	
Flow sum	$\dot{p} = \sum_i \dot{p}_i$	
	Flow for crystal deformation	
Exponential crystal	$\dot{p} = \gamma_0 \exp\left[-\frac{F_0}{RT}\left[1 - \langle f/k \rangle^{n_1}\right]^{n_2}\right]$	
	sinh function	
Miller	$\dot{p} = B\left[\sinh\left(f/K\right)^{1.5}\right]^n$	[MIL 76]
Delobelle	$\dot{p} = A \sinh\left(f/K\right)^n$	[DEL 84]
	exp function	
Bodner	$\dot{p} = A \exp\left(-K/\sigma\right)^n$	[BOD 87]
Onera	$\dot{p} = \langle f/K \rangle^n \exp(\alpha \langle f/K \rangle^{n+1})$	[NOU 89]
Koch	$\dot{p}_\infty \exp\left[-\frac{1}{C}\langle 1 - (\sigma/D)^p \rangle^q\right]$	[KOC 76]
	Other flow rules	
	$\dot{p} = A \exp\left(-\frac{p + p_0}{\alpha\sigma^n}\right)$	
	$\dot{p} = \frac{1}{d^2}\frac{k_1\sigma}{1 + k_2/(d\sigma^m)}$	

Table 1.5. *Flow rules*

1.3. Scale transition rules

1.3.1. *General remarks on scale transition rules*

Phases in the polycrystalline aggregate are defined according to crystal orientation, that is, all the grains falling in the same class of Euler angles are in the same phase. The models used to predict polycrystal behavior in plasticity differ essentially by the transition rule linking local and global stresses and strains.

Besides the simplest and most widely used models (uniform plastic strain or uniform stress), the most popular concept is the self-consistent framework proposed by [HIL 65] and revisited by many authors (see, for instance, [MOL 99]). In terms of rates, the local stress $\underset{\sim}{\sigma}^g$ is expressed according to the global stress $\underset{\sim}{\sigma}$, the global strain $\underset{\sim}{\varepsilon}$ and the local strain $\underset{\sim}{\varepsilon}^g$:

$$\dot{\underset{\sim}{\sigma}}^g = \dot{\underset{\sim}{\sigma}} + \underset{\approx}{\mathbf{L}}^\star : (\dot{\underset{\sim}{\varepsilon}} - \dot{\underset{\sim}{\varepsilon}}^g) \qquad [1.57]$$

The fourth-order tensor $\underset{\approx}{\mathbf{L}}^\star$ takes into account the behavior of the equivalent medium of each grain. $\underset{\approx}{\mathbf{L}}^\star$, also called the accommodation tensor, is defined by:

$$\underset{\approx}{\mathbf{L}}^\star = \underset{\approx}{\mathbf{L}}^{\mathrm{eff}} : \left(\underset{\approx}{\mathbf{S}}^{-1} - \underset{\approx}{\mathbf{I}}\right) \qquad [1.58]$$

where $\underset{\approx}{\mathbf{S}}$ is the well-known Eshelby tensor and $\underset{\approx}{\mathbf{I}}$ the fourth-order identity tensor. The tensor $\underset{\approx}{\mathbf{S}}$ is written in terms of effective modulus $\underset{\approx}{\mathbf{L}}^{\mathrm{eff}}$. Thus,

$$\begin{cases} \left(\underset{\approx}{\mathbf{L}} + \underset{\approx}{\mathbf{L}}^\star\right) : \dot{\underset{\sim}{\varepsilon}}^g = \left(\underset{\approx}{\mathbf{L}}^{\mathrm{eff}} + \underset{\approx}{\mathbf{L}}^\star\right) : \dot{\underset{\sim}{\varepsilon}} \\ \text{and} \\ \dot{\underset{\sim}{\sigma}}^g = \underset{\approx}{\mathbf{L}} : \left(\underset{\approx}{\mathbf{L}} + \underset{\approx}{\mathbf{L}}^\star\right)^{-1} : \left(\underset{\approx}{\mathbf{L}}^{\mathrm{eff}} + \underset{\approx}{\mathbf{L}}^\star\right) : \dot{\underset{\sim}{\varepsilon}} \end{cases} \qquad [1.59]$$

Finally, the following expression can be obtained:

$$\underset{\approx}{\mathbf{L}}^{\mathrm{eff}} = \left\langle \underset{\approx}{\mathbf{L}} : \left(\underset{\approx}{\mathbf{L}} + \underset{\approx}{\mathbf{L}}^\star\right)^{-1} : \left(\underset{\approx}{\mathbf{L}}^{\mathrm{eff}} + \underset{\approx}{\mathbf{L}}^\star\right) \right\rangle \qquad [1.60]$$

Berveiller and Zaoui [BER 79] deduced an explicit transition rule from the previous approach, for the case of a radial monotonic loading path applied to a globally isotropic behavior and isochoric plastic flow:

$$\underset{\sim}{\sigma}^g = \underset{\sim}{\sigma} + \mu\alpha\left(\underset{\sim}{\sigma}, \underset{\sim}{\varepsilon}^p\right)\left(\underset{\sim}{\varepsilon}^p - \underset{\sim}{\varepsilon}^g\right) \quad \text{with} \quad \frac{1}{\alpha} \simeq 1 + \frac{3\mu\varepsilon^p}{2\sigma} \qquad [1.61]$$

where σ and ε^p are, respectively, the overall equivalent stress and the plastic part of overall strain in the uniaxial tension test. They also showed that equation [1.61] allows plastic accommodation in the polycrystal, meanwhile Kröner's rule (obtained with $\alpha = 1$ [KRÖ 61]) produces only elastic accommodation and too large residual stresses. The idea behind all the approaches is finally to introduce a corrective term depending on plastic strains, to compute local stresses. Nevertheless, a linear dependency of this term with respect to plastic strains gives stresses that are too large. The nonlinear accommodation can also be obtained by means of a phenomenological approach, where the difference of the local and global plastic strains present in Kröner's model is replaced by the difference between a local and a global nonlinear hardening variable. The concept of β-rule introduced a variable $\underset{\sim}{\beta}^g$ and its average on the whole aggregate $\underset{\sim}{\beta}$. When applied to a material presenting uniform isotropic elasticity with a shear modulus μ, the expression of the local stress $\underset{\sim}{\sigma}^g$ is:

$$\underset{\sim}{\sigma}^g = \underset{\sim}{\sigma} + \mu\left(\underset{\sim}{\beta} - \underset{\sim}{\beta}^g\right) \quad \text{with} \quad \underset{\sim}{\beta} = \left\langle \underset{\sim}{\beta}^g \right\rangle \qquad [1.62]$$

The symbol $\langle . \rangle$ denotes the volume average. The following rule can be used:

$$\underset{\sim}{\sigma}^g = \underset{\sim}{\sigma} + C\left(\underset{\sim}{\beta} - \underset{\sim}{\beta}^g\right) \quad \text{with} \quad \underset{\sim}{\beta} = \left\langle \underset{\sim}{\beta}^g \right\rangle \qquad [1.63]$$

where C is a material parameter. The variable $\underset{\sim}{\beta}^g$ presents a nonlinear evolution with respect to plastic strain according to the following rule [CAI 94]:

$$\dot{\underset{\sim}{\beta}}^g = \dot{\underset{\sim}{\varepsilon}}^g - D(\underset{\sim}{\beta}^g - \delta\underset{\sim}{\varepsilon}^g)\|\dot{\underset{\sim}{\varepsilon}}^g\| \qquad [1.64]$$

The von Mises equivalent strain rate $\|\dot{\underset{\sim}{\varepsilon}}^g\|$ can also be replaced by the sum of the slip rate absolute values on all slip systems $|\dot{\gamma}^s|$ [CAI 87, SAÏ 06a]

$$\dot{\underset{\sim}{\beta}}^g = \dot{\underset{\sim}{\varepsilon}}^g - D(\underset{\sim}{\beta}^g - \delta\underset{\sim}{\varepsilon}^g)\sum_{s\in\varphi}\left|\dot{\gamma}^s\right| \qquad [1.65]$$

1.3.2. *Scale transition rules for the MM model*

According to Kröner, the physical idea behind the self-consistent framework is the assumption of perfect disorder. It means that the mixture of the two constituents is such that, in a given realization, the probability of finding phase A and phase B in a given place is totally random. As a result, phase A can be seen as an inclusion in the homogeneous equivalent medium made of A and B, and phase B can also be seen as an inclusion in the homogeneous equivalent medium made of A and B. This is why the localization rules for both phases are totally symmetric. The β_rule is generally selected for its simplicity and its versatility. It can be used for a material with N crystallographic phases, and in the case of a 2-phase material as well.

In the MM models, each mechanism I is associated with a local stress tensor $\underset{\sim}{\sigma}^I$ calculated from a stress concentration rule. Several approaches were used to obtain local stresses and strains through different transition rules. They are recalled in the following. The thermodynamic approach of MM models has first been investigated by [CAI 95] for the case of uniform elasticity. In this framework, the model is defined from the free energy $\rho\Psi$, which defines stress and hardening variables knowing elastic strain and internal variables. The plastic strain $\underset{\sim}{\varepsilon}^p$ is decomposed into the two parts $\underset{\sim}{\varepsilon}^1$, $\underset{\sim}{\varepsilon}^2$:

$$\underset{\sim}{\varepsilon}^p = A_1\underset{\sim}{\varepsilon}^1 + A_2\underset{\sim}{\varepsilon}^2 \qquad [1.66]$$

where A_1 and A_2 can be fourth-order tensors or scalars. From a thermodynamic point of view, equation [1.66] leads *de facto* to the following form of the local stresses:

$$\begin{cases} \underset{\sim}{\sigma}^1 = A_1\underset{\sim}{\sigma} \\ \underset{\sim}{\sigma}^2 = A_2\underset{\sim}{\sigma} \end{cases} \qquad [1.67]$$

The simplest assumption involves considering $A_1 = A_2 = 1$. This assumption is widely used by Taleb and co-workers.

For the specific case of two-phase alloys (Figure 1.10), A_1 and A_2 can be seen as the volume fraction of the two phases, so that $A_1 + A_2 = 1$. This kind of consideration includes the works related to the semi-crystalline polymer behavior. The contribution of each phase (amorphous and crystalline) is taken into account through the degree of crystallinity.

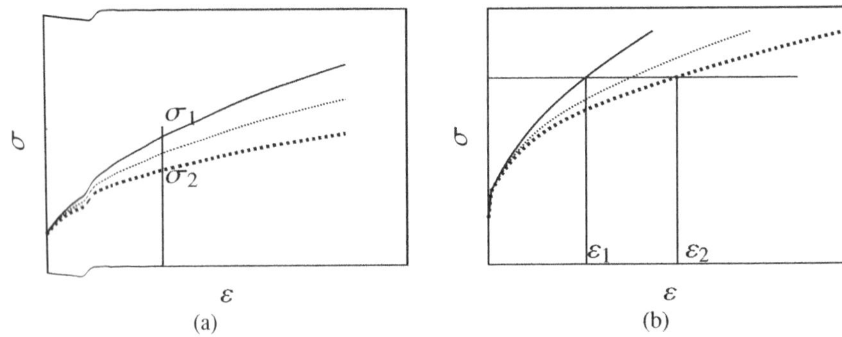

Figure 1.10. *Estimation of the macroscopic behavior according to (a) uniform strain and (b) uniform stress assumption*

Alternatively, the self-consistent schemes can be applied to the case of the MM model. The explicit transition rule of equation [1.61] leads to

$$\begin{cases} \underset{\sim}{\sigma}^1 = \underset{\sim}{\sigma} + A(\underset{\sim}{\varepsilon}^p - \underset{\sim}{\varepsilon}^1) \\ \underset{\sim}{\sigma}^2 = \underset{\sim}{\sigma} + A(\underset{\sim}{\varepsilon}^p - \underset{\sim}{\varepsilon}^2) \end{cases} \quad \text{with} \quad \underset{\sim}{\varepsilon}^p = A_1\underset{\sim}{\varepsilon}^1 + A_2\underset{\sim}{\varepsilon}^2 \qquad [1.68]$$

whereas the "β_rule", equation [1.63], for 2-phase materials can be written as follows:

$$\begin{cases} \underset{\sim}{\sigma}^1 = \underset{\sim}{\sigma} + A\left(\underset{\sim}{\beta} - \underset{\sim}{\beta}_1\right) \\ \underset{\sim}{\sigma}^2 = \underset{\sim}{\sigma} + A\left(\underset{\sim}{\beta} - \underset{\sim}{\beta}_2\right) \end{cases} \quad \text{with} \quad \underset{\sim}{\beta} = A_1\underset{\sim}{\beta}_1 + A_2\underset{\sim}{\beta}_2 \qquad [1.69]$$

The basic idea behind these localization rules is that phases can be hard or soft. Indeed, the softer phase shows a plastic strain $\underset{\sim}{\varepsilon}^1$ larger than the average plastic strain. Therefore, the stress level $\underset{\sim}{\sigma}^1$ supported by the soft phase will be smaller than the macroscopic stress. The hard phase will then be applied a higher stress level, $\underset{\sim}{\sigma}^2$. Accordingly, the plastic strain $\underset{\sim}{\varepsilon}^2$ will be smaller than the average plastic strain.

1.[illegible] Large deformation

MM models are able to describe the mechanical behavior for both metallic materials and polymeric materials. For the latter, a finite strain formulation is mandatory to correctly represent the large strains applied to the material element. Large strain formalism introduces the following standard expressions:

$$\underset{\sim}{\mathbf{L}} = \dot{\underset{\sim}{\mathbf{F}}}\underset{\sim}{\mathbf{F}}^{-1} \qquad \underset{\sim}{\mathbf{D}} = \frac{1}{2}\left(\underset{\sim}{\mathbf{L}} + \underset{\sim}{\mathbf{L}}^T\right) \qquad \underset{\sim}{\mathbf{\Omega}} = \frac{1}{2}\left(\underset{\sim}{\mathbf{L}} - \underset{\sim}{\mathbf{L}}^T\right) \qquad [1.70]$$

where $\underset{\sim}{\mathbf{F}}$ is the deformation gradient, $\underset{\sim}{\mathbf{L}}$ is the rate of deformation, $\underset{\sim}{\mathbf{D}}$ is the stretch rate and $\underset{\sim}{\mathbf{\Omega}}$ is the rotation rate.

The critical part of the algorithm is the definition of the derivatives of the stress and strain tensors. Several possibilities are known to introduce an objective derivation process. Convective transport rules of a given tensor consist of a transformation of this tensor in a rotating frame, where the derivative is applied, then applying the reverse rotation to obtain the value in the laboratory frame. Several candidates have been used in the literature:

– For the case of Green–Naghdi derivative, the rotation is defined by means of the $\underset{\sim}{\mathbf{R}}$ tensor extracted from the polar decomposition $\underset{\sim}{\mathbf{F}} = \underset{\sim}{\mathbf{R}}\underset{\sim}{\mathbf{U}}$:

$$\dot{\underset{\sim}{\mathbf{T}}}^{GN} = \dot{\underset{\sim}{\mathbf{T}}} + \underset{\sim}{\mathbf{T}}\underset{\sim}{\mathbf{R}} - \underset{\sim}{\mathbf{R}}\underset{\sim}{\mathbf{T}} \qquad [1.71]$$

– The Jauman derivative uses the corotational frame, which is defined by a tensor Ω, the spin tensor that comes from the material rotation:

$$\dot{\underset{\sim}{\mathbf{T}}}^{J} = \dot{\underset{\sim}{\mathbf{T}}} + \underset{\sim}{\mathbf{T}}\underset{\sim}{\mathbf{\Omega}} - \underset{\sim}{\mathbf{\Omega}}\underset{\sim}{\mathbf{T}} \qquad [1.72]$$

The spin tensor is found by integrating the relation $\underset{\sim}{\mathbf{\Omega}} = -\underset{\sim}{\mathbf{Q}}^{-1}\dot{\underset{\sim}{\mathbf{Q}}}$, ensuring that the rotating frame follows the material elements:

$$\dot{\underset{\sim}{\mathbf{Q}}} = -\underset{\sim}{\mathbf{Q}}\underset{\sim}{\mathbf{\Omega}} \qquad [1.73]$$

In these cases, the formulation of the material models with tensorial internal variables and even anisotropic response is the same for small perturbations or large strain assumptions [LAD 99].

Specifically, the additional partition of the strain into elastic and inelastic parts is still valid for large strains. The resulting response is an hypoelastic formulation, in agreement with a Green–Naghdi stress rate, that can be problematic, but only for very large strain values. The stress measure is the

Cauchy stress $\underset{\sim}{\sigma}$ obtained using the relevant transformation, with either $\underset{\sim}{\mathbf{R}}$ or $\underset{\sim}{\mathbf{\Omega}}$:

$$\underset{\sim}{\sigma} = det^{-1}\left(\underset{\sim}{\mathbf{F}}\right)\underset{\sim}{\mathbf{R}}\underset{\sim}{\mathbf{S}}\underset{\sim}{\mathbf{R}}^T \qquad \underset{\sim}{\sigma} = det^{-1}\left(\underset{\sim}{\mathbf{F}}\right)\underset{\sim}{\mathbf{Q}}\underset{\sim}{\mathbf{S}}\underset{\sim}{\mathbf{Q}}^T \qquad [1.74]$$

1.5. Brief history of the MM models

MM models have emerged following the development of the crystallographic approaches. Taylor [TAY 38] and Bishop and Hill [BIS 51] were among the pioneers who contributed to the extension of the elasto-plastic macroscopic theory in order to study polycrystal behavior. Sanders [SAN 54], Budiansky and Wu [BUD 62] and Koiter [KOI 60] have considered that the plastic deformation is the result of slip that occurs along parallel planes or more generally involves several "independent" mechanisms. Mandel [MAN 65] has generalized this concept to the case of dependent mechanisms to investigate the uniqueness and existence of the plastic multipliers. Following his work, two kinds of approaches have been developed: (i) the polycrystalline plasticity models, see, for instance, [BUI 69, ZAO 70] and (ii) the MM models as seen by Zarka and co-workers [ZAR 72, ZAR 78, ZAR 79, ZAR 80]. These two approaches take into account, as a common root, the microstructural heterogeneity of the material that leads to internal stresses. Indeed, both approaches account for plastic accommodation induced by the difference between local (grain or mechanism) and global plastic strain. Analogy was carried out between the MM model types and the polycrystalline plasticity approach [CAI 08].

The MM approach belongs to a large class of models, which also includes "*multi-phase*" models. A non-exhaustive list of studies devoted to the "*multi-mechanism*" models type includes the works of [ZAR 79, CON 89, SAÏ 93, CAI 95, ROU 00, SAÏ 04, VEL 06, TAL 06, SAÏ 07, WOL 08, HAS 08, REE 09b, BES 09a, TAL 11, SAÏ 11a]. Investigations of the "*multi-phase*" models can be cited as: [KAR 74, VID 94, GAU 04, GAL 07, BRU 08, DEB 08, BAU 09, GAL 10, HOS 10, CLE 12, ELS 12].

Stress and strains can be expressed for the considered number of mechanisms n through m criteria. n can be equal or not to m. Nevertheless, the most common models relate to one of the following categories:

– "*Model with various mechanisms and various criteria*": The crystal plasticity models for which the behavior of each slip system is represented individually by the resolved shear stress belong to this family. It is worth noting that the single crystal formulation is accounted for by the same set of material parameters for all the slip systems. However, if the mechanisms do not particularly refer to slip systems, the number n can reasonably be set at two or three. Otherwise, the large number of material parameters are difficult to identify. This kind of approach allows introducing several types of plastic mechanisms or phase behavior.

– "*Model with various mechanisms and one criterion*": In this class of models, each mechanism has its own set of variables (i.e. stress, strain, hardening variables, etc.). However, a unique yield function is used to describe the different flow rules similarly to the unified approach. Even if all the mechanisms are simultaneously activated, differentiation between the level of variables can be allowed by the material parameters. The number of the isotropic material parameters is reduced compared with the first case.

2

Model Formulation

In this chapter, a summary of the thermodynamic framework used for deriving the constitutive equations for the MM models is first recalled in section 2.1. The distinction is then made between models with various mechanisms and various criteria (section 2.2) and models with various mechanisms and one criterion (section 2.3). The MM approach is compared with unified models in section 2.4. An emphasis on the nonlinear isotropic hardening rules and kinematic hardening rules specific to the MM models is made in sections 2.5 and 2.6. Finally, the most significant variants of the MM models are proposed. These models differ essentially by the scale transition rule, the strain homogenization scheme, the kinematic hardening rules and the isotropic rules. Special attention is paid to the computation of the inelastic multipliers depending on the plastic or viscoplastic nature of the mechanisms.

2.1. Thermodynamic framework

The basic aspects of the constitutive equations of the MM models are now presented. It has been previously shown [CAI 95] that the form of the constitutive equations of these models is compatible with the general thermodynamic framework developed by [GER 83]. The purpose of this book is not to deliver a comprehensive description of the framework. Only the main operational results are given here. The interested reader can refer, for instance, to a basic textbook like [BES 09b] for a complete introduction.

Thermodynamics of irreversible processes introduces two types of variables, which are linked by the definition of a non-negative convex thermodynamic potential. The specific Gibbs free energy, denoted by Ψ, will be chosen in the following. Knowing this potential is enough to describe any reversible process. The most simple case is the linear elastic behavior, where the stress tensor $\underset{\sim}{\sigma}$ is obtained by a partial derivative of Ψ with respect to the elastic strain $\underset{\sim}{\varepsilon}^e$ (with ρ being the mass per volume unit):

$$\underset{\sim}{\sigma} = \rho \frac{\partial \Psi}{\partial \underset{\sim}{\varepsilon}^e} \quad [2.1]$$

Elastic strain is the state variable that allows us to define the status of the material.

Additional variables are needed if dissipative mechanisms come into play in the material. The common assumption that elastic and plastic mechanisms are not coupled is adopted here, and the new hardening variables A_I are defined according to the preceding scheme, as functions of the state variables α_I, so that

$$\Psi(\underset{\sim}{\varepsilon}^e, \alpha^I) = \Psi_e(\underset{\sim}{\varepsilon}^e) + \Psi_p(\alpha^I) \quad [2.2]$$

and

$$A^I = \rho \frac{\partial \Psi_p}{\partial \alpha^I} \quad [2.3]$$

In the frame of a generalized standard material [HAL 75], the plastic strain rate and the evolution rules for the dissipative variables are elaborated by means of a unique potential Ω:

$$\dot{\underset{\sim}{\varepsilon}}^p = \frac{\partial \Omega}{\partial \underset{\sim}{\sigma}} \qquad \dot{\alpha}^I = -\frac{\partial \Omega}{\partial A^I} \quad [2.4]$$

Using the first and second principles of thermodynamics, it can also be shown [BES 09b] that the so-called "intrinsic dissipation" is the difference between the plastic power and the fraction of power that is temporarily stored by the hardening mechanisms:

$$\mathcal{D} = \underset{\sim}{\sigma} : \dot{\underset{\sim}{\varepsilon}}^p - \rho\dot{\Psi} = \underset{\sim}{\sigma} : \dot{\underset{\sim}{\varepsilon}}^p - A^I \dot{\alpha}^I \quad [2.5]$$

According to the second principle, this dissipation is either null (elastic regime) or positive. The shape of equation [2.5] demonstrates that the condition is automatically fulfilled if and only if Ω is a convex function and $\Omega(0) = 0$.

From a thermodynamic point of view, the starting point of the MM models is now a collection of potentials $\Omega^I, I = 1..n$, where n is the number of the considered mechanisms. For each mechanism I, a local stress $\underset{\sim}{\sigma}^I$ is obtained through a concentration tensor $\underset{\approx}{\mathbf{B}}^I = \partial\underset{\sim}{\sigma}^I / \partial\underset{\sim}{\sigma}$. Note that for the initial version of the models, $\underset{\approx}{\mathbf{B}}^I = \underset{\approx}{\mathbf{I}}$ has been chosen. Two cases have been distinguished in the MM models:

– each $\underset{\sim}{\sigma}^I$ is involved in a different yield function f^I, defining a series of criteria:

$$\dot{\underset{\sim}{\varepsilon}}^p = \sum_I \frac{\partial \Omega^I}{\partial \underset{\sim}{\sigma}} = \sum_I \frac{\partial \Omega^I}{\partial f^I} \frac{\partial f^I}{\partial \underset{\sim}{\sigma}} = \sum_I \frac{\partial \Omega^I}{\partial f^I} \frac{\partial f^I}{\partial \underset{\sim}{\sigma}^I} : \frac{\partial \underset{\sim}{\sigma}^I}{\partial \underset{\sim}{\sigma}} = \sum_I \frac{\partial \Omega^I}{\partial f^I} \underset{\sim}{\mathbf{n}}^I : \underset{\approx}{\mathbf{B}}^I \qquad [2.6]$$

This includes the 2M2C model and the crystal plasticity models. For these models, each mechanism has its own plastic multiplier in time independent plasticity;

– all $\underset{\sim}{\sigma}^I$ are combined into an unique global criterion f:

$$\dot{\underset{\sim}{\varepsilon}}^p = \frac{\partial \Omega}{\partial \underset{\sim}{\sigma}} = \frac{\partial \Omega}{\partial f} \frac{\partial f}{\partial \underset{\sim}{\sigma}} = \frac{\partial \Omega}{\partial f} \sum_I \frac{\partial f}{\partial \underset{\sim}{\sigma}^I} : \frac{\partial \underset{\sim}{\sigma}^I}{\partial \underset{\sim}{\sigma}} = \frac{\partial \Omega}{\partial f} \sum_I \underset{\sim}{\mathbf{n}}^I : \underset{\approx}{\mathbf{B}}^I \qquad [2.7]$$

2M1C model belongs to this second class of model, for which only one plastic multiplier has to be determined in time independent plasticity.

2.2. Model with various mechanisms and various criteria: the 2M2C model

The 2M2C model is assumed to depend on two mechanisms and two criteria (yield functions). The inelastic part of the free energy function can be expressed as a function of the internal variables $\underset{\sim}{\alpha}^1$, $\underset{\sim}{\alpha}^2$, r^1 and r^2 as follows:

$$\rho\Psi_p(\underset{\sim}{\alpha}^I, r^I) = \frac{1}{3} \sum_I \sum_J C_{IJ} \underset{\sim}{\alpha}^I : \underset{\sim}{\alpha}^J + \frac{1}{2} \sum_I b_I Q_I \left(r^I\right)^2 \qquad [2.8]$$

The potential is a quadratic function of the state variables $\underset{\sim}{\alpha}^I$ and r^I. The non-negativity of the free energy will be ensured, provided that the material parameters fulfill the following conditions:

$$C_{11} \geq 0; \quad C_{22} \geq 0; \quad C_{11}C_{22} \geq C_{12}^2; \quad Q_1 \geq 0; \quad Q_2 \geq 0 \qquad [2.9]$$

The hardening variables are then

$$\underset{\sim}{\mathbf{X}}^I = \rho\frac{\partial \Psi_p}{\partial \underset{\sim}{\alpha}^I} = \frac{2}{3}\sum_J C_{IJ}\underset{\sim}{\alpha}^J \qquad R^I = \rho\frac{\partial \Psi_p}{\partial r^I} = b_I\, Q_I\, r^I \qquad [2.10]$$

For the sake of simplicity, $\underset{\sim}{\sigma}^I$ are equal to $\underset{\sim}{\sigma}$ in this section. The flow rule is generated by a potential, which is the sum of two terms:

$$f^I = J(\underset{\sim}{\sigma} - \underset{\sim}{\mathbf{X}}^I) - R^I - R_{0I} \qquad \Omega = \Omega^1(f^1) + \Omega^2(f^2)$$
$$\text{where} \quad J(\underset{\sim}{\sigma} - \underset{\sim}{\mathbf{X}}^I) = \sqrt{\frac{3}{2}\left(\underset{\sim}{\mathbf{s}} - \underset{\sim}{\mathbf{X}}^I\right) : \left(\underset{\sim}{\mathbf{s}} - \underset{\sim}{\mathbf{X}}^I\right)} \qquad [2.11]$$

so that

$$\dot{\underset{\sim}{\varepsilon}}^p = \dot{p}^1\underset{\sim}{\mathbf{n}}^1 + \dot{p}^2\underset{\sim}{\mathbf{n}}^2 \quad \text{with} \quad \underset{\sim}{\mathbf{n}}^I = \frac{\partial f^I}{\partial \underset{\sim}{\sigma}} = \frac{3}{2}\frac{\underset{\sim}{\mathbf{s}} - \underset{\sim}{\mathbf{X}}^I}{J(\underset{\sim}{\sigma} - \underset{\sim}{\mathbf{X}}^I)} \qquad [2.12]$$

At this point, nothing has been specified on $\dot{p}^I$, that can either derive from a viscoplastic potential $\dot{p}^I = \partial\Omega^I/\partial f^I$ or be the plastic multiplier of a time-independent model, where Ω^I is a pseudo potential obtained as the indicator function of the yield surface:

$$\Omega\left(f^I\right) = 0 \quad if \quad f^I < 0$$
$$\Omega\left(f^I\right) \to \infty \quad if \quad f^I = 0 \qquad [2.13]$$

The model would be a "generalized standard" model, by taking $D_I = 0$ and $Q \to \infty$. In the present form, it can be easily checked that the dissipation (equation [2.5]) remains positive. The hardening rules are then expressed as follows:

$$\dot{\underset{\sim}{\alpha}}^I = \left(\underset{\sim}{\mathbf{n}}^I - \frac{3D_I}{2C_{II}}\underset{\sim}{\mathbf{X}}^I\right)\dot{p}^I \qquad \dot{r}^I = \left(1 - \frac{R^I}{Q_I}\right)\dot{p}^I \qquad [2.14]$$

$$\mathcal{D} = \underset{\sim}{\sigma} : \underset{\sim}{\dot{\varepsilon}}^p - \underset{\sim}{\mathbf{X}}^1 : \underset{\sim}{\dot{\alpha}}^1 - \underset{\sim}{\mathbf{X}}^2 : \underset{\sim}{\dot{\alpha}}^2 - R^1 \dot{r}^1 - R^2 \dot{r}^2 \quad [2.15]$$

$$= \sum \left[(\underset{\sim}{\sigma} - \underset{\sim}{\mathbf{X}}^I) : \underset{\sim}{\mathbf{n}}^I - R^I \right] \dot{p}^I + \sum \frac{3D_I}{2C_{II}} \underset{\sim}{\mathbf{X}}^I : \underset{\sim}{\mathbf{X}}^I \dot{p}^I + \sum \frac{b_I}{Q_I} (R^I)^2 \dot{p}^I \quad [2.16]$$

$$= \sum f^I \dot{p}^I + \sum_I R_{0I} \dot{p}^I + \sum \frac{3D_I}{2C_{II}} \underset{\sim}{\mathbf{X}}^I : \underset{\sim}{\mathbf{X}}^I \dot{p}^I + \sum \frac{b_I}{Q_I} (R^I)^2 \dot{p}^I \quad [2.17]$$

In this expression, the first term remains null for time-independent plasticity ($f^I = 0$), and comes into play for the case of viscous dissipation. The second term is a dissipation due to the friction mechanisms. The third term comes from the nonlinear character of kinematic hardening, and the fourth one from the nonlinear character of isotropic hardening.

The 2M2C model type makes it possible to perform the simultaneous treatment of plasticity and viscoplasticity. The model is then able to discriminate between the increase in hardening produced by plasticity or creep. This may be quite important to model complex behaviors like 316 stainless steel at 650°C [CON 89] or N-18 alloy in the temperature range 600–700°C [SAÏ 04].

The 2M2C model type can also be applied to study phase transformation. In this class of models, a stress tensor and a strain tensor are defined in each phase of the material inside the representative volume element. Videau *et al.* [VID 94] were the first who applied the MM models to phase transformation. In the work of Gautier *et al.* [GAU 04] and Saï *et al.* [SAÏ 06b], the transformation-induced plasticity of a 304 stainless steel is carried out using a MM model in which the influence of each phase is balanced by its volume fraction which is calculated by a kinetics transformation rule.

2.3. Model with various mechanisms and one criterion: the 2M1C model

The 2M1C model is assumed to depend on two mechanisms and one criterion (yield function). The inelastic part of the free energy function can be expressed as a function of the internal variables $\underset{\sim}{\alpha}^1$, $\underset{\sim}{\alpha}^2$ and r as follows:

$$\rho\Psi_p = \frac{1}{3}\sum_I\sum_J C_{IJ}\underset{\sim}{\alpha}^I : \underset{\sim}{\alpha}^J + \frac{1}{2}bQr^2 \qquad [2.18]$$

The relations between the internal variables and their associated forces are

$$\underset{\sim}{\mathbf{X}}^I = \rho\frac{\partial\Psi_p}{\partial\underset{\sim}{\alpha}^I} = \frac{2}{3}\sum_J C_{IJ}\underset{\sim}{\alpha}^J \qquad R = \rho\frac{\partial\Psi_p}{\partial r} = bQr \qquad [2.19]$$

The evolution laws of these variables are generated by a potential that introduces a quadratic combination of the two mechanisms:

$$f = \left(J(\underset{\sim}{\sigma} - \underset{\sim}{\mathbf{X}}^1)^2 + J(\underset{\sim}{\sigma} - \underset{\sim}{\mathbf{X}}^2)^2\right)^{1/2} - R - R_0 \qquad \Omega \equiv \Omega(f) \qquad [2.20]$$

This form generates a coupling between the two mechanisms, which is not considered in the 2M2C model. Using the notation $\dot{p} = d\Omega/df$ for the viscoplastic case, or $\dot{p}$ plastic multiplier for time-independent plasticity, we obtain:

$$\begin{cases} \dot{\underset{\sim}{\varepsilon}}^p = \dot{p}\underset{\sim}{\mathbf{n}} = \dfrac{J_1\underset{\sim}{\mathbf{n}}^1 + J_2\underset{\sim}{\mathbf{n}}^2}{(J_1^2 + J_2^2)^{1/2}}\dot{p} \\ \text{with} \quad J_I = J(\underset{\sim}{\sigma} - \underset{\sim}{\mathbf{X}}^I) \quad \text{and} \quad \underset{\sim}{\mathbf{n}}^I = \dfrac{3}{2}\dfrac{\underset{\sim}{\mathbf{s}} - \underset{\sim}{\mathbf{X}}^I}{J_I} \end{cases} \qquad [2.21]$$

Finally, the hardening rules of the 2M1C model are

$$\dot{\underset{\sim}{\alpha}}^I = \left(\underset{\sim}{\mathbf{n}}^I - \frac{3D_I}{2C_{II}}\underset{\sim}{\mathbf{X}}^I\right)\frac{J_I}{J}\dot{p} \qquad \dot{r} = \left(1 - \frac{R}{Q}\right)\dot{p} \qquad [2.22]$$

2.4. Comparison with the unified model

The MM models have inherited the know-how acquired by the unified models during several decades. A review of the building bricks describing the elementary material behaviors (isotropic hardening rules, kinematic hardening rules, flow rules, etc.) was made in section 1.2.

– in the unified model, all the deformation sources are taken into account through a single plastic strain $\underset{\sim}{\varepsilon}^p$. The plastic strain rate $\dot{\underset{\sim}{\varepsilon}}^p$ is normal to the boundary of the elastic domain. This domain is described by a yield function f that evolves from an initial size R_0 through a tensorial variable $\underset{\sim}{\mathbf{X}}$ (the back stress) and a scalar variable R (size change of the elastic domain). $\underset{\sim}{\mathbf{X}}$ and R are the associated forces of the kinematic internal variable $\underset{\sim}{\alpha}$ and the isotropic internal variable r, respectively;

– the 2M2C model can be schematically seen as the collection of two unified models. A scale transition rule can be used to deduce $\underset{\sim}{\sigma}^I$ for the Ith mechanism with respect to $\underset{\sim}{\sigma}$. In the present chapter, the simple assumption $\underset{\sim}{\sigma}^I = \underset{\sim}{\sigma}$ has been used. For each mechanism I, a back stress $\underset{\sim}{\mathbf{X}}^I$ and a size change of the elastic domain R^I are respectively associated with a local kinematic internal variable $\underset{\sim}{\alpha}^I$ and a local isotropic internal variable r^I. Having in hand $\underset{\sim}{\sigma}^I$, $\underset{\sim}{\mathbf{X}}^I$, R^I and the initial size of the local elastic domain R_{0I}, the yield function and then the plastic strain rate of each mechanism can be obtained. The overall plastic strain rate $\dot{\underset{\sim}{\varepsilon}}^p$ is finally calculated as the sum of the plastic strain rate of each mechanism;

– as in the 2M2C model, two local stresses $\underset{\sim}{\sigma}^I$ and two back stresses $\underset{\sim}{\mathbf{X}}^I$ are needed for the 2M1C model. However, only one yield function is used to define the limit of the elastic domain related to R_0 and a size change R. Indeed, a complementary coupling is introduced in the 2M1C model, according to the fact that the effects of the two mechanisms may be inseparable. The variables are then assembled in one criterion, and only one inelastic multiplier is needed.

A schematic view, intended to situate the MM approach with respect to the unified one, is shown in the flowcharts of Figure 2.1.

2.5. Isotropic hardening rules

2.5.1. *Isotropic hardening for models with various mechanisms and one criterion*

If only one criterion is considered, isotropic hardening plays the same role as in the unified model type. It characterizes the expansion of the yield surface through a scalar variable that defines the radius of the elastic domain.

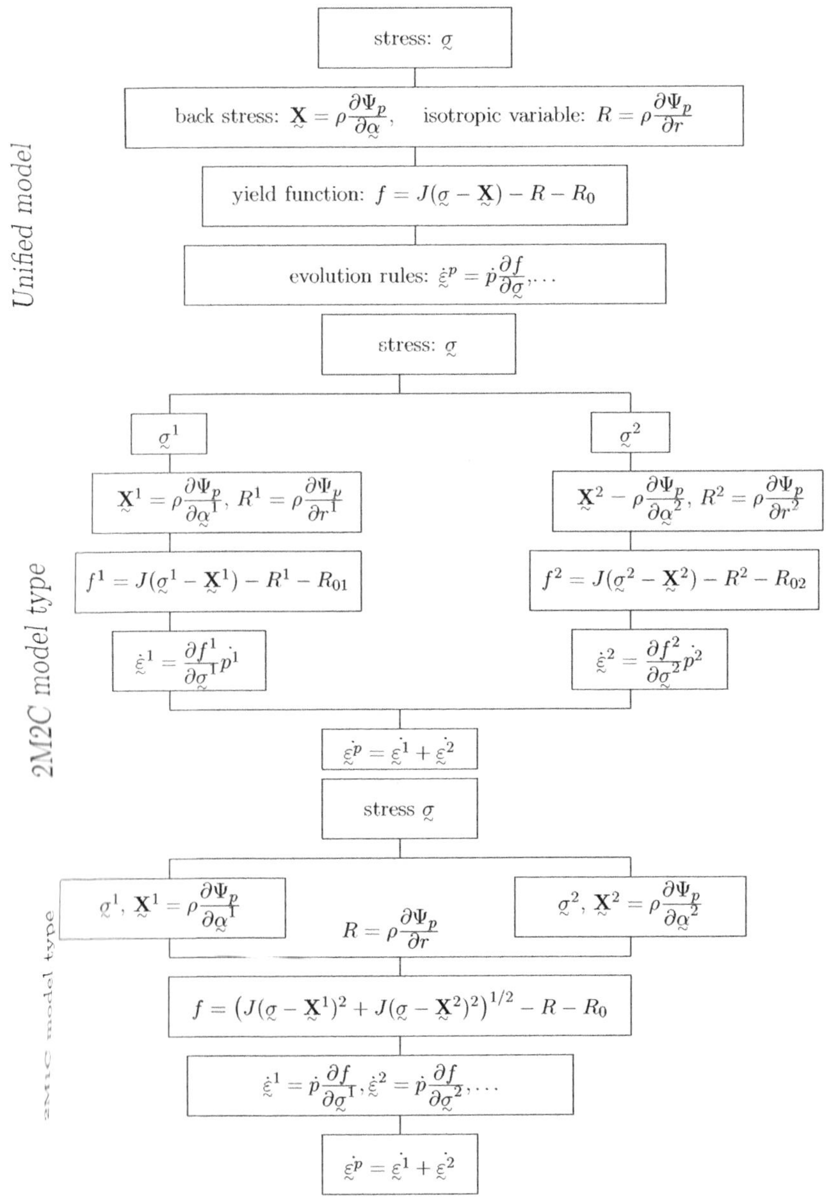

Figure 2.1. *Comparative flowcharts of the unified model, the 2M2C and the 2M1C model types*

Following the formalism of section 2.3, the threshold function is expressed as a quadratic combination of the two mechanisms:

$$f = \left(J(\underset{\sim}{\sigma} - \underset{\sim}{\mathbf{X}}^1)^2 + J(\underset{\sim}{\sigma} - \underset{\sim}{\mathbf{X}}^2)^2\right)^{1/2} - R - R_0 \qquad [2.23]$$

The rate of the isotropic hardening variables is thus

$$\dot{r} = \left(1 - \frac{R}{Q}\right)\dot{p} \qquad [2.24]$$

According to the form of the free energy $\rho\Psi_p = \ldots + \frac{1}{2}\,bQr^2$, the size change of the elastic domain $R = bQr$ can be written in the explicit form

$$R = Q(1 - e^{-bp}) \qquad [2.25]$$

For a better description of the cyclic (softening/hardening), two isotropic variables are generally introduced to describe different regimes of stress evolution in the first cycles and until the steady state is reached. The strain memory effect can be incorporated into either one isotropic variable or both variables. Velay *et al.* [VEL 06, TAL 10] described the yield function of the "two mechanisms and one criterion" model by means of two additive isotropic variables. The strain memory effect was introduced into the first isotropic variable only.

2.5.2. *Isotropic hardening for models with various mechanisms and various criteria*

Following the formalism of section 2.2, each mechanism has a separate yield surface, which evolves according to the local stress $\underset{\sim}{\sigma}^I$, the back stress $\underset{\sim}{\mathbf{X}}^I$ and the size change of the elastic domain R^I:

$$f^I = J(\underset{\sim}{\sigma}^I - \underset{\sim}{\mathbf{X}}^I) - R^I - R_{0I} \qquad [2.26]$$

R_{0I} denotes the initial size of the yield surface of the considered mechanism.

If no coupling is considered between the hardening variables, the size change of the elastic domain is given in integrated form using the isotropic material parameters Q_I and b_I:

$$R^I = Q_I(1 - \exp(-b_I p^I)) \qquad [2.27]$$

The extra-hardening effect and the strain memory effect described previously can be adapted for one or both mechanisms.

A coupling between the isotropic variables can improve the model's capabilities. For the sake of illustration, a model with three mechanisms and three criteria is considered.

Each stress tensor ($\underset{\sim}{\sigma}_1, \underset{\sim}{\sigma}_2$ and $\underset{\sim}{\sigma}_3$) is involved in a yield function (f_1, f_2 and f_3) to define the elastic domain related to each mechanism. If the kinematic hardening variables are not considered, the yield functions can be written as follows:

$$f^I = J\left(\underset{\sim}{\sigma}^I\right) - R^I - R_{0I} \quad I = 1 \ldots 3 \qquad [2.28]$$

R_{01}, R_{02} and R_{03} denote the initial size of the elastic domain for mechanisms 1, 2 and 3, respectively. The isotropic hardening variables (R_1, R_2 and R_3) that characterize the size change of the elastic domain related to each yield surface are described by three internal variables (r_1, r_2 and r_3), respectively. In this formulation, an isotropic–isotropic coupling has been introduced between the hardening variables to account for the interaction between the different mechanisms:

$$\begin{pmatrix} R^1 \\ R^2 \\ R^3 \end{pmatrix} = \begin{pmatrix} Q_{11} & Q_{12} & Q_{13} \\ Q_{21} & Q_{22} & Q_{23} \\ Q_{31} & Q_{32} & Q_{33} \end{pmatrix} \begin{pmatrix} b_1 r^1 \\ b_2 r^2 \\ b_3 r^3 \end{pmatrix} \qquad [2.29]$$

Q_{11}, Q_{22} and Q_{33} are the isotropic hardening moduli. ($Q_{IJ}, I \neq J$) are coupling parameters that play an essential role in the material behavior. For example, a positive value of Q_{12} allows mechanism 2 (if active) to attenuate mechanism 1. Increasing Q_{12} leads to deactivate mechanism 1. Inversely, a negative value of Q_{21} induces a premature activation of mechanism 2 caused by mechanism 1. This feature can be helpful, for instance, to activate the third mechanism only after the softening stage.

2.6. Kinematic hardening rules

Unlike the expression used for KHR in the unified models (Table 1.3), the evolution rules of the back stresses introduce a coupling of the kinematic internal variables. In the following, they are discussed separately for (i) models with various mechanisms and various criteria and then for (ii) models with various mechanisms and one criterion.

2.6.1. *KHR: models with various mechanisms and various criteria*

For the model with two mechanisms and two criteria, the KHR are expressed by

$$\dot{\underset{\sim}{\alpha}}^1 = \dot{p}^1 \left(\underset{\sim}{\mathbf{n}}^1 - \frac{3D_1}{2C_{11}} \left((1-\delta_1)\underset{\sim}{\mathbf{X}}^1 + (2/3)\delta_1 \left(\underset{\sim}{\mathbf{X}}^1 : \underset{\sim}{\mathbf{n}}^1 \right) \underset{\sim}{\mathbf{n}}^1 \right) \right) \quad [2.30]$$

$$\dot{\underset{\sim}{\alpha}}^2 = \dot{p}^2 \left(\underset{\sim}{\mathbf{n}}^2 - \frac{3D_2}{2C_{22}} \left((1-\delta_2)\underset{\sim}{\mathbf{X}}^2 + (2/3)\delta_2 \left(\underset{\sim}{\mathbf{X}}^2 : \underset{\sim}{\mathbf{n}}^2 \right) \underset{\sim}{\mathbf{n}}^2 \right) \right) \quad [2.31]$$

The previous two equations lead to the following scalar expression of the KHR when tension–torsion loading is considered:

$$\dot{\alpha}_{11}^1 = \dot{p}^1 \left(n_{11}^1 - \frac{D_1}{C_{11}} \left((1-\delta_1)X_{11}^1 + \delta_1 \left(X_{11}^1 n_{11}^1 + (4/3)X_{12}^1 n_{12}^1 \right) n_{11}^1 \right) \right) \quad [2.32]$$

$$\dot{\alpha}_{12}^1 = \dot{p}^1 \left(n_{12}^1 - \frac{D_1}{C_{11}} \left((1-\delta_1)X_{12}^1 + \delta_1 \left(X_{12}^1 n_{12}^1 + (4/3)X_{12}^1 n_{12}^1 \right) n_{12}^1 \right) \right) \quad [2.33]$$

$$\dot{\alpha}_{11}^2 = \dot{p}^2 \left(n_{11}^2 - \frac{D_2}{C_{22}} \left((1-\delta_2)X_{11}^2 + \delta_2 \left(X_{11}^2 n_{11}^2 + (4/3)X_{12}^2 n_{12}^2 \right) n_{11}^2 \right) \right) \quad [2.34]$$

$$\dot{\alpha}_{12}^2 = \dot{p}^2 \left(n_{12}^2 - \frac{D_2}{C_{22}} \left((1-\delta_2)X_{12}^2 + \delta_2 \left(X_{11}^2 n_{11}^2 + (4/3)X_{12}^2 n_{12}^2 \right) n_{12}^2 \right) \right) \quad [2.35]$$

In the case of uniaxial loading (e.g. pure tensile test), $n_{11}^i n_{11}^i = 1$. Therefore, the previous KHR degenerates into the classical KHR, even the parameters δ_1 and δ_2 are not null (these parameters do not affect one-dimensional loading):

$$\dot{\alpha}^1_{11} = \dot{p}^1 \left(n^1_{11} - (D_1/C_{11})X^1_{11}\right) \quad \dot{\alpha}^2_{11} = \dot{p}^2 \left(n^2_{11} - (D_2/C_{22})X^2_{11}\right) \qquad [2.36]$$

In the same way, for a pure shear test, the KHR reads

$$\dot{\alpha}^1_{12} = \dot{p}^1 \left(n^1_{12} - (D_1/C_{11})X^1_{12}\right) \quad \dot{\alpha}^2_{12} = \dot{p}^2 \left(n^2_{12} - (D_2/C_{22})X^2_{12}\right) \qquad [2.37]$$

2.6.2. *KHR: models with various mechanisms and one criterion*

In the case of two mechanisms and one criterion, the product of each normal tensor by itself is no longer equal to 3/2. Instead of that, $\underset{\sim}{\mathbf{n}}^1 : \underset{\sim}{\mathbf{n}}^1 + \underset{\sim}{\mathbf{n}}^2 : \underset{\sim}{\mathbf{n}}^2 = 3/2$. Hence, the KHR described above are not able to ensure decoupling between the behavior under uniaxial loading and the behavior under biaxial loading. The parameters δ_1 and δ_2 are still present in the case of uniaxial loading. To obtain this decoupling effect, a new "normal" tensor is introduced in [TAL 10]: $\underset{\sim}{\mathbf{m}}^i = (\underset{\sim}{\sigma}^i - \underset{\sim}{\mathbf{X}}^i)/(\|\underset{\sim}{\sigma}^i - \underset{\sim}{\mathbf{X}}^i\|)$. The KHR are thus:

$$\dot{\underset{\sim}{\alpha}}^1 = \dot{p}\left(\underset{\sim}{\mathbf{n}}^1 - \frac{3D_1}{2C_{11}}\left((1-\delta_1)\underset{\sim}{\mathbf{X}}^1 + (2/3)\delta_1\left(\underset{\sim}{\mathbf{X}}^1 : \underset{\sim}{\mathbf{m}}^1\right)\underset{\sim}{\mathbf{m}}^1\right)\right) \qquad [2.38]$$

$$\dot{\underset{\sim}{\alpha}}^2 = \dot{p}\left(\underset{\sim}{\mathbf{n}}^2 - \frac{3D_2}{2C_{22}}\left((1-\delta_2)\underset{\sim}{\mathbf{X}}^2 + (2/3)\delta_2\left(\underset{\sim}{\mathbf{X}}^2 : \underset{\sim}{\mathbf{m}}^2\right)\underset{\sim}{\mathbf{m}}^2\right)\right) \qquad [2.39]$$

The previous KHR are key features in the description of ratcheting phenomenon, as shown later in Chapter 3.

2.7. Computation of the inelastic multipliers

In the MM models, the direction of the inelastic strain rate of each mechanism is given by the gradient of the yield function and its magnitude by the inelastic multiplier. In the following, the computation of these inelastic multipliers is established for the initial version of the 2M1C and 2M2C models summarized in Tables 3.2 and 3.1.

2.7.1. *Flow rate for the 2M1C model*

For the 2M1C model, there is only one type of flow:

– if the mechanism is viscoplastic, the viscoplastic strain rate is defined by the partial derivative of the potential Ω with respect to the yield function f. Assuming, for instance, a power law

$$\Omega = \frac{K}{n+1}\left\langle \frac{f}{K} \right\rangle^{n+1} \qquad [2.40]$$

We obtain

$$\dot{p} = \left\langle \frac{f}{K} \right\rangle^{n} \qquad [2.41]$$

Alternative expressions of the flow rules, as shown in Table 1.5, can also be used;

– if the mechanism is time independent, the consistency condition is written to obtain the inelastic (plastic) multiplier:

$$\frac{\partial f}{\partial \underset{\sim}{\sigma}^1} : \dot{\underset{\sim}{\sigma}}^1 + \frac{\partial f}{\partial \underset{\sim}{\sigma}^2} : \dot{\underset{\sim}{\sigma}}^2 + \frac{\partial f}{\partial \underset{\sim}{\mathbf{X}}^1} : \dot{\underset{\sim}{\mathbf{X}}}^1 + \frac{\partial f}{\partial \underset{\sim}{\mathbf{X}}^2} : \dot{\underset{\sim}{\mathbf{X}}}^2 + \frac{\partial f}{\partial R} : \dot{R} = 0 \qquad [2.42]$$

$$\dot{p} = \frac{\left(\underset{\sim}{\mathbf{n}}^1 : \dot{\underset{\sim}{\sigma}}^1 + \underset{\sim}{\mathbf{n}}^2 : \dot{\underset{\sim}{\sigma}}^2\right)}{H_R + H_{X1} + H_{X2}} \qquad [2.43]$$

$$H_R = \left(1 - \frac{R}{Q}\right) Q \qquad [2.44]$$

$$H_{X1} = (2/3)\left(\underset{\sim}{\mathbf{n}}^1 C_{11} + \underset{\sim}{\mathbf{n}}^2 C_{12}\right)\left(\underset{\sim}{\mathbf{n}}^1 - \frac{3D_1}{2C_{11}}\underset{\sim}{\mathbf{X}}^1\right) \qquad [2.45]$$

$$H_{X2} = (2/3)\left(\underset{\sim}{\mathbf{n}}^1 C_{12} + \underset{\sim}{\mathbf{n}}^2 C_{22}\right)\left(\underset{\sim}{\mathbf{n}}^2 - \frac{3D_2}{2C_{22}}\underset{\sim}{\mathbf{X}}^2\right) \qquad [2.46]$$

2.7.2. *Flow rates for the 2M2C model*

For the model 2M2C, inelastic flow can be computed by distinguishing three cases.

– The two mechanisms are time dependent;

Assuming a power law for the expression of the potentials:

$$\Omega^I = \frac{K_I}{n_I + 1}\left\langle \frac{f^I}{K_I} \right\rangle^{n_I} \qquad [2.47]$$

the two viscoplastic flow rates are simply:

$$\dot{p}^1 = \left\langle \frac{f^1}{K_1} \right\rangle^{n_1} \quad \text{if} \quad f^1 > 0; \qquad \dot{p}^2 = \left\langle \frac{f^2}{K_2} \right\rangle^{n_2} \quad \text{if} \quad f^2 > 0 \qquad [2.48]$$

– The two mechanisms are time independent. In this case, two consistency conditions can be written:

$$\frac{\partial f^1}{\partial \underset{\sim}{\sigma}^1} : \dot{\underset{\sim}{\sigma}}^1 + \frac{\partial f^1}{\partial \underset{\sim}{\mathbf{X}}^1} : \dot{\underset{\sim}{\mathbf{X}}}^1 + \frac{\partial f^1}{\partial R^1} : \dot{R}^1 = 0 \qquad [2.49]$$

$$\frac{\partial f^2}{\partial \underset{\sim}{\sigma}^2} : \dot{\underset{\sim}{\sigma}}^2 + \frac{\partial f^2}{\partial \underset{\sim}{\mathbf{X}}^2} : \dot{\underset{\sim}{\mathbf{X}}}^2 + \frac{\partial f^2}{\partial R^2} : \dot{R}^2 = 0 \qquad [2.50]$$

The previous equations lead to the following linear system:

$$\begin{bmatrix} M_{11} & M_{12} \\ M_{21} & M_{22} \end{bmatrix} \begin{bmatrix} \dot{p}^1 \\ \dot{p}^2 \end{bmatrix} = \begin{bmatrix} Q_1 \\ Q_2 \end{bmatrix} \qquad [2.51]$$

where the known coefficients M_{11}, M_{12}, M_{21}, M_{22} depend on the values of the state variables. The other coefficients, Q_1 and Q_2, are obtained from the loading increment. The plastic multipliers are

$$\dot{p}^1 = \left\langle \frac{Q_1 M_{22} - Q_2 M_{12}}{M_{11} M_{22} - M_{21} M_{12}} \right\rangle \qquad \dot{p}^2 = \left\langle \frac{Q_2 M_{11} - Q_1 M_{21}}{M_{11} M_{22} - M_{21} M_{12}} \right\rangle \qquad [2.52]$$

where

$$Q_1 = \underset{\sim}{\mathbf{n}}^1 : \dot{\underset{\sim}{\sigma}}^1 \quad Q_2 = \underset{\sim}{\mathbf{n}}^2 : \dot{\underset{\sim}{\sigma}}^2 \qquad [2.53]$$

$$M_{11} = \frac{2}{3} C_{11} \underset{\sim}{\mathbf{n}}^1 \left(\underset{\sim}{\mathbf{n}}^1 - \frac{3D_1}{2C_{11}} \underset{\sim}{\mathbf{X}}^1 \right) + Q_1 \left(1 - \frac{R^1}{Q_{11}} \right) \qquad [2.54]$$

$$M_{12} = \frac{2}{3} C_{12} \underset{\sim}{\mathbf{n}}^1 \left(\underset{\sim}{\mathbf{n}}^2 - \frac{3D_2}{2C_{22}} \underset{\sim}{\mathbf{X}}^2 \right) \qquad [2.55]$$

$$M_{21} = \frac{2}{3} C_{12} \underset{\sim}{\mathbf{n}}^1 \left(\underset{\sim}{\mathbf{n}}^1 - \frac{3D_1}{2C_{11}} \underset{\sim}{\mathbf{X}}^1 \right) \qquad [2.56]$$

$$M_{22} = \frac{2}{3}C_{22}\underset{\sim}{\mathbf{n}}^2\left(\underset{\sim}{\mathbf{n}}^2 - \frac{3D_2}{2C_{22}}\underset{\sim}{\mathbf{X}}^2\right) + Q_{22}\left(1 - \frac{R^2}{Q_{22}}\right) \quad [2.57]$$

The preceding equations are not valid if the determinant $(M_{11}M_{22} - M_{21}M_{12})$ is null. This specific condition is studied below, for the simple case of one-dimensional loading and linear kinematic hardening. Using $f^1 = 0$ and $f^2 = 0$ leads to

$$\begin{bmatrix} C_{11} & C_{12} \\ C_{12} & C_{22} \end{bmatrix}\begin{bmatrix} \alpha^1 \\ \alpha^2 \end{bmatrix} = \begin{bmatrix} A_1\sigma - n^1R^1 \\ A_2\sigma - n^2R^2 \end{bmatrix} \quad [2.58]$$

To avoid the indetermination linked to $(C_{11}C_{22} - C_{12}C_{12}=0)$, the two secondary determinants should also be assigned to zero. It leads to a constant value of the stress:

$$\sigma^* = \frac{C_{12}n^1R^1 - C_{11}n^2R^2}{A_1C_{12} - A_2C_{11}} = \frac{C_{22}n^1R^1 - C_{12}n^2R^2}{A_1C_{22} - A_2C12} \quad [2.59]$$

The model degenerates then to give a perfect plastic behavior as soon as the second mechanism becomes active. The same result can also be obtained from the system [2.51]. For the sake of clarity, the model is restricted to the case of linear kinematic hardening, without isotropic hardening. In these conditions, system [2.51] writes for stress controlled loading:

$$\begin{bmatrix} C_{11} & n^1C_{12}n^2 \\ n^1C_{12}n^2 & C_{22} \end{bmatrix}\begin{bmatrix} \dot{p}^1 \\ \dot{p}^2 \end{bmatrix} = \begin{bmatrix} n^1A_1\dot{\sigma} \\ n^2A_2\dot{\sigma} \end{bmatrix} \quad [2.60]$$

whereas for strain controlled loading case:

$$\begin{bmatrix} 3\mu A_1^2 + C_{11} & 3\mu A_1A_2n^1n^2 + n^1C_{12}n^2 \\ 3\mu A_1A_2n^1n^2 + n^1C_{12}n^2 & 3\mu A_2^2 + C_{22} \end{bmatrix}\begin{bmatrix} \dot{p}^1 \\ \dot{p}^2 \end{bmatrix} = \begin{bmatrix} 2\mu n^1A_1\dot{\varepsilon} \\ 2\mu n^2A_2\dot{\varepsilon} \end{bmatrix} \quad [2.61]$$

where μ is the elastic shear modulus. The determinant of the system [2.60] vanishes when $C_{11}C_{22} - C_{12}C_{12}=0$ and leads to a perfect plastic behavior. However, if the determinant of the system [2.61] is equal to zero, the plastic multiplier can be obtained as

$$\dot{p}^1 = K\dot{\varepsilon} \quad \text{and} \quad \dot{p}^2 = K\frac{C_{11}}{C_{12}}\dot{\varepsilon} \quad [2.62]$$

with

$$K = \frac{2(A_1C_{22} - A_2C_{12})}{3\mu(A_1^2C_{22} + A_2^2C_{11} - 2A_1A_2C_{12})}n^1 \qquad [2.63]$$

– The first mechanism is viscoplastic, whereas the second one is plastic. Assuming a power law

$$\Omega^1 = \frac{K_1}{n_1 + 1}\left\langle\frac{f^1}{K_1}\right\rangle^{n_1} \qquad [2.64]$$

the viscoplastic flow rate for the first mechanism is

$$\dot{p}^1 = \left(\frac{f^1}{K_1}\right)^{n_1} \quad \text{if} \quad f^1 > 0; \qquad [2.65]$$

The plastic multiplier for the second mechanism is given by the local consistency condition

$$\frac{\partial f^2}{\partial \underset{\sim}{\sigma}^2} : \dot{\underset{\sim}{\sigma}}^2 + \frac{\partial f^2}{\partial \underset{\sim}{\mathbf{X}}^2} : \dot{\underset{\sim}{\mathbf{X}}}^2 + \frac{\partial f^2}{\partial R^2} : \dot{R}^2 = 0 \qquad [2.66]$$

This leads to

$$\dot{p}^2 = \left\langle\frac{Q_2 - M_{21}\dot{p}^1}{M_{22}}\right\rangle \qquad [2.67]$$

The plastic mechanism can be stopped by viscoplastic flow. The reason is that the model offers the material two opportunities to accommodate a certain amount of inelastic strain. For a fast global strain rate, there is "not enough time" for the development of viscoplasticity: $\dot{p}^1$ is small, so that the inelastic strain rate is almost equal to $\dot{p}^2$ (time independent). For a low global strain rate, the viscoplastic part may become predominant, and eventually stop time-independent plastic flow if $M_{21}\dot{p}^1 > Q_2$.

3

Typical MM Responses

The thermodynamic framework described in section 2.1 led to the 2M2C and 2M1C models detailed in sections 2.2 and 2.3, respectively. The typical stress/strain responses discussed in this chapter deal with the first version of the MM models. Indeed, the main effects including creep–plasticity interaction, rate sensitivity and ratcheting, which are governed by the kinematic hardening matrix, will be presented in sections 3.2, 3.3 and 3.6, respectively.

In the first versions of the MM models, extra-hardening due to non-proportional loading was represented by means of a specific isotropic variable. An alternative solution would consist of extending the yield criteria (equation [2.23]) proposed by Taleb *et al.* [TAL 06] who have modified the flow function using a new parameter N. The effect of this parameter on the extra-hardening under non-proportional loading is investigated in section 3.8. The static recovery effect is investigated in section 3.9.

3.1. Some MM model variants

3.1.1. *Initial MM models*

Tables 3.1 and 3.2 summarize the initial 2M2C and 2M1C versions according to [CAI 95]. The models are developed in a thermodynamically consistent framework. The inelastic strain is split into two contributions between the two mechanisms through the material parameters A_1 and A_2. From a thermodynamic point of view, this "homogenization" process leads *de facto* to a form of local stresses expressed by means of A_1 and A_2. It is worth noting that A_1 and A_2 can be seen as weighting parameters so that $A_1+A_2=1$. Alternatively, $A_1=A_2=1$ leads to a "static" model type that is also consistent from a thermodynamic point of view.

$$\dot{\underset{\sim}{\varepsilon}} = \dot{\underset{\sim}{\varepsilon}}^e + A_1\dot{\underset{\sim}{\varepsilon}}^1 + A_2\dot{\underset{\sim}{\varepsilon}}^2$$

$$f^1 = J(\underset{\sim}{\sigma}^1 - \underset{\sim}{\mathbf{X}}^1) - R^1 - R_{01} \quad f^2 = J(\underset{\sim}{\sigma}^2 - \underset{\sim}{\mathbf{X}}^2) - R^2 - R_{02}$$

$$\underset{\sim}{\sigma}^1 = A_1\underset{\sim}{\sigma} \quad \underset{\sim}{\sigma}^2 = A_2\underset{\sim}{\sigma}$$

$$\begin{pmatrix}\underset{\sim}{\mathbf{X}}^1\\ \underset{\sim}{\mathbf{X}}^2\end{pmatrix} = (2/3)\begin{pmatrix}C_{11} & C_{12}\\ C_{12} & C_{22}\end{pmatrix}\begin{pmatrix}\underset{\sim}{\alpha}^1\\ \underset{\sim}{\alpha}^2\end{pmatrix}$$

$$\begin{pmatrix}R^1\\ R^2\end{pmatrix} = \begin{pmatrix}Q_{11} & Q_{12}\\ Q_{12} & Q_{22}\end{pmatrix}\begin{pmatrix}b_1 r^1\\ b_2 r^2\end{pmatrix}$$

$$\dot{\underset{\sim}{\varepsilon}}^1 = \dot{\lambda}^1\underset{\sim}{\mathbf{n}}^1 \quad \dot{\underset{\sim}{\varepsilon}}^2 = \dot{\lambda}^2\underset{\sim}{\mathbf{n}}^2$$

$$\dot{\underset{\sim}{\alpha}}^1 = \dot{\lambda}^1\left(\underset{\sim}{\mathbf{n}}^1 - (3D_1/2C_{11})\underset{\sim}{\mathbf{X}}^1\right) \quad \dot{\underset{\sim}{\alpha}}^2 = \dot{\lambda}^2\left(\underset{\sim}{\mathbf{n}}^2 - (3D_2/2C_{22})\underset{\sim}{\mathbf{X}}^2\right)$$

$$\dot{r}^1 = \dot{\lambda}^1(1 - (R_1/Q_{11})) \quad \dot{r}^2 = \dot{\lambda}^2(1 - (R_2/Q_{22}))$$

Table 3.1. *Constitutive equations of the initial 2M2C model according to [CAI 95]*

$$\dot{\underset{\sim}{\varepsilon}} = \dot{\underset{\sim}{\varepsilon}}^e + A_1\dot{\underset{\sim}{\varepsilon}}^1 + A_2\dot{\underset{\sim}{\varepsilon}}^2$$

$$f = \left(J(\underset{\sim}{\sigma}^1 - \underset{\sim}{\mathbf{X}}^1)^2 + J(\underset{\sim}{\sigma}^2 - \underset{\sim}{\mathbf{X}}^2)^2\right)^{1/2} - R - R_0$$

$$\underset{\sim}{\sigma}^1 = A_1\underset{\sim}{\sigma} \quad \underset{\sim}{\sigma}^2 = A_2\underset{\sim}{\sigma}$$

$$\begin{pmatrix}\underset{\sim}{\mathbf{X}}^1\\ \underset{\sim}{\mathbf{X}}^2\end{pmatrix} = (2/3)\begin{pmatrix}C_{11} & C_{12}\\ C_{12} & C_{22}\end{pmatrix}\begin{pmatrix}\underset{\sim}{\alpha}^1\\ \underset{\sim}{\alpha}^2\end{pmatrix}$$

$$R = bQr$$

$$\dot{\underset{\sim}{\varepsilon}}^1 = \dot{\lambda}\underset{\sim}{\mathbf{n}}^1 \quad \dot{\underset{\sim}{\varepsilon}}^2 = \dot{\lambda}\underset{\sim}{\mathbf{n}}^2$$

$$\dot{\underset{\sim}{\alpha}}^1 = \dot{\lambda}\left(\underset{\sim}{\mathbf{n}}^1 - (3D_1/2C_{11})\underset{\sim}{\mathbf{X}}^1\right) \quad \dot{\underset{\sim}{\alpha}}^2 = \dot{\lambda}\left(\underset{\sim}{\mathbf{n}}^2 - (3D_2/2C_{22})\underset{\sim}{\mathbf{X}}^2\right)$$

$$\dot{r} = \dot{\lambda}(1 - R/Q)$$

Table 3.2. *Constitutive equations of the initial 2M1C model according to [CAI 95]*

3.1.2. *Updated 2M1C models after [TAL 06]*

Two modifications of the 2M1C model have been proposed in order to improve the description of the ratcheting behavior for the one-dimensional case and for two-dimensional cases as well [TAL 06]: (i) the flow function is modified by the introduction of a material parameter N and (ii) a combination of classical fading memory and radial fading memory is introduced in the dynamic recovery term following [BUR 87] and [DEL 95]. Table 3.3 summarizes the equations.

$$\dot{\underset{\sim}{\varepsilon}} = \dot{\underset{\sim}{\varepsilon}}^e + A_1\dot{\underset{\sim}{\varepsilon}}^1 + A_2\dot{\underset{\sim}{\varepsilon}}^2$$

$$f = \left(J(\underset{\sim}{\sigma}^1 - \underset{\sim}{\mathbf{X}}^1)^N + J(\underset{\sim}{\sigma}^2 - \underset{\sim}{\mathbf{X}}^2)^N\right)^{1/N} - R - R_0$$

$$R = Q(1 - e^{-bv})$$

$$\underset{\sim}{\sigma}^1 = A_1\underset{\sim}{\sigma} \qquad \underset{\sim}{\sigma}^2 = A_2\underset{\sim}{\sigma}$$

$$\begin{pmatrix} \underset{\sim}{\mathbf{X}}^1 \\ \underset{\sim}{\mathbf{X}}^2 \end{pmatrix} = (2/3)\begin{pmatrix} C_{11} & C_{12} \\ C_{12} & C_{22} \end{pmatrix}\begin{pmatrix} \underset{\sim}{\alpha}^1 \\ \underset{\sim}{\alpha}^2 \end{pmatrix}$$

$$\dot{v} = (f/K)^n \quad \dot{\underset{\sim}{\varepsilon}}^1 = \dot{v}\underset{\sim}{\mathbf{n}}^1 \quad \dot{\underset{\sim}{\varepsilon}}^2 = \dot{v}\underset{\sim}{\mathbf{n}}^2$$

$$\dot{\underset{\sim}{\alpha}}^1 = \dot{v}\underset{\sim}{\mathbf{n}}^1 - D_1\left((1-\eta)\underset{\sim}{\alpha}^1 + \eta(\underset{\sim}{\alpha}^1 : \underset{\sim}{\mathbf{m}}^1)\underset{\sim}{\mathbf{m}}^1\right)\dot{v}$$

$$\dot{\underset{\sim}{\alpha}}^2 = \dot{v}\underset{\sim}{\mathbf{n}}^2 - D_2\left((1-\eta)\underset{\sim}{\alpha}^2 + \eta(\underset{\sim}{\alpha}^2 : \underset{\sim}{\mathbf{m}}^2)\underset{\sim}{\mathbf{m}}^2\right)\dot{v}$$

$$\underset{\sim}{\mathbf{m}}^i = (\underset{\sim}{\sigma}^i - \underset{\sim}{\mathbf{X}}^i)/(\|\underset{\sim}{\sigma}^i - \underset{\sim}{\mathbf{X}}^i\|)$$

Table 3.3. *Constitutive equations of the 2M1C model according to [TAL 06]*

3.1.3. *Updated MM models after [SAÏ 07]*

An alternative situation for evaluating the local stresses on each mechanism is to see them as a representative of phase behavior in heterogeneous materials. In such a case, a scale transition rule can be implemented, instead of using $\underset{\sim}{\sigma}^i = A_i\underset{\sim}{\sigma}$ [SAÏ 07]. The β-rule [CAI 94] is selected for that purpose. It makes it possible to control the level of the stress redistribution between the phases, starting from an elastic accommodation rule and then progressively switching toward a plastic accommodation, in order to lower the residual stresses originating from phase heterogeneities. The weighting parameters A_1 and A_2 are then assigned to $1 - z$ and z in [SAÏ 07], figuring phase volume fractions. The resulting models are also used to represent the semi-crystalline polymer behavior [REG 09b]. In this case, the contributions of the amorphous phase and the crystalline phase to the inelastic behavior of the semi-crystalline polymer are characterized by their crystallinity ratio z. Moreover, the kinematic hardening rules are modified for the fine-tuning of multi-axial ratcheting. Table 3.4 summarizes the corresponding MM versions.

3.1.4. A general nMnC model

The 2M2C model can be extended by using any number of mechanisms. The resulting nMnC can be written as

$$\underset{\sim}{\varepsilon}^p = \sum_{j=1}^{n} \underset{\sim}{\varepsilon}^j \qquad [3.1]$$

$$f^i = J\left(\underset{\sim}{\sigma}^i - \underset{\sim}{\mathbf{X}}^i\right) - R^i - R_{0i} \qquad [3.2]$$

$$\underset{\sim}{\mathbf{X}}^i = (2/3) \sum_{j=1}^{n} C_{ij} \underset{\sim}{\alpha}^j \qquad [3.3]$$

$$R^i = \sum_{j=1}^{n} Q_{ij} b_j r^j \qquad [3.4]$$

$$\dot{\underset{\sim}{\varepsilon}}^i = \dot{\lambda}^i \underset{\sim}{\mathbf{n}}^i \qquad [3.5]$$

$$\dot{\underset{\sim}{\alpha}}^i = \dot{\lambda}^i \left(\underset{\sim}{\mathbf{n}}^i - (3D_i/C_{ii})\underset{\sim}{\mathbf{X}}^i\right) \qquad [3.6]$$

$$\dot{r}^i = \dot{\lambda}^i \left(1 - R^i/Q_{ii}\right) \qquad [3.7]$$

The main drawback of this formulation is the large number of material parameters. For example, for n=10, 100 kinematic hardening moduli are needed in addition to the rest of the material parameters.

2M2C model
$\dot{\underset{\sim}{\varepsilon}} = \dot{\underset{\sim}{\varepsilon}}^e + (1-z)\dot{\underset{\sim}{\varepsilon}}^1 + z\dot{\underset{\sim}{\varepsilon}}^2$
$f^1 = J(\underset{\sim}{\sigma}^1 - \underset{\sim}{\mathbf{X}}^1) - R^1 - R_{01} \quad f^2 = J(\underset{\sim}{\sigma}^2 - \underset{\sim}{\mathbf{X}}^2) - R^2 - R_{02}$
$\underset{\sim}{\sigma}^1 = \underset{\sim}{\sigma} + \mu'\left(\underset{\sim}{\beta} - \underset{\sim}{\beta}^1\right) \quad \underset{\sim}{\sigma}^2 = \underset{\sim}{\sigma} + \mu'\left(\underset{\sim}{\beta} - \underset{\sim}{\beta}^2\right)$
$\underset{\sim}{\beta} = (1-z)\underset{\sim}{\beta}^1 + z\underset{\sim}{\beta}^2$
$\begin{pmatrix}\underset{\sim}{\mathbf{X}}^1 \\ \underset{\sim}{\mathbf{X}}^2\end{pmatrix} = (2/3)\begin{pmatrix}C_{11} & C_{12} \\ C_{12} & C_{22}\end{pmatrix}\begin{pmatrix}\underset{\sim}{\alpha}^1 \\ \underset{\sim}{\alpha}^2\end{pmatrix}$
$R^1 = b_1 Q_1 r^1 \quad R^2 = b_2 Q_2 r^2$
$\dot{\underset{\sim}{\varepsilon}}^1 = \dot{\lambda}^1 \underset{\sim}{\mathbf{n}}^1 \quad \dot{\underset{\sim}{\varepsilon}}^2 = \dot{\lambda}^2 \underset{\sim}{\mathbf{n}}^2$
$\dot{\underset{\sim}{\boldsymbol{\beta}}}^1 = \dot{\underset{\sim}{\boldsymbol{\varepsilon}}}^1 - d_1\underset{\sim}{\boldsymbol{\beta}}^1\Vert\dot{\varepsilon}^1\Vert \quad \dot{\underset{\sim}{\boldsymbol{\beta}}}^2 = \dot{\underset{\sim}{\boldsymbol{\varepsilon}}}^2 - d_2\underset{\sim}{\boldsymbol{\beta}}^2\Vert\dot{\varepsilon}^2\Vert$
$\dot{\underset{\sim}{\alpha}}^1 = \dot{\lambda}\left(\underset{\sim}{\mathbf{n}}^1 - \frac{3D_1}{2C_{11}}\left((1-\delta_1)\underset{\sim}{\mathbf{X}}^1 + \delta_1\left(\underset{\sim}{\mathbf{X}}^1 : \underset{\sim}{\mathbf{n}}^1\right)\underset{\sim}{\mathbf{n}}^1\right)\right)$
$\dot{\underset{\sim}{\alpha}}^2 = \dot{\lambda}\left(\underset{\sim}{\mathbf{n}}^2 - \frac{3D_2}{2C_{22}}\left((1-\delta_2)\underset{\sim}{\mathbf{X}}^2 + \delta_2\left(\underset{\sim}{\mathbf{X}}^2 : \underset{\sim}{\mathbf{n}}^2\right)\underset{\sim}{\mathbf{n}}^2\right)\right)$
$\dot{r}^1 = \dot{\lambda}^1\left(1 - R^1/Q_1\right) \quad \dot{r}^2 = \dot{\lambda}^2\left(1 - R^2/Q_2\right)$
2M1C model
$\dot{\underset{\sim}{\varepsilon}} = \dot{\underset{\sim}{\varepsilon}}^e + (1-z)\dot{\underset{\sim}{\varepsilon}}^1 + z\dot{\underset{\sim}{\varepsilon}}^2$
$f = \left(J(\underset{\sim}{\sigma}^1 - \underset{\sim}{\mathbf{X}}^1)^2 + J(\underset{\sim}{\sigma}^2 - \underset{\sim}{\mathbf{X}}^2)^2\right)^{1/2} - R - R_0$
$\underset{\sim}{\sigma}^1 = \underset{\sim}{\sigma} + \mu'\left(\underset{\sim}{\beta} - \underset{\sim}{\beta}^1\right) \quad \underset{\sim}{\sigma}^2 = \underset{\sim}{\sigma} + \mu'\left(\underset{\sim}{\beta} - \underset{\sim}{\beta}^2\right)$
$\underset{\sim}{\beta} = (1-z)\underset{\sim}{\beta}^1 + z\underset{\sim}{\beta}^2$
$\begin{pmatrix}\underset{\sim}{\mathbf{X}}^1 \\ \underset{\sim}{\mathbf{X}}^2\end{pmatrix} = (2/3)\begin{pmatrix}C_{11} & C_{12} \\ C_{12} & C_{22}\end{pmatrix}\begin{pmatrix}\underset{\sim}{\alpha}^1 \\ \underset{\sim}{\alpha}^2\end{pmatrix}$
$R = bQr$
$\dot{\underset{\sim}{\varepsilon}}^1 = \dot{\lambda}\underset{\sim}{\mathbf{n}}^1 \quad \dot{\underset{\sim}{\varepsilon}}^2 = \dot{\lambda}\underset{\sim}{\mathbf{n}}^2$
$\dot{\underset{\sim}{\boldsymbol{\beta}}}^1 = \dot{\underset{\sim}{\boldsymbol{\varepsilon}}}^1 - d_1\underset{\sim}{\boldsymbol{\beta}}^1\Vert\dot{\varepsilon}^1\Vert \quad \dot{\underset{\sim}{\boldsymbol{\beta}}}^2 = \dot{\underset{\sim}{\boldsymbol{\varepsilon}}}^2 - d_2\underset{\sim}{\boldsymbol{\beta}}^2\Vert\dot{\varepsilon}^2\Vert$
$\dot{\underset{\sim}{\alpha}}^1 = \dot{\lambda}\left(\underset{\sim}{\mathbf{n}}^1 - \frac{3D_1}{2C_{11}}\left((1-\delta_1)\underset{\sim}{\mathbf{X}}^1 + \delta_1\left(\underset{\sim}{\mathbf{X}}^1 : \underset{\sim}{\mathbf{n}}^1\right)\underset{\sim}{\mathbf{n}}^1\right)\right)$
$\dot{\underset{\sim}{\alpha}}^2 = \dot{\lambda}\left(\underset{\sim}{\mathbf{n}}^2 - \frac{3D_2}{2C_{22}}\left((1-\delta_2)\underset{\sim}{\mathbf{X}}^2 + \delta_2\left(\underset{\sim}{\mathbf{X}}^2 : \underset{\sim}{\mathbf{n}}^2\right)\underset{\sim}{\mathbf{n}}^2\right)\right)$
$\dot{r} = \dot{\lambda}(1 - R/Q)$

Table 3.4. *Constitutive equations of the modified 2M2C model according to [SAÏ 07]*

3.1.5. Generalization of the 2M1C model

More than two mechanisms can be involved in the construction of the final criterion, leading to the so-called nM1C model for n mechanisms. As an example, the main lines of a 3M1C model are defined below:

– the yield criterion used is based on a nonlinear combination of the three mechanisms:

$$f = \left[\left(J(\underset{\sim}{\sigma}^1 - \underset{\sim}{\mathbf{X}}^1) \right)^2 + \left(J(\underset{\sim}{\sigma}^2 - \underset{\sim}{\mathbf{X}}^2) \right)^2 + \left(J(\underset{\sim}{\sigma}^3 - \underset{\sim}{\mathbf{X}}^3) \right)^2 \right]^{1/2} - R - R_0 \quad [3.8]$$

– the modification of the size of the yield surface is obtained by:

$$R = bQr \quad [3.9]$$

– the back stresses of each mechanism are related to the kinematic internal variables by means of an interaction matrix implementing a kinematic–kinematic coupling:

$$\begin{pmatrix} \underset{\sim}{\mathbf{X}}^1 \\ \underset{\sim}{\mathbf{X}}^2 \\ \underset{\sim}{\mathbf{X}}^3 \end{pmatrix} = \frac{2}{3} \begin{pmatrix} C_{11} & C_{12} & C_{13} \\ C_{12} & C_{22} & C_{23} \\ C_{13} & C_{23} & C_{33} \end{pmatrix} \begin{pmatrix} \underset{\sim}{\alpha}^1 \\ \underset{\sim}{\alpha}^2 \\ \underset{\sim}{\alpha}^3 \end{pmatrix} \quad [3.10]$$

The local stresses are deduced from a selected transition rule. Following the classical 2M1C model, the evolution equations are

$$\dot{\underset{\sim}{\varepsilon}}^I = \dot{p}\underset{\sim}{\mathbf{n}}^{\mathbf{I}} \quad \text{with} \quad (I = 1, 3) \quad [3.11]$$

$$\dot{\underset{\sim}{\alpha}}^I = \dot{p}\left(\underset{\sim}{\mathbf{n}}^{\mathbf{I}} - \frac{3D_I}{2C_{II}} \underset{\sim}{\mathbf{X}}^I \right) \quad \text{with} \quad (I = 1, 3) \quad [3.12]$$

$$\dot{r} = \dot{p}\left(1 - \frac{R}{Q} \right) \quad [3.13]$$

3.2. Creep–plasticity interaction

The interaction between plasticity and creep for tests involving both monotonous loading and hold time under stress control is naturally taken into account in the unified viscoplastic approach. This is no longer true if the type of deformation during the two sections of the test is not the same.

Nevertheless, the material parameter C_{12} offers the opportunity to reintroduce this interaction. This has to be investigated, as there is experimental evidence of creep–plasticity interaction [OHA 83, OHA 86]. On the other hand, according to the chosen value, the interaction may produce various mechanical effects. Consider a tensile test A where the load is controlled as follows: (i) loading under strain control from t_0 to t_1, (ii) constant stress (creep state) from t_1 to t_2 and finally (iii) a new step at the same prescribed strain rate as the first one. The response can be compared to a simple reference test B under a prescribed strain rate.

If no coupling is considered, as in the classical creep–plasticity models, the load level in the third part of the test A is lower than the reference level in test B. However, if the parameter C_{12} is not zero, the following three different effects can be obtained (Figure 3.1):

- if $C_{12} < C_{22}$, the stress level in test A is lower than that in test B;
- if $C_{12} = C_{22}$, the stress in test A has the same level as in test B;
- if $C_{12} > C_{22}$, the stress level in test A is larger than that in test B.

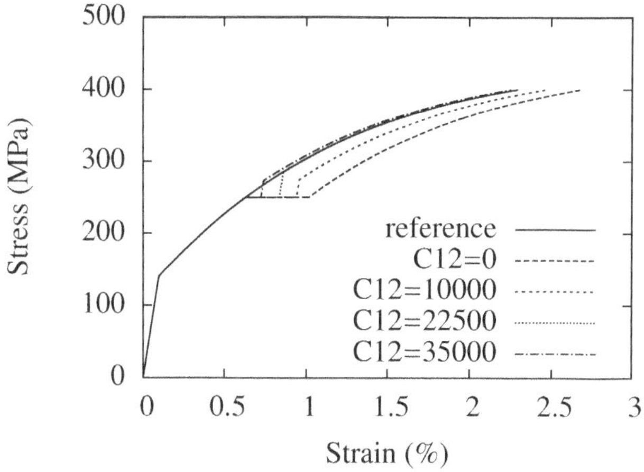

Figure 3.1. *Creep–plasticity interaction, 2M2C model: $A_1 = A_2 = 1$, $E = 144\ 000$, $n = 7.2$, $K = 1200$, $D_1 = D_2 = 100$, $R_{01} = 0$, $R_{02} = 140$ $C_{11} = 50\ 000$, $C_{22} = 30\ 000$ (MPa, s)*

3.3. Rate sensitivity for the 2M2C model

A material exhibits a "normal rate sensitivity" if the stress level increases with strain rate. In many materials (e.g. austenitic stainless steels), an inverse rate sensitivity is observed: in this case, increasing strain rate leads to a decrease in the stress level for a given strain. This inverse rate sensitivity is linked to jerky flow (Portevin–Le Chatelier instability). This was studied by many authors after the landmark paper [KUB 85], where it was proposed to introduce an additional variable representing aging time. The model has been used in finite-element calculations to represent jerky flow [MAZ 10]. Two deformation regimes are present: the first involves drag of impurities and produces large hardening rates, and in the second one, dislocations are able to escape from the array of impurities, so that the deformation process is much easier. In fact, the MM model family with two mechanisms and two criteria can reproduce these two regimes if a large value is affected to C_{11} (assuming that mechanism "1" is viscoplastic, whereas mechanism "2" follows a time-independent behavior). To illustrate these two possibilities, a first set of tests with a high strain rate ($\dot{\varepsilon}=10^{-3}$ s^{-1}) is compared to a second one with a low strain rate ($\dot{\varepsilon}=10^{-6}$ s^{-1}). Figure 3.2 shows that an inverse rate sensitivity is obtained with a high value of C_{11} compared to C_{12}. This is no longer true if C_{11} is smaller than C_{12}. In this last case, a normal rate sensitivity is obtained. For the inverse rate sensitivity, kinematic hardening is more pronounced at a lower strain rate because the ratio of viscoplastic strain rate to plastic strain rate becomes larger at a lower strain rate. On the other hand, as C_{12} is larger than C_{22}, the plastic threshold at low strain is fixed by the value of $C_{12}\alpha^1$ more than $C_{22}\alpha^2$. It is then higher for a low strain rate than for a high strain rate. It follows that the stress–strain curve driven by viscoplasticity at a low strain rate may become higher than the stress–strain curve given by the plastic model at a high strain rate. Note that, in the present case, the model does not deliver an unstable response, as with the Kubin–Estrin model, so that it cannot be claimed that it fully captures the effective physical effect. Nevertheless, it is able to reproduce the abnormal relative values of stress that is associated with the phenomenon.

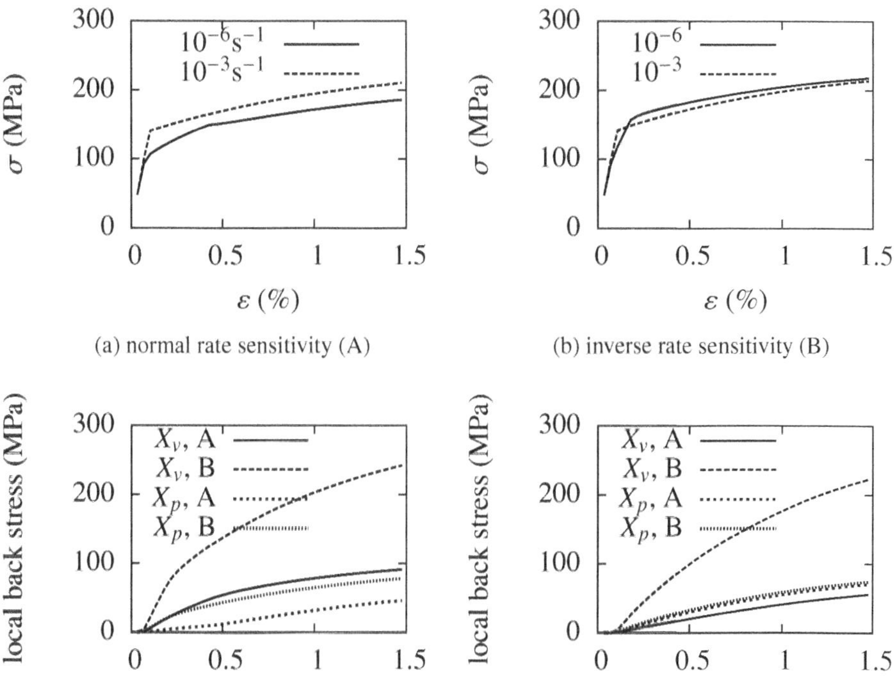

(a) normal rate sensitivity (A)

(b) inverse rate sensitivity (B)

(c) $10^{-6}s^{-1}$

(d) $10^{-3}s^{-1}$

Figure 3.2. *Study of rate sensitivity, 2M2C model:* $A_1 = A_2=1$, E*=20 0000,* n*=7.2,* K*=70,* D_2*=100,* R_{01}*=0,* R_{02}*=140,* C_{12}*=40 000,* C_{22}*=1000. Normal rate sensitivity:* C_{11}*=20000,* D_1*=200. Inverse rate sensitivity:* C_{11}*=100 000,* D_1*=400. a, b stress–strain curves. c, d back stresses (units: MPa, s)*

3.4. Stabilized behavior of viscoplastic 2M1C model

For the case of a one-dimensional loading in a time-dependent framework, the 2M1C model of Table 3.2 reduces to the following equations in the case of linear kinematic hardening:

$$
\begin{aligned}
\dot{p} &= \left\langle \frac{f}{K} \right\rangle^n \\
\sigma^1 &= A_1\sigma\sigma^2 = A_2\sigma \\
x^1 &= C_{11}\alpha^1 + C_{12}\alpha^2 \\
x^2 &= C_{12}\alpha^1 + C_{22}\alpha^2 \\
\dot{\alpha}^1 &= \dot{\varepsilon}^1\dot{\alpha}^2 = \dot{\varepsilon}^2
\end{aligned}
$$

For a strain-controlled test, the following expression is obtained once a mechanical steady state is reached [VEL 06]:

$$R_v = R_0 + K\dot{p}^{1/n} = \sqrt{(A_1\sigma - x^1)^2 + (A_2\sigma - x^2)^2} \qquad [3.14]$$

Rearranging the previous equation, we find

$$\left(A_1^2 + A_2^2\right)\sigma^2 - 2\sigma\left(A_1 x^1 + A_2 x^2\right) + (x^1)^2 + (x^2)^2 = R_v^2 \qquad [3.15]$$

The solution of this equation provides the value of the maximum stress for the strain-controlled test:

$$\sigma = \frac{\left(A_1 x^1 + A_2 x^2\right) \pm \sqrt{-R_v^2\left(A_1^2 + A_2^2\right) + (A_1 x^1 + A_2 x^2)^2}}{A_1^2 + A_2^2} \qquad [3.16]$$

3.5. Closed-form solution for ratcheting behavior of the 2M2C model: case of linear kinematic hardening rules

If linear kinematic rules are used and if $A_1 = A_2 = 1$, the 2M2C model reduces to the following equations:

$$\begin{aligned}
\varepsilon^{in} &= \varepsilon^1 + \varepsilon^2 \\
f^1 &= |\sigma - C_{11}\varepsilon^1 - C_{12}\varepsilon^2| - R_{01} \\
f^2 &= |\sigma - C_{12}\varepsilon^1 - C_{22}\varepsilon^2| - R_{02} \\
\dot{\varepsilon}^1 &= \dot{p}^1 n^1 \\
\dot{\varepsilon}^2 &= \dot{p}^2 n^2 \\
n^1 &= sign(\sigma - C_{11}\varepsilon^1 - C_{12}\varepsilon^2) \\
n^2 &= sign(\sigma - C_{12}\varepsilon^1 - C_{22}\varepsilon^2)
\end{aligned}$$

In a cyclic stress-controlled test (σ_{min},σ_{max}), assuming that the stress σ_{max} is large enough to activate the two mechanisms, the local inelastic strains ε^1 and ε^2 can be deduced from the conditions $f^1 = 0$ and $f^2 = 0$, so that

$$\begin{pmatrix} C_{11} & C_{12} \\ C_{12} & C_{22} \end{pmatrix}\begin{pmatrix} \varepsilon^1 \\ \varepsilon^2 \end{pmatrix} = \begin{pmatrix} \sigma_{max} - R_{01} \\ \sigma_{max} - R_{02} \end{pmatrix} \qquad [3.17]$$

The ratchet strain per cycle $\delta\varepsilon_{in}$ is the strain shift between corresponding points on successive cycles (i.e. A and G).

The expression of the inelastic strain at points A and G in Figure 3.3 can be obtained from equation [3.17] provided the determinant $C_{11}C_{22} - C_{12}^2$ is non-zero. It can then be deduced that $\delta\varepsilon^{in} = 0$ since the stress level is the same at points A and G.

However, if the determinant $C_{11}C_{22} - C_{12}^2$ is null, equation [3.17] does not provide the inelastic strains. In this case, the behavior is illustrated in Figure 3.3 according to the following path:

– from A to B, the behavior is purely elastic;

– from B to C, only the first mechanism is active;

– from C to D, mechanism 2 is active and the coupling effect extinguishes mechanism 1. In fact, the two consistency conditions $\dot{f}^1 = \dot{f}^2 = 0$ cannot be satisfied simultaneously. Using only the consistency condition $\dot{f}^2 = 0$ leads to $C_{12}\dot{p}_1 + C_{22}\dot{p}_2 = n^2\dot{\sigma}$, so that any positive value of $\dot{p}^2$ gives a negative solution of $\dot{p}^1$ due to the high value of C_{12} with respect to C_{22}. The only way to be consistent is to assign $\dot{p}^1$ to zero;

– from D to E, the behavior is purely elastic;

– from E to F, only mechanism 1 is active;

– from F to G, mechanism 2 is active and extinguishes mechanism 1 as previously explained.

Figure 3.3(b) shows the evolution of the inelastic local strain of the two mechanisms according to the stress–strain curve of Figure 3.3(a).

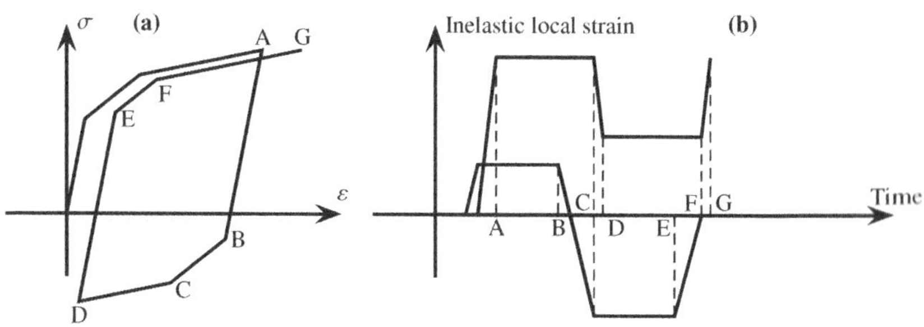

Figure 3.3. *Analytical study of the ratcheting behavior of the 2M2C model: (a) distinction of the different branches in the stress-strain loop; (b) activation of the mechanisms according to the different branches*

In the one-dimensional case, the yield surfaces are, respectively,

$$C_{11}\varepsilon^1 + C_{12}\varepsilon^2 = \sigma - n^1 R_{01}, \quad C_{12}\varepsilon^1 + C_{22}\varepsilon^2 = \sigma - n^2 R_{02} \qquad [3.18]$$

During the loading phase ($n^1 = n^2 = 1$), σ reaches σ_{max}. Using $C_{11}C_{22} - C_{12}^2 = 0$ in equation [3.18], the following expressions are obtained:

$$C_{11}\varepsilon^1 + C_{12}\varepsilon^2 = \sigma_{max} - R_{01} = \frac{C_{11}}{C_{12}}(\sigma_{max} - R_{02}) \qquad [3.19]$$

$$C_{11}\dot{\varepsilon}^1 + C_{12}\dot{\varepsilon}^2 = \dot{\sigma}, \quad C_{12}\dot{\varepsilon}^1 + C_{22}\dot{\varepsilon}^2 = \dot{\sigma} \qquad [3.20]$$

Considering the above behaviors in different branches and using the property $C_{11}C_{22} - C_{12}{}^2 = 0$, it comes

$$\varepsilon_F^1 = \varepsilon_B^1 + \frac{\sigma_F + \sigma_C - \sigma_E - \sigma_B}{C_{11}}, \quad \varepsilon_G^2 = \varepsilon_B^2 + \frac{\sigma_{max} + \sigma_{min} - \sigma_C - \sigma_F}{C_{22}} \qquad [3.21]$$

with

$$\sigma_F = \frac{C_{11}R_{02} - C_{12}R_{01}}{C_{11} - C_{12}} \quad \text{and} \quad \sigma_C = \frac{C_{12}R_{01} - C_{11}R_{02}}{C_{11} - C_{12}} \qquad [3.22]$$

The stress state at B and E (end of branches with elastic behavior) is such that

$$\sigma_B = -R_{01} + \frac{C_{12}}{C_{22}}(\sigma_{max} - R_{02}) \qquad [3.23]$$

$$\sigma_E = R_{01} + \frac{C_{12}}{C_{22}}(\sigma_{min} + R_{02}) \qquad [3.24]$$

Considering the expressions of these stresses, the strain increase between two successive cycles at the tensile peak can be deduced as

$$\delta\varepsilon^{in} = \left(\frac{1}{C_{22}} - \frac{1}{C_{12}}\right)(\sigma_{max} + \sigma_{min}) \quad \text{with} \quad C_{11}C_{22} - C_{12}^2 = 0 \qquad [3.25]$$

The variation of the tensile peak strain between two successive cycles can be obtained analytically for the kinematic hardening rule:

$$\dot{\alpha}^I = \left(\underset{\sim}{\mathbf{n}}^I - \frac{3D_I}{2C_{II}}\underset{\sim}{\mathbf{X}}^{\mathbf{I}}\right)\frac{\partial\Omega^I}{\partial f^I} \qquad \dot{r}^I = \left(1 - \frac{R^I}{Q_I}\right)\frac{\partial\Omega^I}{\partial f^I} \qquad [3.26]$$

However, an analytic solution can be obtained if the dynamic recovery term $D_I \underset{\sim}{\alpha}^I$ is used instead of the term $(3/2)(D_I/C_{II})\underset{\sim}{\mathbf{X}}^I$:

$$\dot{\underset{\sim}{\alpha}}^I = \dot{p}\left(\underset{\sim}{\mathbf{n}}^I - D_I \underset{\sim}{\alpha}^I\right) \quad (I = 1, 2) \qquad [3.27]$$

The branches of Figure 3.3(a) are now nonlinear. The following behaviors are considered:

– from A to B, the behavior is purely elastic;

– from B to C, only the first mechanism is active;

– from C to D, the two mechanisms are active;

– from D to E, the behavior is purely elastic;

– from E to F, only the first mechanism is active;

– from F to G, the two mechanisms are active.

The integration of the state variables along the different branches gives the expression of the variation of the tensile peak strain:

$$\delta\varepsilon^{in} = \ln\left[\left(\frac{1 - D_1^2(\alpha_D^1)^2}{1 - D_1^2(\alpha_A^1)^2}\right)^{1/D_1} \left(\frac{1 - D_2^2(\alpha_D^2)^2}{1 - D_2^2(\alpha_A^2)^2}\right)^{1/D_2}\right] \qquad [3.28]$$

where α_A^1, α_D^1, α_A^2 and α_D^2 are the intermediate values of the kinematic hardening variables:

– $\alpha_A^1 = [(C_{22} - C_{12})\sigma_{max} + R_{02}C_{12} - R_{01}C_{22}]/(C_{22}C_{11} - C_{12}^2)$

– $\alpha_A^2 = [(C_{11} - C_{12})\sigma_{max} + R_{01}C_{12} - R_{02}C_{11}]/(C_{22}C_{11} - C_{12}^2)$

– $\alpha_D^1 = [(C_{22} - C_{12})\sigma_{min} - R_{02}C_{12} + R_{01}C_{22}]/(C_{22}C_{11} - C_{12}^2)$

– $\alpha_D^2 = [(C_{11} - C_{12})\sigma_{min} - R_{01}C_{12} + R_{02}C_{11}]/(C_{22}C_{11} - C_{12}^2)$

The strain variation at peak stress between two successive cycles (equation [3.28]) is constant, as in the classical unified model proposed by Chaboche [CHA 86].

3.6. Ratcheting for 2M1C model

Even if the yield criteria are basically different, the two classes of MM models have common characteristics with respect to ratcheting behavior. Ratcheting behavior of the 2M1C model is closely related to the hardening matrix (Figure 3.4). It was shown analytically for the case of a one-dimensional loading that:

– when linear kinematic hardening rules are considered, the ratcheting behavior is controlled by the character of the hardening matrix. If the determinant of this matrix is equal to zero, then ratcheting behavior is observed. Conversely, a regular matrix leads to shakedown;

– in the case of nonlinear kinematic hardening rules, a constant evolution of the tensile peak strain is obtained in the case of one-dimensional unbalanced loading in the plastic range. The variation of the tensile peak strain between two successive cycles is expressed analytically according to the components of the matrix $[C]$, the fading memory parameters and the applied cyclic stresses $(\sigma_{min}, \sigma_{max})$.

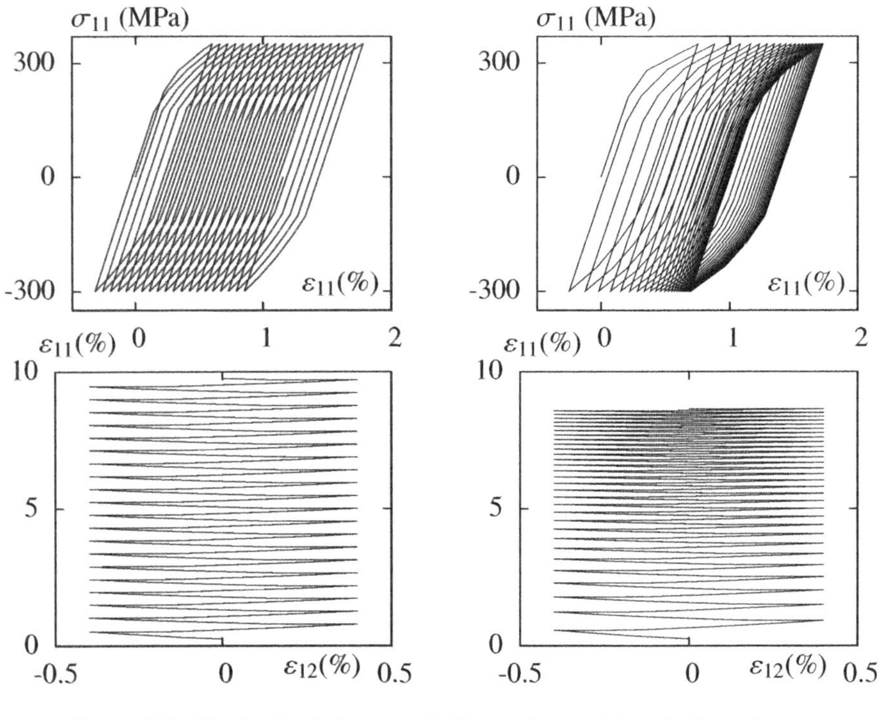

Figure 3.4. *Ratcheting behavior with linear kinematic hardening rules (2M1C model). Left: 1D and 2D ratcheting allowed by a singular matrix. Right: 1D and 2D ratcheting stopped by a regular matrix*

In the so-called β-MM models, the effective stresses for each mechanism can be rewritten as $\underset{\sim}{\sigma} - \underset{\sim}{\mathbf{Y}}^I$, with

$$\begin{pmatrix} \underset{\sim}{\mathbf{Y}}^1 \\ \underset{\sim}{\mathbf{Y}}^2 \end{pmatrix} = \frac{2}{3}\begin{pmatrix} C_{11} & C_{12} \\ C_{12} & C_{22} \end{pmatrix}\begin{pmatrix} \underset{\sim}{\alpha}^1 \\ \underset{\sim}{\alpha}^2 \end{pmatrix} + \mu' \begin{pmatrix} z & -z \\ -(1-z) & (1-z) \end{pmatrix}\begin{pmatrix} \underset{\sim}{\beta}^1 \\ \underset{\sim}{\beta}^2 \end{pmatrix}$$

This illustrates the difference between the intermechanism and intramechanism corrective terms for computing internal stresses. When both $\underset{\sim}{\beta}^I$ and $\underset{\sim}{\mathbf{X}}^I$ have linear evolution rules, it can be easily shown that the modified models are reduced to the initial form with the following determinant of the hardening matrix:

$$\Delta = (4/9)(C_{11}C_{22} - C_{12}^2) + (2/3)\mu' \left[(1-z)C_{11} + zC_{22} + C_{12}\right]$$

As a result, ratcheting behavior depends on both the initial determinant $C_{11}C_{22} - C_{12}^2$ and the localization parameter μ'. A systematic study is then proposed to illustrate the uniaxial and multi-axial ratcheting behaviors with respect to (i) the evolution rule of the accommodation variables $\underset{\sim}{\beta}^I$, (ii) the evolution rule of the back stresses $\underset{\sim}{\mathbf{X}}^I$ and (iii) the characteristics of the hardening matrix [**C**]. In order to easily identify the various tests, their names are defined by five letters. The first one characterizes the regular (R)/ singular (S) matrix. The next two capital letters are devoted to the description of the kinematic hardening evolution rules (L for linear and N for nonlinear hardening rules). The last two small letters are reserved for describing the evolution rules of the accommodation variables (l for linear and n for nonlinear). For instance:

– the model with a singular matrix, linear evolution rules for the kinematic variables and linear evolution rules for the accommodation variables will be referred to as model SLLll;

– the model with a regular matrix, linear evolution rule for the first kinematic variable, nonlinear evolution rule for the second kinematic variable and nonlinear evolution rules for the accommodation variables will be referred to as model RLNnn.

The list of the parameters used for each model is given in Table 3.5, and the corresponding predicted uniaxial ratcheting response to applied stresses is shown in Figure 3.5.

Model parameters								
	SLLll	RLLll	RLLnn	RLLln	SLLnn	RNNll	RNLll	RNNnn
C_{11}	30	30	50	50	50	50	50	100
C_{22}	3	3	5	5	5	5	5	10
C_{12}	9.486	9.486	10	10	15.81	10	10	20
μ'	30	0	40	40	40	40	40	80
D_1	0	0	0	0	0	100	300	100
D_2	0	0	0	0	0	10	0	10
d_1	0	0	20	0	20	0	0	20
d_2	0	0	200	200	200	0	0	200

(z=0.79, R_0=250 MPa, Q=50 MPa, b=30)

Table 3.5. *2M1C_β model: study of ratcheting behavior with respect to the different material parameters (GPa, s)*

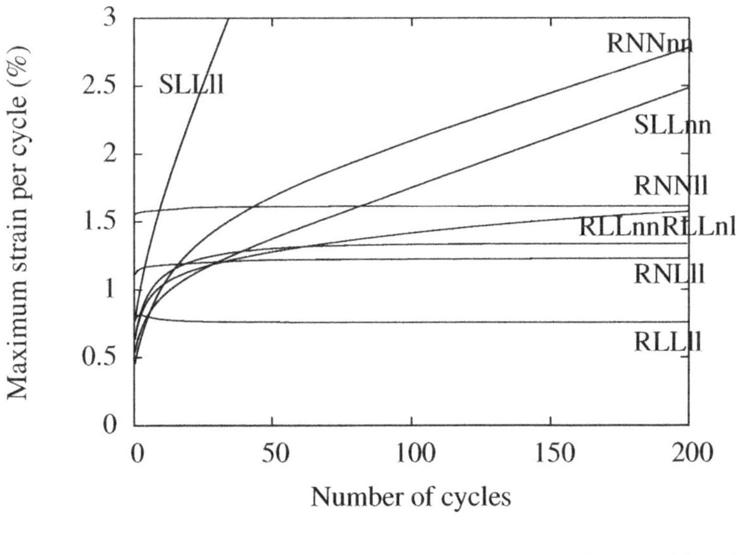

Figure 3.5. *Simulation of the 1D ratcheting test using the 2M1C_β model under one-dimensional loading at a prescribed axial stress (–150 MPa, +300 MPa): Systematic study of the effect of (i) the hardening matrix and (ii) the evolution rules of the kinematic hardening variables and the accommodation variables. (a) 200 cycles and (b) 2000 cycles*

The main results can be summarized as follows:

– ratcheting behavior with a constant increase in a tensile peak strain is obtained in the following cases:

- for a singular matrix ($\Delta = 0$) and linear evolution rules for both kinematic hardening variables regardless of the evolution rules of the accommodation variables (models SLLll and SLLnn),

- for a regular matrix ($\Delta \neq 0$) and nonlinear evolution rules for the two kinematic hardening variables and the two accommodation variables (model RNNnn).

– ratcheting behavior also occurs but at a lower rate with a regular matrix, two linear kinematic hardening variables and two nonlinear accommodation variables (RLLnn);

– a shakedown behavior is obtained with a regular matrix if one (at least) of the kinematic hardening variables is linear and one (at least) of the accommodation variables is linear (models RLLll, RNLll and RLLnl);

– ratcheting can also be stopped with two nonlinear kinematic variables if a regular matrix and two linear accommodation variables are used (model RNNll). However, the asymptotic tensile peak strain reached at a steady state is large by comparison with the models RLLll, RNLll and RLLnl.

3.7. Ratcheting behavior of the 10M10C model

The generalized model 10M10C is investigated here to study its ratcheting behavior. The mechanical response is governed by the numerical value of the determinant of the matrix $[C]$. If this matrix is singular, ratcheting is observed. A regular matrix produces shakedown. In order to illustrate this property, two ratcheting tests are selected. A viscoplastic framework is chosen in order to avoid the problem of determining the plastic multipliers:

– for the first test, the loading is such that $\sigma_{max} = 530\,\text{MPa}$ and $\sigma_{min} = -400\,\text{MPa}$. A singular matrix is obtained if, for example, all C_{ij} with $i \neq j$ are set to -100,000 and all C_{ij} with $i = j$ are set to 900,000. We can easily check the singularity of this matrix by adding the components of the same column or the same row. The isotropic hardening variables are constant (R_{01}=50, R_{02}=100, R_{03}=200, R_{04}=300 R_{05}=400, R_{06}=500, R_{07}=600, R_{08}=700, R_{07}=800, R_{010}=900) and the Norton parameters are K=200 and n=10;

– for the second test, the loading is such that $\sigma_{max} = 900\,\text{MPa}$ and $\sigma_{min} = -700\,\text{MPa}$. A regular matrix is obtained if all C_{ij} with $i \neq j$ are set to 1,000 and all C_{ij} with $i = j$ are set to 80,000. The isotropic hardening variables are constant (R_{01}=50, R_{02}=150, R_{03}=200, R_{04}=250 R_{05}=300, R_{06}=350, R_{07}=400, R_{08}=500, R_{07}=550, R_{010}=600). The Norton parameters are K=100 and n=10.

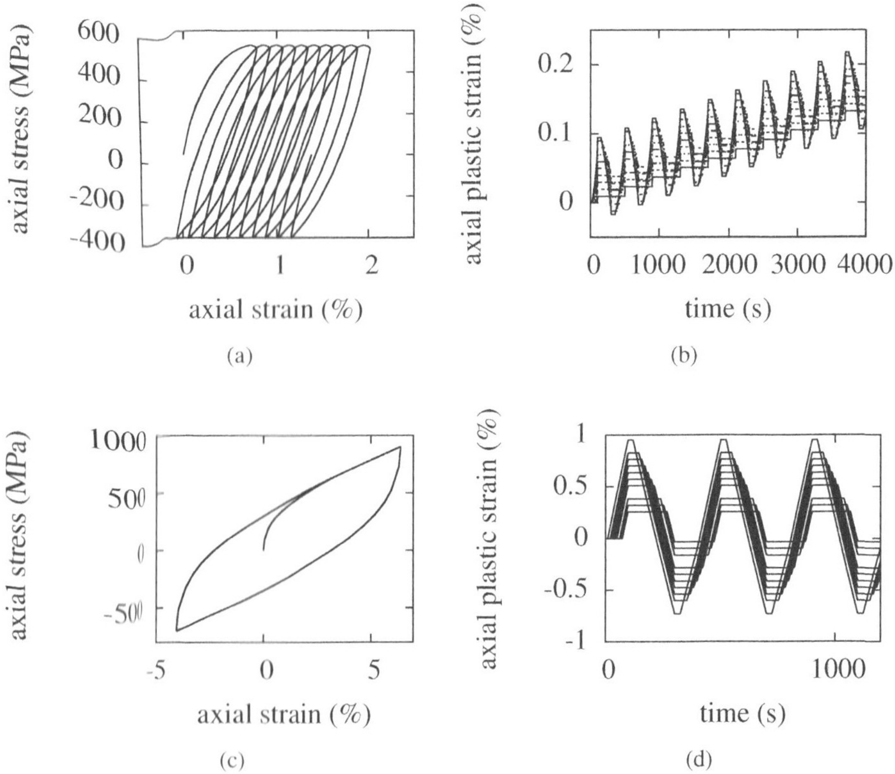

Figure 3.6. *Simulation of uniaxial ratcheting tests using the 10M10C model. Singular matrix: (a) stress–strain response, (b) plastic strain for the different mechanisms. Regular matrix: (c) stress–strain response, (d) plastic strain for the different mechanisms*

All the previous parameters are expressed in the system unit (MPa, s). Figure 3.6 shows the simulations of the two cases described above. As shown in Figure 3.6(c) and Figure 3.6(d), all the mechanisms are active for the applied loading. However, the link between the character of the hardening matrix and the ratcheting behavior could vanish in other configurations. Indeed, if only a subset of mechanisms is active, the ratcheting behavior will now be linked to the corresponding sub-determinant. In other words, a regular matrix and a singular matrix could lead to ratcheting and shakedown, respectively. Only the "active" part of the matrix must be considered.

3.8. Extra-hardening under non-proportional loading

In combined axial–torsional fatigue tests, "out-of-phase" tests refer to sinusoidal signals with a 90° phase lag (Figure 3.7), meanwhile, for "in-phase" loadings, the phase lag is zero. A material exhibits an "additional hardening" if the equivalent stress range obtained in an "out-of-phase" test is larger than that obtained in any "in-phase" test having the same equivalent strain range.

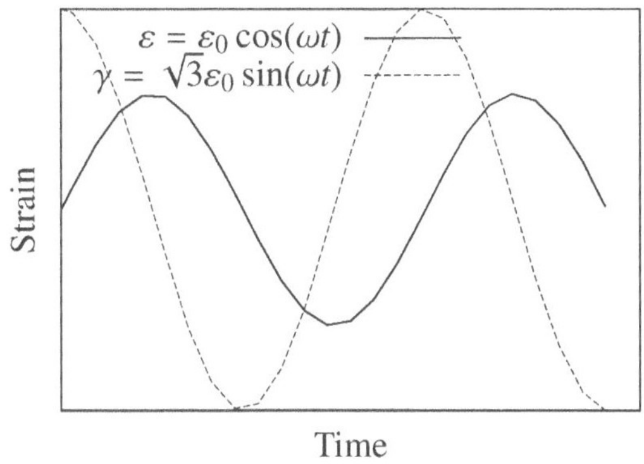

Figure 3.7. *Typical loading for "out-of-phase" tests*

Materials like austenitic stainless steel or copper are well known to be prone to additional hardening [CAI 84, LAM 78, BEN 87]. The 2M1C model can reproduce this additional hardening. To illustrate this possibility, an axial fatigue test with a strain range of 1% is compared to a tension–torsion out-of-phase test with the same equivalent axial strain range. As shown in Figure 3.8, the resulting stress range is larger for the out-of-phase test than for the reference axial test. Note that the model does not need any specific material parameter to describe the degree of additional hardening. As shown in [SAÏ 07], the 2M1C model is able to predict with a good accuracy the amount of additional hardening obtained for 316 stainless steel. On the other hand, it is worth noting that this extra-hardening has a pure kinematic source, so that it will vanish after one or two cycles if a "in-phase" loading follows an "out-of-phase" block. This is not the case in real experiments, where the memory of the initial extra-hardening will vanish very slowly in a subsequent in-phase loading.

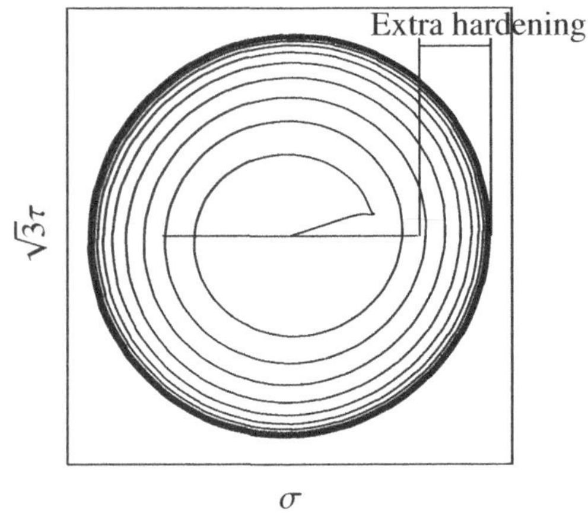

Figure 3.8. *Simulation of an out-of-phase test using the 2M1C model. The material parameters are those of the 316 stainless steel (out-of-phase loading and stabilized cycle of axial loading)*

The 2M1C and 2M2C models can be merged into a unique formulation. Intermediate solutions can then be found by considering the following yield criterion:

$$f = \left[\left(\frac{J(\underset{\sim}{\sigma} - \underset{\sim}{\mathbf{X}}^1)}{K_1} \right)^N + \left(\frac{J(\underset{\sim}{\sigma} - \underset{\sim}{\mathbf{X}}^2)}{K_2} \right)^N \right]^{1/N} - R \qquad [3.29]$$

where N is a new material parameter. The 2M1C model is recovered by using $N = 2$; meanwhile, the 2M2C model is the limit case when $N \to \infty$. K_1 and K_2 are not new material parameters. They are equal for the 2M1C model. For the 2M2C model, K_1 and K_2 are related to the ratio of the initial yield surfaces R_{01} and R_{02}. The effect of the new parameter N is shown in Figure 3.9(a). For $N = 1$, the curve obtained is a rhomboid, for the $N = 2$ (2M1C model), it is a circle, while for $N \to \infty$, it is a rectangle (2M2C model). Figure 3.9(b) shows the resulting additional hardening for different values of N. It is worth noting that for the sake of normalization, the initial size of the elastic domain R_0 is a function of N (R_0=160 MPa for N=2). Accordingly,

$$R_0(N) = 2^{1/N}\sigma_y \quad [3.30]$$

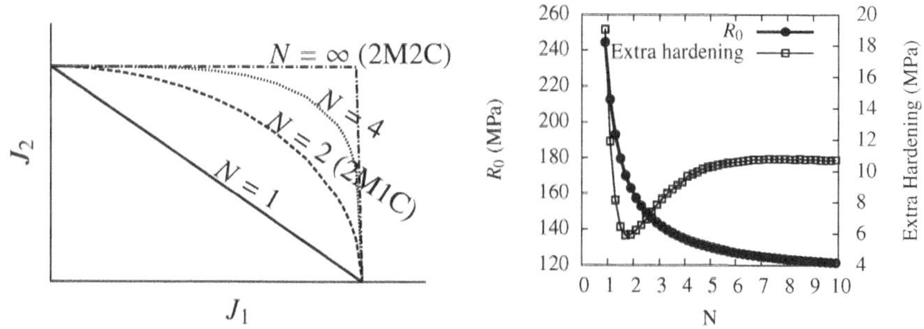

(a) Generalization of the 2M1C and the 2M2C models

(b) Additional hardening and R_0

Figure 3.9. *Influence of the parameter N on the out-of-phase test*

It can be noted that the additional hardening is high for small values of N and decreases to reach a minimum value for a value of N slightly smaller than 2. Thereafter, the amount of additional hardening increases to reach a steady state for high values of N. In this unique formulation, choosing N=1 produces a version, which may degenerate into a unified model with two back stresses for some particular loadings. For instance, in one-dimensional tensile loading, the yield function of equation [3.29] is such that

$$f = |\sigma - X^1| + |\sigma - X^2| = 2\sigma - X^1 - X^2 \quad [3.31]$$

On the other hand, taking a yield function as in equation [3.29] does not preserve the opportunity to introduce two different mechanisms, namely viscoplastic and plastic, in the constitutive equations. Further simulations are needed to explore the effect of the parameter N on describing the whole range of experimental data.

3.9. Static recovery effect

The introduction of the static recovery term in the kinematic hardening rules was initially proposed by Lemaitre and Chaboche [LEM 90]. Static recovery processes occur mainly at high temperatures. It is characterized by a slow, partial restoration of the kinematic variable and by a progressive recovery of the hardening. In the case of isothermal conditions, Chaboche's kinematic hardening law can be written for an unified model type:

$$\dot{\underset{\sim}{\alpha}} = \dot{p}\left(\underset{\sim}{\mathbf{n}} - \frac{3D}{2C}\underset{\sim}{\mathbf{X}}\right) - \frac{3}{2}\left(\frac{J(\underset{\sim}{\mathbf{X}})}{M}\right)^m \frac{\underset{\sim}{\mathbf{X}}}{J(\underset{\sim}{\mathbf{X}})} \quad [3.32]$$

It means that, once the plastic strain rate is null, a total recovery may be obtained. In practical applications, the parameter values are such that hardening remains present until failure is obtained. On the other hand, the balance between the driving strain rate-dependent term and the recovery generates an asymptotic value of the hardening that decreases for smaller strain rates or smaller stresses. This can be illustrated for one-dimensional loading cases, where the asymptotic value α_s can be deduced from the following equation:

$$\dot{\alpha} = \dot{\varepsilon}_s^p(1 - D\alpha_s) - \left(\frac{\alpha_s}{\alpha_0}\right)^m = 0 \quad [3.33]$$

As $\underset{\sim}{\mathbf{X}} = (2/3)C\underset{\sim}{\alpha}$, it can be easily checked that $\alpha_0 = M/C$. If the viscous effect is described by a power function with the material parameters K and n, a model with an initial yield of R_0 and no isotropic hardening will produce an analytic solution for a creep test at σ_0:

$$\sigma_0 = K\left[\frac{(\frac{\alpha_s}{\alpha_0})^m}{1 - D\alpha_s}\right]^{(1/n)} + C\alpha_s + R_0 \quad [3.34]$$

Figure 3.10 shows the variation of the variable α_s with the stress level σ. For high values of σ_0, the asymptotic value of α_s is equal to $1/D = 0.005$; meanwhile, for smaller values, it tends to zero.

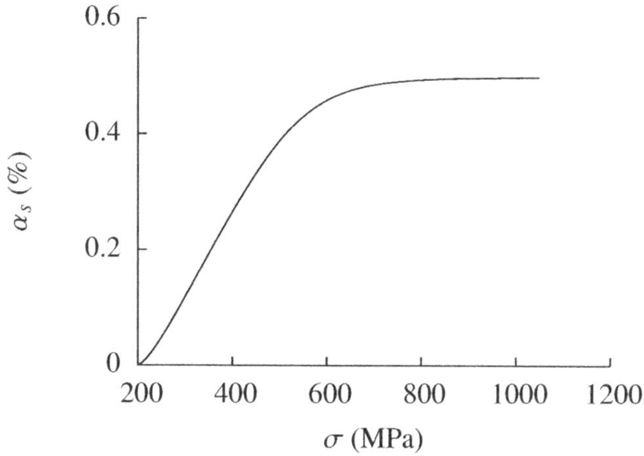

Figure 3.10. *Recovery effect, unified model, one-dimensional case: R_0=200, n=4, K=100, m=2.5, M=80, C=30000, D=200 (MPa, s)*

For the MM models, a specific recovery term may be introduced in each kinematic variable. The evolution rules of the kinematic internal variables can be written for the 2M2C model:

$$\dot{\underset{\sim}{\alpha}}^1 = \dot{p}^1\left(\underset{\sim}{\mathbf{n}}^1 - \frac{3D_1}{2C_{11}}\underset{\sim}{\mathbf{X}}^1\right) - \frac{3}{2}\left(\frac{J(\underset{\sim}{\mathbf{X}}^1)}{M_1}\right)^{m_1}\frac{\underset{\sim}{\mathbf{X}}^1}{J(\underset{\sim}{\mathbf{X}}^1)} \quad [3.35]$$

$$\dot{\underset{\sim}{\alpha}}^2 = \dot{p}^2\left(\underset{\sim}{\mathbf{n}}^2 - \frac{3D_2}{2C_{22}}\underset{\sim}{\mathbf{X}}^2\right) - \frac{3}{2}\left(\frac{J(\underset{\sim}{\mathbf{X}}^2)}{M_2}\right)^{m_2}\frac{\underset{\sim}{\mathbf{X}}^2}{J(\underset{\sim}{\mathbf{X}}^2)} \quad [3.36]$$

Similarly, they can be written for the 2M1C model:

$$\dot{\underset{\sim}{\alpha}}^1 = \dot{p}\left(\underset{\sim}{\mathbf{n}}^1 - \frac{3D_1}{2C_{11}}\underset{\sim}{\mathbf{X}}^1\right) - \frac{3}{2}\left(\frac{J(\underset{\sim}{\mathbf{X}}^1)}{M_1}\right)^{m_1}\frac{\underset{\sim}{\mathbf{X}}^1}{J(\underset{\sim}{\mathbf{X}}^1)} \quad [3.37]$$

$$\dot{\underset{\sim}{\alpha}}^2 = \dot{p}\left(\underset{\sim}{\mathbf{n}}^2 - \frac{3D_2}{2C_{22}}\underset{\sim}{\mathbf{X}}^2\right) - \frac{3}{2}\left(\frac{J(\underset{\sim}{\mathbf{X}}^2)}{M_2}\right)^{m_2}\frac{\underset{\sim}{\mathbf{X}}^2}{J(\underset{\sim}{\mathbf{X}}^2)} \quad [3.38]$$

where m_I and M_I (I = 1..2) are four additive material parameters. The way each mechanism is affected by static recovery depends on material behavior. If only one mechanism is affected, namely the first one, the evolution laws of the kinematic variables can be expressed in one-dimensional case (with $M_1 = M$ and $m_1 = m$) as

$$\dot{\alpha}^1 = \dot{\varepsilon}_s^1(1 - \frac{D_1}{C_{11}}x_s^1) - \left(\frac{x_s^1}{M}\right)^m = 0 \qquad [3.39]$$

$$\dot{\alpha}^2 = \dot{\varepsilon}_s^2(1 - \frac{D_2}{C_{22}}x_s^2) = 0 \qquad [3.40]$$

If the viscous effect for the first mechanism is described by a power function, the creep stress σ_0 is linked to the back stress x^1 by

$$\sigma = K_1\left[\frac{(\frac{x_s^1}{M})^m}{1 - \frac{D_1}{C_{11}}x_s^1}\right]^{(1/n_1)} + x^1 + R_0^1 \quad \text{for the 2M2C model} \qquad [3.41]$$

$$\sigma = K\left[\frac{(\frac{x_s^1}{M})^m}{1 - \frac{D_1}{C_{11}}x_s^1}\right]^{(1/n)} + x^1 + R_0 \quad \text{for the 2M1C model} \qquad [3.42]$$

After a few algebraic manipulations and using the kinematic–kinematic coupling equations,

$$x^1 = C_{11}\alpha^1 + C_{12}\alpha^2 \qquad [3.43]$$

$$x^2 = C_{12}\alpha^1 + C_{22}\alpha^2 \qquad [3.44]$$

The back stress x^1 can be expressed only as a function of α^1:

$$x^1 = \frac{C_{11}C_{22} - C_{12}^2}{C_{22}}\alpha^1 + \frac{C_{12}}{D_2} \qquad [3.45]$$

The numerical results are the same for the 2M2C and 2M1C models (Figure 3.11). It can be easily demonstrated that

$$\alpha_{lim}^1 = \frac{C_{22}}{C_{11}C_{22} - C_{12}^2}\left(\frac{C_{11}}{D_1} - \frac{C_{12}}{D_2}\right) \qquad [3.46]$$

where α_{lim}^1 is positive for tensile loading. Indeed, according to [WOL 08], the non-negativity of the free energy is guaranteed by the condition $C_{12}^2 \leq C_{11}C_{22}$.

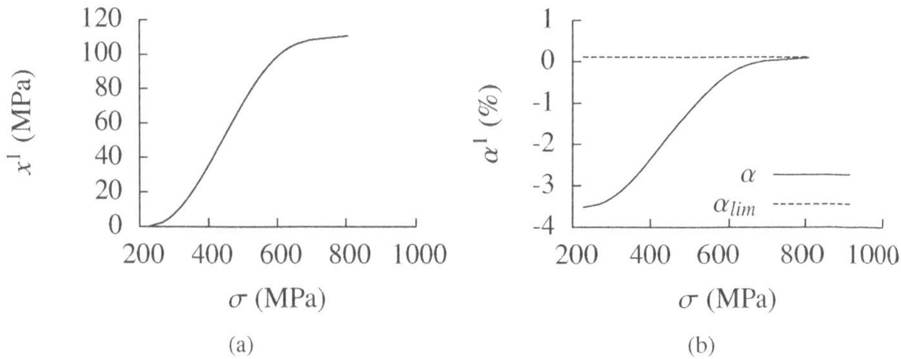

Figure 3.11. *Recovery effect, MM model, one-dimensional case: (a) back stress and (b) kinematic internal variable.* m*=2.5,* M*=82,* C_{11}*=26,000,* C_{21}*=52,800,* C_{12}*=34,807,* D_1*=234,* D_2*=323. 2M2C model:* R_0^1*=208,* n_1*=6.9,* K_1*=200. 2M1C model:* R_0*=208,* n*=6.9,* K*=200 (MPa, s)*

4

Comparison with Experimental Databases

This chapter is devoted to illustrative examples dealing with the application of MM models to various materials, including metals, semi-crystalline polymers and amorphous polymers. Each section shows a brief description of the context of the study in which the experimental database of the material is given and the main features of the model are selected. The identification strategy is also discussed. Most of the cases refer to model variants summarized in section 3.1, with an emphasis on specific modifications related to scale transition scheme, kinematic hardening rules and isotropic variables. The different materials investigated by means of the MM models are recalled in Table 4.1. The models have as a common root the plastic strain decomposition into several parts ($\underset{\sim}{\varepsilon}^p = A_I \sum \underset{\sim}{\varepsilon}^I$). They differ by their type (number of mechanisms and criteria), the scale transition rules and the hardening laws, according to the following remarks:

– the material parameters A_I are most of the time assigned to 1. For the case where mechanisms are attributed to particular phases, the coefficients are such that $\sum A_I = 1$;

– the local stresses are obtained through a static scheme ($\underset{\sim}{\sigma}^I = \underset{\sim}{\sigma}$) or using the β-scale transition rule;

– the fading memory terms in the kinematic hardening rule of each mechanism may be expressed by means of the back stresses $\underset{\sim}{\mathbf{X}}^I$ or the internal variables $\underset{\sim}{\alpha}^I$;

– to obtain a fine description of isotropic hardening, two- or three-term evolution rules are used in models with various mechanisms and one criterion.

Material	Model	A_I	σ^I	$\dot{\alpha}^I$	Isotropic hardening
Inconel 718 [SAÏ 93]	2M2C	A_1=0.65 A_2=0.32	$A_I\sigma$	$\left(\mathbf{n}^I - (3D_I/2C_{II})\mathbf{X}^I\right)\dot{p}^I$	$R_I = Q_I(1 - exp(-b_I p^I)$
N18 alloy [SAÏ 04]	2M2C	A_I=1	σ	$\left(\mathbf{n}^I - \frac{3D_I}{2C_{II}}\mathbf{X}^I\right)\dot{p}^I - \frac{3}{2}\left(\frac{J(\mathbf{X}^I)}{M}\right)^m \frac{\mathbf{X}^I}{J(\mathbf{X}^I)}$	$R_I = Q_I(1 - exp(-b_I p^I)$
CS1026 [TAL 06]	2M1C	A_I=1	σ	$\left(\mathbf{n}^I - D_I\left((1-\eta)\alpha^I + \eta(\alpha^I : \mathbf{m}^I)\mathbf{m}^I\right)\right)\dot{p}$	$R_I = Q_I(1 - exp(-b_I p^I)$
316L [SAÏ 07]	2M1C 2M2C	$A_1 = (1 - z)$ $A_2 = z$	$\sigma + \mu'(\beta - \beta^I)$	$\left(\mathbf{n}^I - \frac{3D_1}{2C_{11}}\left((1-\eta_I)\mathbf{X}^I + \eta_I\left(\mathbf{X}^I : \mathbf{n}^I\right)\mathbf{n}^I\right)\right)\dot{p}$ $\left(\mathbf{n}^I - \frac{3D_1}{2C_{11}}\left((1-\eta_I)\mathbf{X}^I + \eta_I\left(\mathbf{X}^I : \mathbf{n}^I\right)\mathbf{n}^I\right)\right)\dot{p}^I$	$R_I = Q_I(1 - exp(-b_I p^I)$
316L [TAL 14]	2M1C	A_I=1	σ	$\left(\mathbf{n}^I - D_I\left((1-\eta)\alpha^I + \eta(\alpha^I : \mathbf{m}^I)\mathbf{m}^I\right)\right)\dot{p}$	$R = R^1 + R^2 + R^3$
316L [SAÏ 11a]	3M1C	$A_I = z_I$	$\sigma + \mu'(\beta - \beta^I)$	$\left(\mathbf{n}^I - \frac{3D_1}{2C_{11}}\left((1-\eta_I)\mathbf{X}^I + \eta_I\left(\mathbf{X}^I : \mathbf{n}^I\right)\mathbf{n}^I\right)\right)\dot{p}$	$R_I = Q_I(1 - exp(-b_I p^I)$
304L [HAS 08] [TAL 10] [TAL 11] [TAL 14] [SAÏ 14]	2M1C	A_I=1	σ	$\left(\mathbf{n}^I - D_I\left((1-\eta)\alpha^I + \eta(\alpha^I : \mathbf{m}^I)\mathbf{m}^I\right)\right)\dot{p}$	$R = R^1 + R^2 + R^2$
304L [SAÏ 14]	3M1C	A_I=1	σ	$\left(\mathbf{n}^I - D_I\left((1-\eta)\alpha^I + \eta(\alpha^I : \mathbf{m}^I)\mathbf{m}^I\right)\right)\dot{p}$	$R = R^1 + R^2 + R^3$
55NiCrMoV7 [VEL 06]	2M1C	A_1=0.78 A_2=0.4	$A_I\sigma$	$\left(\mathbf{n}^I - (3D_I/2C_{II})\mathbf{X}^I\right)\dot{p}^I$	$R = R^1 + R^2$
PA6 [REG 09b]	2M1C	$A_1 = 1 - z\ A_2 = z$	$\sigma + \mu'(\beta - \beta^I)$	$\left(\mathbf{n}^I - (3D_I/2C_{II})\mathbf{X}^I\right)\dot{p}^I$	$R_I = Q_I(1 - exp(-b_I p^I)$
PMMA, PC [JER 14]	3M3C	A_I=1	σ	-	$R_I = Q_I(1 - exp(-b_I p^I)$
CuZn27 [TAL 15]	2M1C	A_I=1	σ	$\left(\mathbf{n}^I - D_I\left((1-\eta)\alpha^I + \eta(\alpha^I : \mathbf{m}^I)\mathbf{m}^I\right)\right)\dot{p}$	$R = R^1 + R^2 + R^3$
35NCD16 [TAL 15]	2M1C	A_I=1	σ	$\left(\mathbf{n}^I - D_I\left((1-\eta)\alpha^I + \eta(\alpha^I : \mathbf{m}^I)\mathbf{m}^I\right)\right)\dot{p}$	$R = R^1 + R^2 + R^3$
2017A [TAL 15] [SAÏ 12]	2M1C	A_I=1	σ	$\left(\mathbf{n}^I - D_I\left((1-\eta)\alpha^I + \eta(\alpha^I : \mathbf{m}^I)\mathbf{m}^I\right)\right)\dot{p}$	$R = R^1 + R^2 + R^3$
XC18 [TAL 15]	2M1C	A_I=1	σ	$\left(\mathbf{n}^I - D_I\left((1-\eta)\alpha^I + \eta(\alpha^I : \mathbf{m}^I)\mathbf{m}^I\right)\right)\dot{p}$	$R = R^1 + R^2 + R^3$
Zircaloy4 [PRI 08]	2M2C	A_I=1	σ	-	$R_I = \alpha_I \mu b\sqrt{\rho_I + H\rho_J}$
Ti6Al4V [LON 09]	2M2C	A_I=1	$\sigma + \mu'(\beta - \beta^I)$	$\left(\mathbf{n}^I - \frac{3D_I}{2C_{II}}\mathbf{X}^I\right)\dot{p}^I - \frac{3}{2}\left(\frac{J(\mathbf{X}^I)}{M}\right)^m \frac{\mathbf{X}^I}{J(\mathbf{X}^I)}$	$R_I = Q_I(1 - exp(-b_I p^I)$

Table 4.1. *Application of the MM models to different classes of materials*

Following these applications, the rest of the chapter provides recommendations to select the proper expression of the model for a given material behavior in order to select the best MM model in each case.

4.1. Inconel 718 [SAÏ 93]

4.1.1. *Context of the case study*

The 2M2C model reliability is investigated by studying the mechanical behavior of a nickel-based superalloy INCO718 at 650℃. Experimental tests from different sources [THA 83, AND 87, GHO 91] were selected. The value of the yield stress seems to be around 650–700 MPa when it is estimated from tensile stress–strain curves using the earliest deviation from the linear regime, as shown in Figure 4.1(a). However, uniaxial creep tests performed at stresses of 600, 650 and 700 MPa generate a creep strain of 2, 5 and 5%, respectively, after 400, 200 or 100 h. The use of a unified approach would need to introduce a smaller yield stress, and a series of kinematic variables to generate a large kinematic effect at the onset of plastic flow. The 2M2C model is selected to account for this behavior by considering two different regimes.

4.1.2. *Particular model features*

The equations used are those presented in Table 3.1. Specifically, the nonlinearity of the kinematic hardening variable of the first mechanism is given by a dynamic recovery term in the Prager rule and a static recovery term (see section 3.9), which corresponds to the rearrangement of dislocations by climb and recrystallization from annealing:

$$\dot{\underset{\sim}{\alpha}}^1 = \dot{p}^1 \left(\underset{\sim}{\mathbf{n}}^1 - \frac{3D_1}{2C_{11}} \underset{\sim}{\mathbf{X}}^1 \right) - \frac{3}{2} \left(\frac{J(\underset{\sim}{\mathbf{X}}^1)}{M} \right)^m \frac{\underset{\sim}{\mathbf{X}}^1}{J(\underset{\sim}{\mathbf{X}}^1)} \quad [4.1]$$

$$\dot{\underset{\sim}{\alpha}}^2 = \dot{p}^2 \left(\underset{\sim}{\mathbf{n}}^2 - \frac{3D_2}{2C_{22}} \underset{\sim}{\mathbf{X}}^2 \right) \quad [4.2]$$

4.1.3. *Numerical results*

The model is able to correctly predict the tensile tests at different strain rates (Figure 4.1(a)), the creep response for the three loading levels (Figure 4.1(b)) and the cyclic uniaxial tension–compression tests for different strain ranges and strain rates (Figure 4.2). The creep tests reveal a predominant tertiary stage. There is no primary creep, and the second creep stage is very short. Microstructural observations [AND 87] confirm that there is no damage in these tests and that this behavior is mainly attributed to localized deformation, producing material softening.

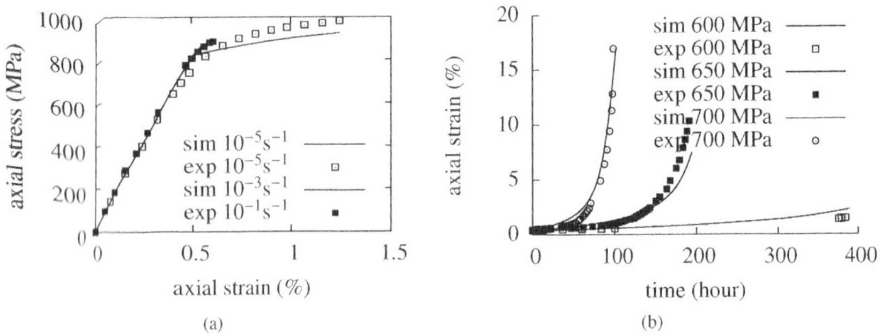

Figure 4.1. *Simulation of the tensile and creep tests of INCO718 at 650°C using the 2M2C model: (a) tensile, (b) creep*

The parameters used to simulate the different tests of the database related to INCO718 at 650°C are shown in Table 4.2. In this application, the two mechanisms did not refer to specific sources of deformation.

4.2. Deformation mechanisms of Ni–Ti shape memory alloy [ROU 00]

4.2.1. *Context of the case study*

Roumagnac *et al.* [ROU 00] investigated the behavior in tension and at room temperature of Ti–Ni shape memory alloys in the martensitic state. Three stages are distinguished in the isothermal mechanical behavior. The first stage (I) corresponds to the reorientation of the martensite under the applied mechanical loading. A linear stress–strain behavior characterizes the second stage (II). In this work, the mechanical response results from two deformation mechanisms: (i) martensite reorientation and (ii) irreversible plasticity. The authors showed that these two deformation processes arise successively at different total strain and stress levels.

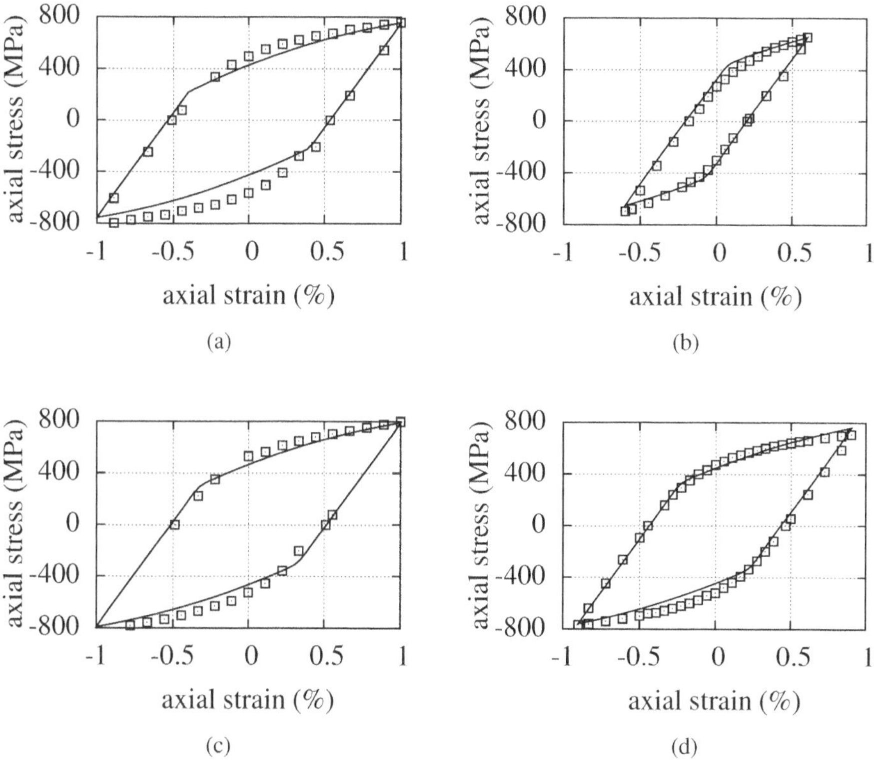

Figure 4.2. *Simulation of the fatigue tests of INCO718 at 650°C using the 2M2C model. Stabilized stress strain loops: (a) $\Delta\varepsilon$=2%, $\dot{\varepsilon}=10^{-6}s^{-1}$, (b) $\Delta\varepsilon$=1.2%, $\dot{\varepsilon}=10^{-3}s^{-1}$, (c) $\Delta\varepsilon$=2%, $\dot{\varepsilon}=10^{-3}s^{-1}$, (d) $\Delta\varepsilon$=1.8%, $\dot{\varepsilon}=10^{-3}s^{-1}$*

Young's modulus	E=162,000
Scale transition rule	A_1=0.65, A_2=0.32
Isotropic hardening	R_{01}=208, Q_{11}=-208, b_1=14 R_{02}=261, Q_{22}=-106, b_2=34, Q_{12}=0
Kinematic hardening	C_{11}=26,033, C_{22}=52,789, C_{12}=34,807 D_1=234, D_2=323
Norton rule	K_1=2,790, n_1=6.9, K_2=57, n_2=3
Recovery parameters	M=82, m=2.5

Table 4.2. *Simulation of the behavior of INCO718 at 650°C using the 2M2C model according to [SAÏ 93] (MPa, s)*

4.2.2. *Particular model features*

The thermodynamically based constitutive equations account for the deformation mechanisms that have been observed. In the present case, kinematic–kinematic coupling is neglected in the inelastic free energy ($C_{12} = 0$). [ROU 00] demonstrated that the material does not exhibit any isotropic hardening. The proposed model is then a simple approach with two loading surfaces. Two plastic potentials are then used to account for the nonlinear behavior, as in the 2M2C model. The parameter D_1 depends on the "reorientation strain" ε^1:

$$\dot{\underset{\sim}{\alpha}}^1 = \dot{p}^1 \left(\underset{\sim}{\mathbf{n}}^1 - \frac{3D_1}{2C_{11}} \underset{\sim}{\mathbf{X}}^1 \right) \quad \text{with} \quad D_1 = k_1 \exp(-k_2 (\varepsilon^1)^{k_3}) \qquad [4.3]$$

4.2.3. *Numerical results*

The behavior in tension and at room temperature of the Ti–Ni shape memory alloy in the martensitic state is simulated by the proposed model. Due to the decrease in the coefficient D_1 with respect to accumulated strain (see Table 4.3), the kinematic hardening becomes linear for large strain in stage II. The two mechanisms then become active simultaneously. The material parameters listed in Table 4.3 are used to simulate the stress–strain behavior of the material, as shown in Figure 4.3(a). Further validation is needed to explore the reliability of the proposed model to successfully describe larger experimental databases. In particular, the kinematic hardening rule proposed in equation 4.3 should be checked by simulating tests under strain or stress control. Figure 4.3(b) compares the local strains calculated at the end of the loading/unloading cycles with the corresponding experimental results.

Elasticity	Reorientation plasticity			Irreversible plasticity		
E	R_{01}	C_{11}	D_1	R_{02}	C_{22}	D_2
57,000	40	84,300	$642\exp(-15750(\varepsilon^1)^{3.09})$	180	46,100	102

Table 4.3. *Material parameters for the Ni–Ti alloy according to [ROU 00] (MPa)*

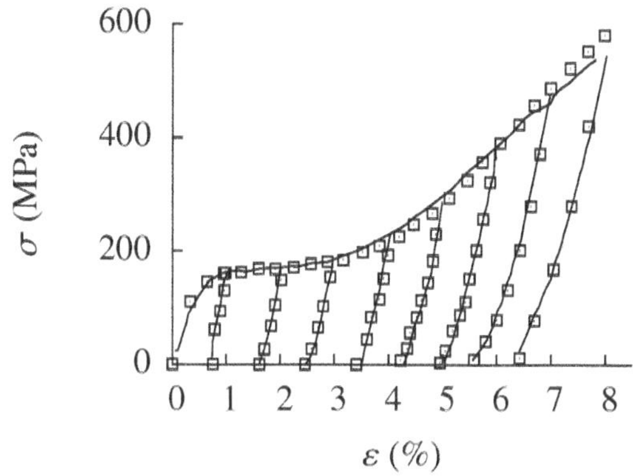

(a)

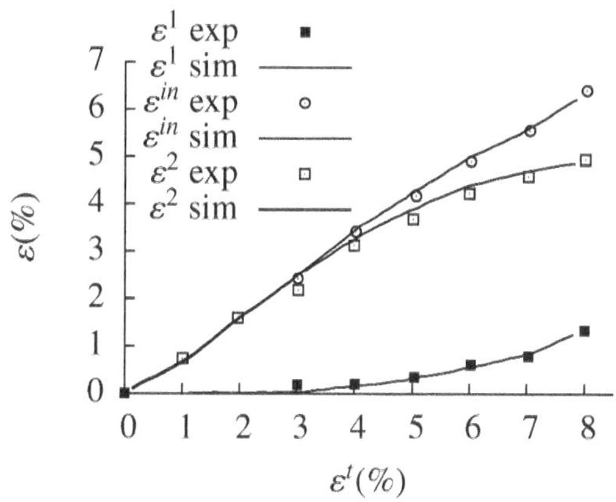

(b)

Figure 4.3. *Study of NiTi alloy (from [ROU 00]): (a) simulation of the loading–unloading behavior, (b) experimental and simulated local strains*

4.3. N18 alloy [SAÏ 04]

4.3.1. *Context of the case study*

A new variant of the 2M2C model was proposed by Saï *et al.* [SAÏ 04]. In this version, the set of constitutive equations is linked to the activation of physical phenomena. The shape of the rules describing the viscoplastic strain rate takes into account the crossing of walls by dislocations that do not interact with each other. The activation energy and the activation volume are estimated by means of mechanical tests. The critical variable measuring material microstructure evolution during the deformation process is the density of dislocations. Three plasticity mechanisms are then identified:

– the first mechanism consists in Orowan's bypassing of γ' precipitates by matrix dislocations. The hardening phase γ' remains elastic and the plastic strain is restrained in the matrix;

– the second mechanism is active during the second stage of relaxation and corresponds to lower strain rates;

– the third mechanism of viscoplastic flow can develop at the lowest stress levels (800–900 MPa at 650°C). It involves local rearrangement of atoms in the ordered phase γ'.

Figure 4.4. *Microstructure of the N18 alloy [SAÏ 04]*

In fact, only the first two mechanisms are taken into account in the final model. The nonlinear behavior is assumed to originate from dislocation slip, and the hardening may be a consequence of the dislocation–dislocation interaction.

4.3.2. *Particular model features*

The main modification of the 2M2C model is based on the Bailey–Orowan transport equation $\dot{\gamma}' = \rho_m b v$, where $\dot{\gamma}'$ is the creep rate produced by the average density of mobile dislocations ρ_m with an average Burgers vector b, moving at an average velocity v. According to Hausselt and Blum [HAU 76], this equation was generalized to take into account the competition between the creation of mobile dislocations and the annihilation process. For the N18 alloy, the density of mobile dislocations increases with the level of viscoplastic strain rate. The multiplication of the dislocations increases with material hardening (ρ_m^+). When the density of mobile dislocations increases, the probability that they meet each other increases and leads to their annihilation (ρ_m^-). The density of mobile dislocations ρ_m can then be written as

$$\rho_m = \rho_{ini} + \rho_m^+ + \rho_m^- \quad \dot{\rho}_m^+ = a_1 J\left(\underset{\sim}{\dot{\varepsilon}}^1\right) + a_2 J\left(\underset{\sim}{\dot{\varepsilon}}^2\right) \quad \dot{\rho}_m^- = -S\,exp\left(\frac{-\Delta H_S}{kT}\right) \quad [4.4]$$

where ΔH_S is the enthalpy related to the recovery phenomenon and S is a material parameter. The equivalent viscoplastic strain rates are derived from an Arrhenius formulation, in agreement with the definition of a thermally activated process:

$$\dot{p}^1 = \rho C_1 \exp\left(\frac{J(\underset{\sim}{\sigma}^1)V_1 - \Delta H_{01}}{kT}\right)$$

$$\dot{p}^2 = \rho C_2 \exp\left(\frac{J(\underset{\sim}{\sigma}^2 - \underset{\sim}{\mathbf{X}}^2)V_2 - R^2 - \Delta H_{01}}{kT}\right) \quad [4.5]$$

In these expressions, ΔH_0^1 and ΔH_0^2 are the activation enthalpies, k is the Boltzmann constant and V_1 and V_2 are the activation volumes. Note that the back stress $\underset{\sim}{\mathbf{X}}^1$ for the first mechanism is equal to zero.

4.3.3. *Numerical results*

The material parameters were separated into three types as reported in Table 4.4. A comparison between the experimental and numerical results is presented in Figures 4.5(a), 4.5(b) and 4.5(c). It can be seen that, besides the good agreement between simulated and experimental curves, the model is able to reproduce the local information related to the competition between the creation of mobile dislocation density and the annihilation process. Figure 4.5(d) shows the evolution of the dislocation densities in a strain-controlled test.

Measured parameters	ΔH_0^1=640 kJ mole^{-1}, V_1=115 J mole^{-1} MPa^{-1} ΔH_0^2=918 kJ mole^{-1}, V_2=840 J mole^{-1} MPa^{-1}
Manually estimated parameters	E=175,000 MPa, ΔH_S=300 J mole^{-1}, $\rho_{ini}=10^{-13}$ m^{-2} $R_{01}=0$, R_{02}=700 MPa
Identified parameters	$C_1 = 55$, a_1=1.2 10^4, $C_2=10^{14}$, $a_2=10^3$ $C_{22} = 3\times10^4$, $D_2 = 900$, $S = 6\times10^{-9}$

Table 4.4. *Classification and values of the material parameters related to the 2M2C model proposed by [SAÏ 04] for the N18 alloy at 650°C*

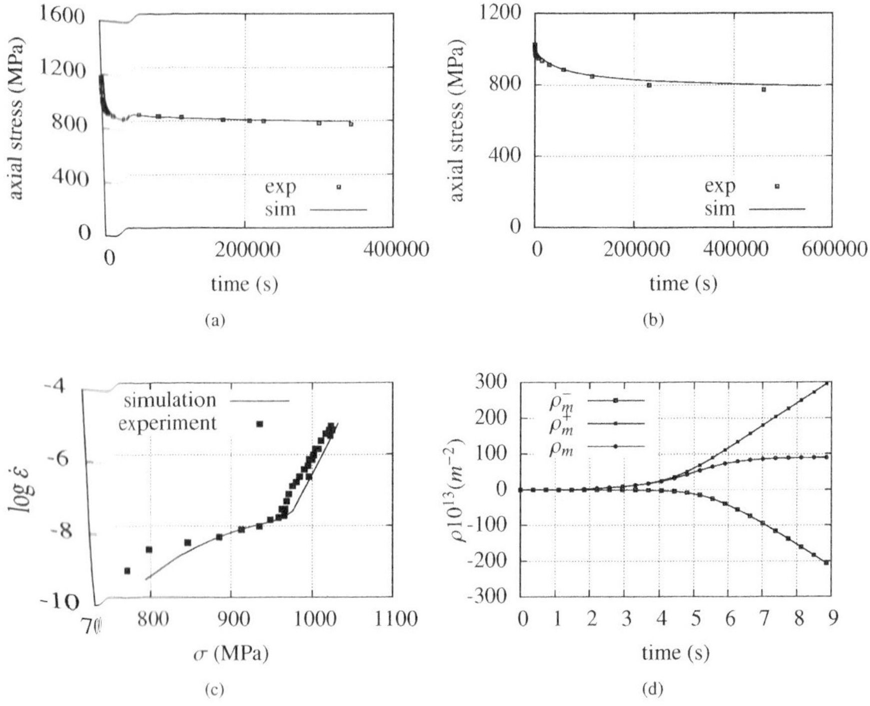

Figure 4.5. *Study of N18 behavior: (a) relaxation test 1.4% × 10^{-5} s^{-1}, (b) relaxation test 0.2% × 10^{-5} s^{-1}, (c) slope discontinuity in the relaxation test 0.2% × 10^{-5} s^{-1} and (d) simulation of the evolution of dislocation densities*

4.4. Carbon steel CS1026 [TAL 06]

4.4.1. *Context of the case study*

This work is devoted to the study of the capabilities of the initial version of the 2M1C model proposed by Cailletaud and Saï [CAI 95]. The predictions of the model are evaluated by considering 1D and 2D ratcheting results taken from [HAS 92a, HAS 92b]. Such an evaluation reveals the difficulty for the 2M1C model to describe simultaneously 1D and 2D ratcheting using the same set of material parameters. This weakness was generally observed for most of the existing phenomenological models. In order to correct the problem, a modification of the kinematic hardening rules is proposed in [TAL 06]. A comparison between the predictions of the new model with the same set of 1D and 2D ratcheting test results, considered above, shows the encouraging results of the updated formulation.

4.4.2. *Particular model features*

The modification proposed by Taleb *et al.* [TAL 06] to improve the description of the ratcheting behavior by the 2M1C model for one-dimensional and two-dimensional cases is summarized in Table 3.3:

– the flow function is modified by the introduction of a material parameter N. The parameter N is taken equal to 2 as in the initial 2M1C model. The effect of this parameter on the simulation of out-of-phase tests is studied in [SAÏ 07, SAÏ 14]. Figure 3.9 shows the resulting additional hardening together with the evolution of R_0 for different values of N;

– following the kinematic hardening rules proposed in [BUR 87, DEL 95], a combination of classical fading memory and radial fading memory is introduced in the dynamic recovery term. The new material parameter η, active under multi-axial loading only, is introduced to control the dynamic recovery term for each mechanism.

On the contrary, Taleb *et al.* [TAL 06] showed analytically the importance of the relative values of the hardening parameters C_{11}, C_{22} and C_{12}. The main difference between the models studied in [TAL 06] and [CAI 95, SAÏ 07] is due to the expression of the fading memory terms of the evolution rule of the kinematic hardening variable, which is a function of the internal variables $\underset{\sim}{\alpha}^i$

instead of the back stresses $\underset{\sim}{\mathbf{X}}^i$. The effect of the material parameters is studied in [WOL 08], which gives the conditions that must be enforced to ensure the thermodynamic consistency of the model:

– a necessary and sufficient condition to ensure the non-negativity of the free energy ($C_{12}^2 \leq C_{11}C_{22}$);

– a sufficient condition to ensure the non-negativity of the dissipation ($C_{12}^2(D_1 - D_2)^2 \leq 4D_1D_2(C_{11}C_{22}(1-\eta)^2 - C_{12}^2)$.

Taleb *et al.* [TAL 06] evaluated their model by comparison to 1D and 2D ratcheting performed on CS1026 steel as recalled above. The predictions are improved compared to the initial model of [CAI 95] thanks to the proposed formulation.

4.4.3. *Numerical results*

The considered set of uniaxial and multi-axial ratcheting tests [HAS 92a, HAS 92b] on 1026 carbon steel (CS1026) at room temperature was composed of:

– classical tension–compression ratcheting tests under different mean stresses (σ_{xm}) and constant amplitude (σ_{xa});

– classical tension–compression ratcheting tests under constant mean stress (σ_{xm}) but increasing amplitude (σ_{xa});

– axial strain cycle (ε_{xc}) combined with constant internal pressure ($\sigma_{\theta m}$): several axial strain amplitudes as well as internal pressure values;

– axial strain cycle (ε_{xc}) combined with variable internal pressure ($\sigma_{\theta m}$) following a positive or negative slope;

– biaxial bow-tie cycle;

– biaxial reverse bow-tie cycle.

The procedure used to obtain the optimal set of material parameters is described in detail in [TAL 06]. The material parameters of the considered 2M1C model have been identified according to the following considerations:

– time effects not being significant for the considered tests, the parameters n and K have been taken to be equal to 11 and 31, respectively, which are two usual values under these conditions;

– R_0 represents $\sqrt{2}\sigma_y$, where σ_y is the yield stress of the material that is determined from the stabilized stress–strain loop (the isotropic hardening was neglected in these simulations);

– the parameters C_{11} and C_{22} have been taken equal to Young's modulus E and $E/10$, respectively;

– the rest of the parameters have been determined using an optimization process considering the whole database (1D and 2D ratcheting tests).

Table 4.5 gives the values obtained for the material parameters. It is worth noting that before ratcheting tests, the specimens were subjected to a number of cycles under 1D strain control in order to obtain cyclically stabilized specimens. This process may justify the fact that the isotropic hardening is neglected in the simulations. A few examples of the numerical simulations (lines) are shown together with the experimental results (open-square symbols) in Figures 4.6(a) and (b) and Figures 4.6(c) and (d).

R_0	K	n	η	C_{11}	C_{22}	C_{12}	D_1	D_2
211	31	11	0.75	181,500	18,150	12,000	600	35

Table 4.5. *Material parameters used for CS1026 (MPa, s)*

4.5. Thermo-mechanical behavior of 55NiCrMoV7 [VEL 06]

4.5.1. *Context of the case study*

The 2M1C model was investigated in the work of Velay *et al.* [VEL 06] to describe the cyclic stress–strain response of 55NiCrMoV7 tempered martensitic steel (AISI L6). The 2M1C model allows the determination of two inelastic strain mechanisms resulting from two stress states, which can be related to the typical continuous cyclic softening of the alloy. This cyclic softening is reproduced through the evolution of the isotropic component (drag stress). A memory effect is introduced to take into account the influence of the plastic strain range on the amount of cyclic softening. The kinematic component (back stress) of the model allows the description of complex loading conditions applied to tool steels. Strain-induced recovery, time recovery (cyclic behavior, including tensile dwell times) and ratcheting effects are considered.

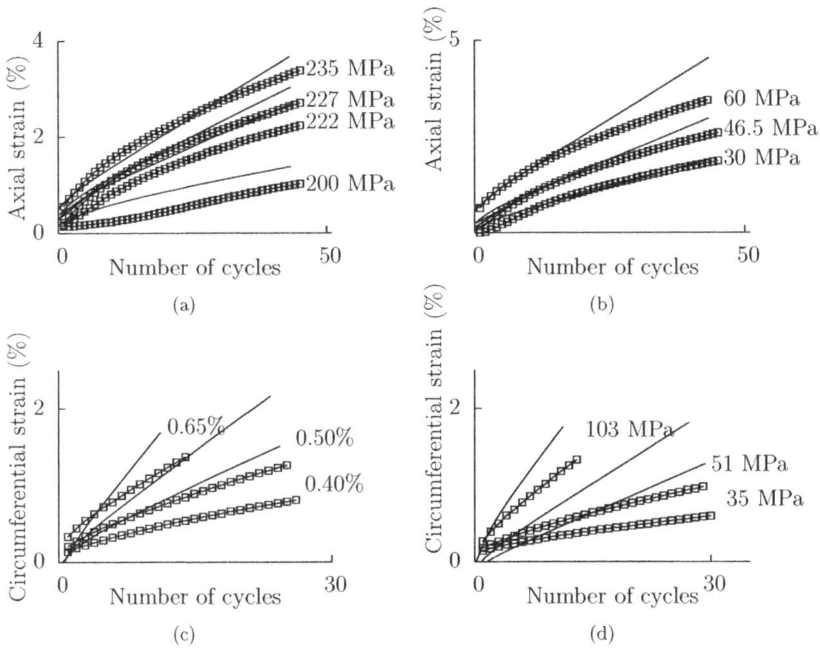

Figure 4.6. *Simulation (—) of experimental tests (□) [HAS 92a, HAS 92b] on 1026 carbon steel using the 2M1C model. One-dimensional ratcheting tests: (a) constant mean stress (σ_{xm} = 46.5 MPa) under different stress amplitudes σ_{xa} and (b) constant stress amplitude (σ_{xa} = 222 MPa) under different mean stresses σ_{xm}. Two-dimensional ratcheting tests: (c) different strain ranges ε_{xc} ($\sigma_{\theta m}$=69.4 MPa) and (d) different internal pressures $\sigma_{\theta m}$ ($\varepsilon_{xc} = 0.5\%$)*

4.5.2. *Particular model features*

A memory effect was introduced to take into account the influence of the plastic strain range on the amount of cyclic softening. The yield function was described using two isotropic variables R^1 and R^2. R^1 describes a softening much more pronounced than R^2. Its asymptotic value changes when the inelastic strain reaches a given threshold [CHA 86]. The formalism used is described in section 1.2.2.

4.5.3. *Numerical results*

Two types of cyclic tests were performed in order to identify the model parameters at temperatures of 20, 300, 400 and 500°C. The first type can be divided into two different steps:

– the first step is a symmetrical strain-controlled low-cycle fatigue test, with a fixed strain range $\Delta\varepsilon_t = 1.6\%$, which itself can be divided into three sub-steps:

- sub-step 1 consists of a number of fatigue cycles to reach a cumulated plastic strain of 400% at a constant strain rate of 10^{-2} s^{-1},

- during sub-step 2, several strain rates were successively applied, i.e. 10^{-2}, 10^{-3} and 10^{-4} s^{-1}, with three cycles being performed at each strain rate,

- finally, fatigue relaxation cycles were performed at a strain rate of 10^{-2} s^{-1} and relaxation times of 30 s (three cycles), then 90 and 600 s (two cycles for each),

– the second step consists in a non-symmetrical stress-controlled fatigue test at a constant stress rate of 100 MPa.s^{-1}. A total of 10 cycles were performed.

Young's modulus	E=15,6935
Localization parameters	A_1=0.78, A_2=0.4
Isotropic parameters	R_0=410, Q_∞=–100, b_1=5.5 Q_2=–75, b_2=0.2
Kinematic parameters	C_{11}=195,655, C_{22}=13,215, C_{12}=–40,500 D_1=0, D_2=0
Norton parameters	K=268, n=15
Recovery parameters	M_1=705, m_1=10.5, M_1=700, m_1=4.3

Table 4.6. *Identified parameters for 55NiCrMoV7 at 500°C (MPa, s)*

Table 4.6 provides the set of identified material parameters. Figure 4.7 shows some identification results, whereas Figure 4.8(a) compares the experimental results and model responses for continuous cycling and several strain ranges. This is an illustration of the memory of the plastic strain range. There is no nonlinear term in the evolution of the kinematic variable, neither strain nor time dependent, so an analytical treatment of the evolution equations can be made. This solution, already detailed in section 3.4, can then be obtained in the one-dimensional case at the stabilized cycle:

$$\sigma = \frac{(A_1 x^1 + A_2 x^2) \pm \sqrt{R^2(A_1^2 + A_2^2) - (A_1 x^1 - A_2 x^2)}}{A_1^2 + A_2^2} \qquad [4.6]$$

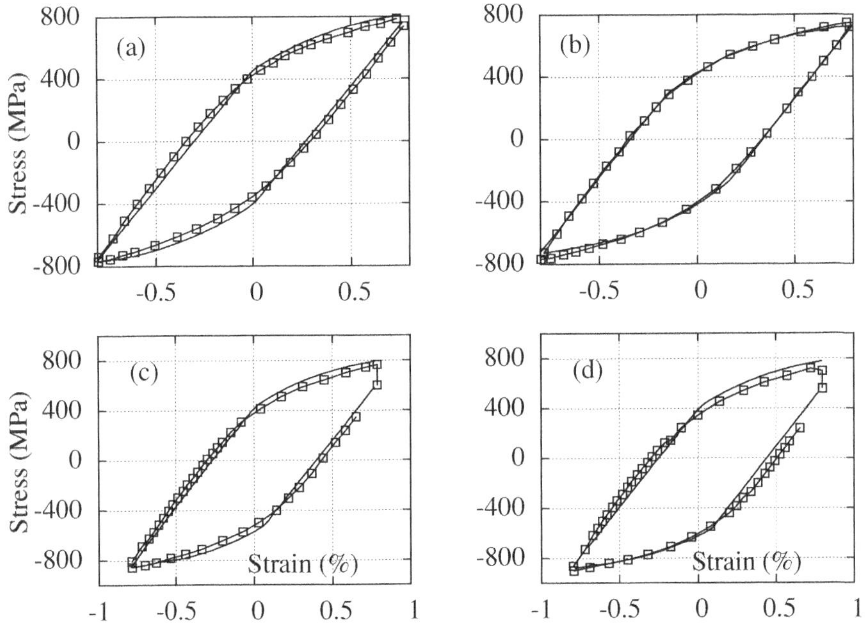

Figure 4.7. *Comparison between experimental results of 55NiCrMoV7 tempered martensitic steels at 500°C and the calculated responses using the 2M1C model type [VEL 06] with (a) a strain rate of 10^{-2} s^{-1}, (b) a strain rate of 10^{-4} s^{-1}, (c) a dwell time of 30 s and (d) a dwell time of 600 s*

Figure 4.8(b) compares the analytical response with experiments for a stabilized cycle at a strain rate of 10^{-2} s^{-1} and a temperature of 500°C.

According to Velay *et al.* [VEL 06], the two mechanisms are due to microstructural evolution. Carbide coarsening and the decrease in the dislocation density are the main mechanisms inducing cyclic softening, as shown by Mebarki [MEB 03]. The quenching generates a high dislocation density within the martensitic laths. The cyclic load leads to an increase in the mean-free motion of the dislocations. The rearrangement mechanism is also critical. In the formulation, it is represented through the isotropic variables R^1 (strong softening during the first 100 cycles) and R^2 (linear softening). Similarly to [MEB 03], the relative influence (one-third/two-thirds) of these mechanisms on the cyclic softening amplitude is illustrated in Figure 4.8(c).

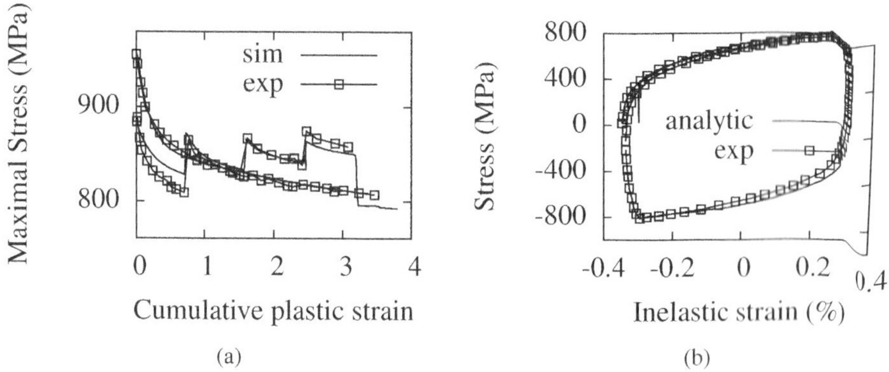

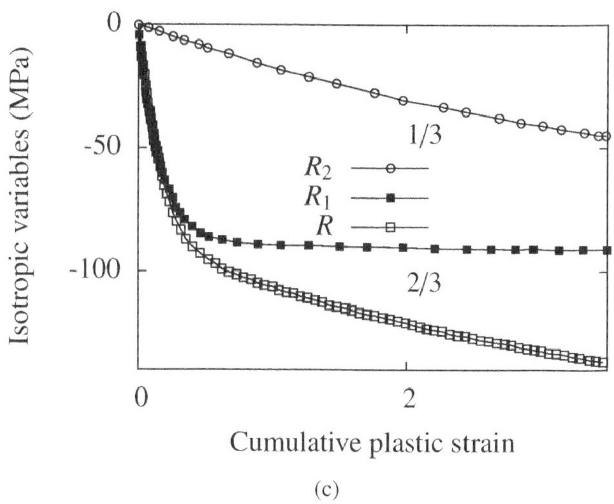

Figure 4.8. *55NiCrMoV7 at 500°C [VEL 06]: (a) comparison between experimental and calculated responses for a continuous cycling and different strain ranges, (b) comparison between the analytical solution and experiments at a strain rate of 10^{-2} s^{-1} for a stabilized cycle and (c) simulation of the relative influence of softening mechanisms*

4.6. 2017 Aluminum alloy

4.6.1. 2017A [SAÏ 12]

4.6.1.1. *Context of the case study*

In this section, the anisotropic mechanical behavior of 2017 aluminum alloy at room temperature is investigated. The specimens are machined from bars parallel to the extrusion axis inducing anisotropic behavior. For the specific case of the aluminum alloy, the two mechanisms do not refer precisely to lower-scale phenomena, but there are a few reasons why a model with two different strain tensors is relevant. The main point is that microstructural features such as intermetallic particles, dispersoids and fine precipitates interact with dislocations and generate a hard phase in the vicinity of the obstacles, and a soft phase in the matrix. The same is true at a higher level when considering the grain size effect and crystallographic textures that play a role in anisotropic mechanical properties. Hardening can be seen as the combined effect of the increase in dislocation density in soft grains (mechanism 1) which interacts with grains that are stronger due to their crystallographic orientation (mechanism 2).

The imposed loading in the tests performed here will involve two levels: the higher level is applied to shed light on the non-proportional effect during the first cycles, whereas the low level aims at describing the mechanical behavior for several hundreds of cycles.

The tested specimens are machined from bars parallel to the extrusion axis. The initial yield surface built by means of the proposed MM model is compared (Figure 4.9) to the experimental yield surface taken from the work of [ROU 85b].

In the previous work, the initial yield surface was not far from the Tresca criterion. In this chapter, the experimental ratio between plastic threshold in shear and axial load is $125/(200\sqrt{3})=0.36$, which is much lower than the value predicted by the Tresca (0.5) or von Mises criterion (0.577). In fact, the low value of the ratio may come from the anisotropy or result from a translation of the yield surface along the positive axis of the axial stress (kinematic hardening), due to a possible preloading during the material or specimen processing. The question has been answered by means of two cyclic tests shown in Figure 4.10. The first test consists of a preloading in tension, followed by a compression; meanwhile, the second test starts in compression and then tension. In both cases, a torsional loop follows the initial tensile cycle. Here, "equivalent strain" is equal to ε and $\gamma/\sqrt{3}$ for the axial and torsional tests, respectively, while the "equivalent stress" is used to denote σ and $\sqrt{3}\tau$ for the axial and torsional tests, respectively. The tension–compression and compression–tension preloadings provide the same stress–strain loop. In the following, anisotropy is introduced in the model in addition to the non-proportional effect.

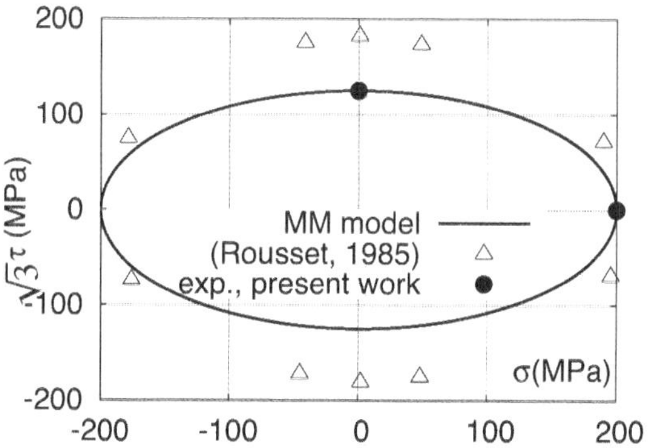

Figure 4.9. *2017A : Comparison between (i) the simulated yield surface obtained by the anisotropic MM model, (ii) experimental yield taken from [ROU 85b], (iii) experimental data [SAÏ 12]*

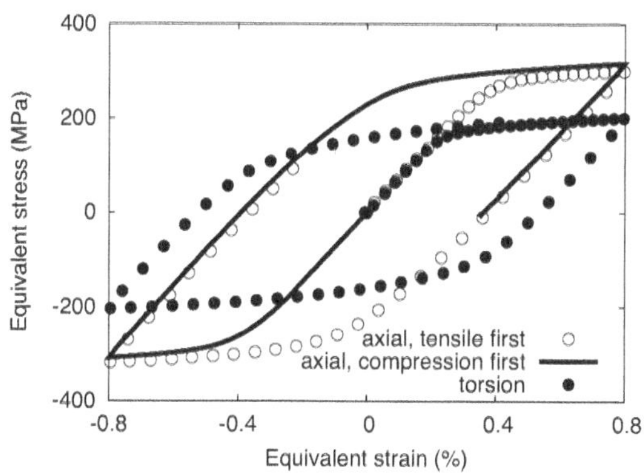

Figure 4.10. *2017A [SAÏ 12]: verification of the anisotropy. Comparison between the tension–compression, compression–tension and torsional first loops (experimental results)*

4.6.1.2. *Particular model features*

The 2M1C model was selected to describe the anisotropic behavior of the 2017 alloy. The anisotropy will be simply represented by changing the former J invariant used in the yield criterion into J_L, which introduces a new fourth-order tensor $\underset{\approx}{\mathbf{L}}$. According to the Voigt notation, $\underset{\approx}{\mathbf{L}}$ is such that $L^{ij} = L^{ji}$ and reduces to six material parameters (L_a, L_b, L_c, L_d, L_e and L_f) in the case of orthotropic symmetry. The 6×6 matrix representing $\underset{\approx}{\mathbf{L}}$ in the orthotropic frame is diagonal:

$$\underset{\approx}{\mathbf{L}} = \begin{pmatrix} L_a & 0 & 0 & 0 & 0 & 0 \\ 0 & L_b & 0 & 0 & 0 & 0 \\ 0 & 0 & L_c & 0 & 0 & 0 \\ 0 & 0 & 0 & L_d & 0 & 0 \\ 0 & 0 & 0 & 0 & L_e & 0 \\ 0 & 0 & 0 & 0 & 0 & L_f \end{pmatrix} \tag{4.7}$$

The anisotropic yield criterion can be written as

$$f = \sqrt{\left(J_L(\underset{\sim}{\sigma} - \underset{\sim}{\mathbf{X}}^1)\right)^2 + \left(J_L(\underset{\sim}{\sigma} - \underset{\sim}{\mathbf{X}}^2)\right)^2} - R - R_0 \tag{4.8}$$

$$J_{LI} = J_L(\underset{\sim}{\sigma} - \underset{\sim}{\mathbf{X}}^I) = \left(\frac{3}{2}(\underset{\sim}{\mathbf{s}} - \underset{\sim}{\mathbf{X}}^I) : \underset{\approx}{\mathbf{L}} : (\underset{\sim}{\mathbf{s}} - \underset{\sim}{\mathbf{X}}^I)\right)^{1/2} \qquad I = 1, 2 \tag{4.9}$$

In addition to the yield function, the normal to the yield surface for each mechanism is modified as

$$\underset{\sim}{\mathbf{n}}^1 = \frac{3}{2}\frac{\underset{\approx}{\mathbf{L}} : \left(\underset{\sim}{\mathbf{s}} - \underset{\sim}{\mathbf{X}}^1\right)}{\left(J_{L1}^2 + J_{L2}^2\right)^{1/2}} \quad \text{and} \quad \underset{\sim}{\mathbf{n}}^2 = \frac{3}{2}\frac{\underset{\approx}{\mathbf{L}} : \left(\underset{\sim}{\mathbf{s}} - \underset{\sim}{\mathbf{X}}^2\right)}{\left(J_{L1}^2 + J_{L2}^2\right)^{1/2}} \tag{4.10}$$

A time-independent framework is considered, so that the plastic multiplier is given by the consistency condition $\dot{f} = 0$:

$$\frac{\partial f}{\partial \underset{\sim}{\sigma}} : \dot{\underset{\sim}{\sigma}} + \frac{\partial f}{\partial \underset{\sim}{\mathbf{X}}^1} : \dot{\underset{\sim}{\mathbf{X}}}^1 + \frac{\partial f}{\partial \underset{\sim}{\mathbf{X}}^2} : \dot{\underset{\sim}{\mathbf{X}}}^2 + \frac{\partial f}{\partial R} : \dot{R} = 0 \tag{4.11}$$

4.6.1.3. *Numerical results*

The model was compared to a series of strain-controlled biaxial tests performed at room temperature. The experimental database involves three loading paths. (i) In the first test, the specimen was subjected to 20 tension-compression cycles and then 20 torsion cycles were applied. (ii) In the second test, a cyclic torsion loading of 20 cycles was followed by 20 tension-compression cycles. (iii) In the third test, the specimen was subjected to 20 cycles. Each cycle is composed of two sequences: during the first sequence, a pure axial strain is imposed, whereas a torsional loading is imposed in the second sequence. A comparison between simulated responses using material parameters of Table 4.7 and experimental data is shown in Figure 4.11.

E=72,000	$\nu = 0.33$	R_0=280	Q_1=92	b_1 =15.8
Q_m=71	d=58.60	b_2=18.4		
C_{11}=7,300	C_{22}=205,300	C_{12}=35,000	D_1=650	D_2=440
L_a = 1	L_b=1	L_c=1	L_d=2.56	

Table 4.7. *Material parameters (MPa, s) of the anisotropic MM model for the 2017 aluminum alloy (L_e and L_f are not identified)*

4.6.2. *2017A [TAL 15]*

The 2017A behavior is also investigated in [TAL 15]. The numerical simulation is performed thanks to the 2M1C model summarized in Table 3.3 and resumed in section 3.1. In order to identify the material parameters of the model, the strain-controlled experiments of Figure 4.12 and the stress-controlled experiments of Figure 4.13 were considered. The tests are the following:

- 2017A_50-250: 100 cycles: σ_{max}=300 MPa, σ_{min}=-200 MPa;
- 2017A_300-0.4: 100 cycles: σ_{max}=300 MPa, $\frac{\gamma_{max}}{\sqrt{3}}$=0.4%;
- 2017A_Ax_0.8: 20 cycles: $\varepsilon_{zz} = \pm 0.8\%$;
- 2017A_Sh_0.8: 20 cycles: $\gamma/\sqrt{3} = \pm 0.8\%$;
- 2017A_Cross_Ax-Sh_0.8: 20 cycles: $\varepsilon_{zz} = \pm 0.8\% \rightarrow \gamma/\sqrt{3} = \pm 0.8\%$.

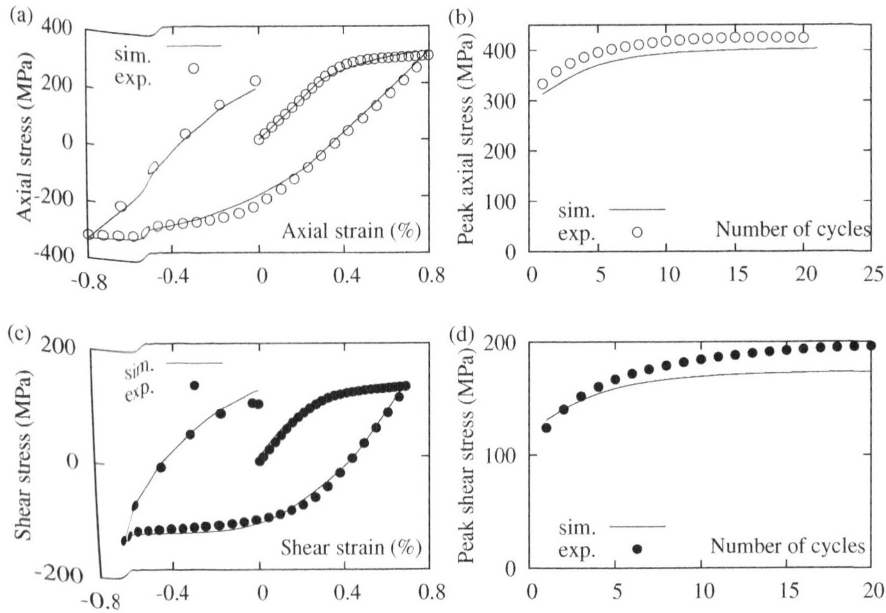

Figure 4.11. *2017A [SAÏ 12]: comparison between simulated and experimental data for the first hysteresis loop. Third test: (a) first axial loop, (b) peak axial stress, (c) first shear loop and (d) peak shear stress*

The following assumptions were considered for this material during the identification process. Some of these assumptions were supported by the tests performed here on the material:

— insignificant memory of the strain hardening;

— time-dependent phenomena were assumed to be small, so that $K = 31$ and $n = 11$.

Furthermore, the optimization process leads to very small values of D_1 and D_2; therefore, it is assumed that $D_1 = D_2 = 0$. The other parameters are given in Table 4.8, and the corresponding comparisons between simulated responses and experimental data are shown in Figures 4.14 and 4.15.

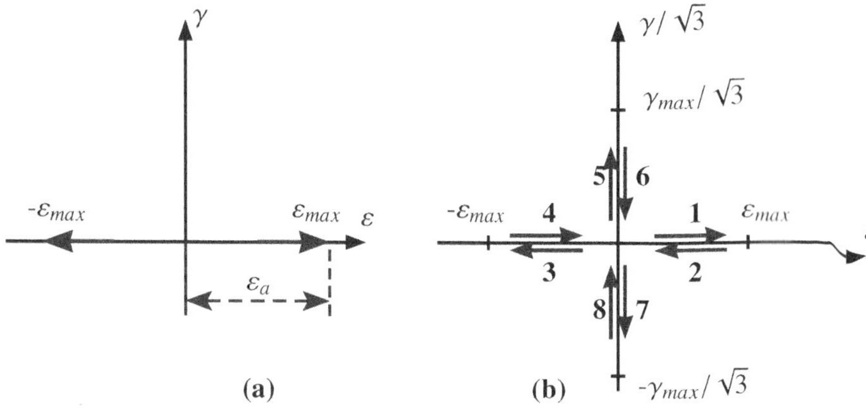

Figure 4.12. *Loading paths applied in the tests performed: (a) cyclic tension-compression under strain control; (b) path "cross" where each cycle is composed of one cycle of tension-compression with given amplitude ε_a followed by one cycle of fully reversed torsion with the same equivalent amplitude. For a color version of the figure, see www.iste.co.uk/cailletaud/multi-mechanism.zip*

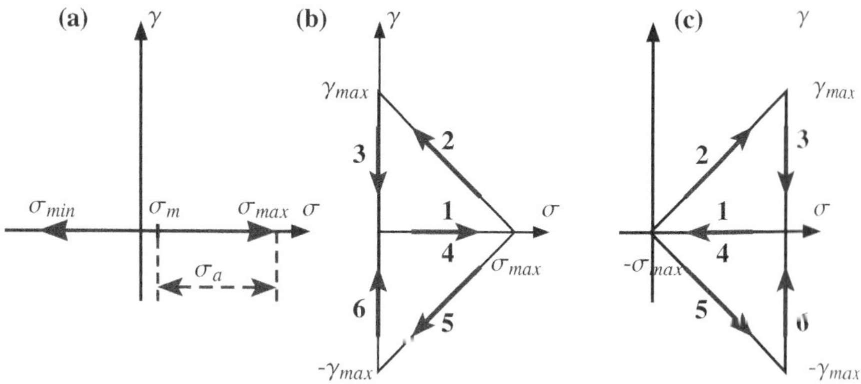

Figure 4.13. *Loading paths applied in the tests performed: (a) cyclic tension-compression under stress control with non-zero mean stress (σ_m); (b) path "triangle" combining prescribed stress between 0 and σ_{max} and torsional strain between γ_{max} and -γ_{max} following a triangular shape; (c) same path as (b) with $-\sigma_{max}$ instead of σ_{max} (inverse triangle). For a color version of the figure, see www.iste.co.uk/cailletaud/multi-mechanism.zip*

R_0	Q_2	b_2	Q_{3m}	b_3	d_3	C_{11}	C_{22}	C_{12}
283	92	12.2	500	18.5	15	5,030	607,000	–35,000
Q_{30}	L_a	L_b	L_c	L_d	L_e	L_f		
0	1	1	1	2.56	1	1		

Table 4.8. *Material parameters for the 2017A alloy (MPa, s)*

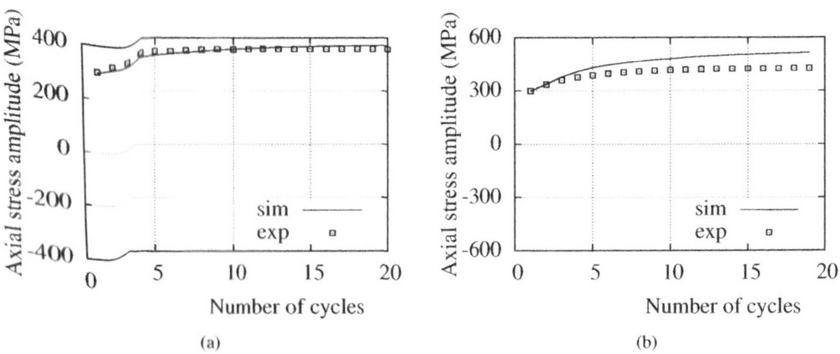

Figure 4.14. *2017A [TAL 15]: simulations of the tests used in the optimization process for the 2017A alloy. (a) axial stress amplitude versus the number of cycles for the test 2017A_Ax_0.8; (b) axial stress amplitude versus the number of cycles for the test 2017A_Cross_0.8*

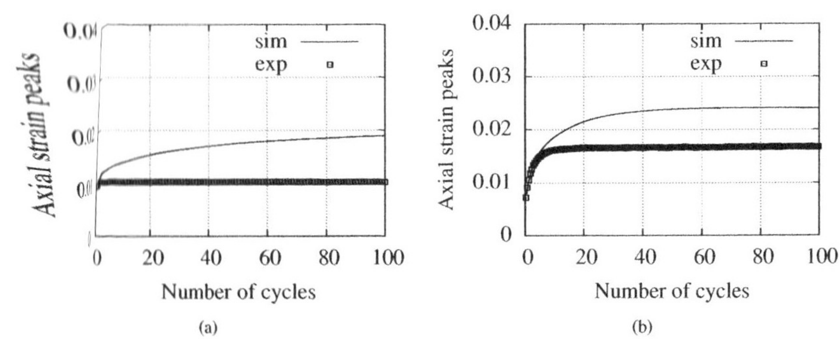

Figure 4.15. *2017A [TAL 15]: simulations of the tests performed on 2017A alloy under stress control. Axial strain peaks versus the number of cycles, comparison between experiments and their simulations: (a) test 2017A_50-250 and (b) test 2017A_300-0.4*

4.7. 304 austenitic stainless steel

This section aims at presenting the simulation of some experimental databases related to 304SS by means of different versions of the 2M1C and 2M2C models. The origin of the tests is as follows:

– the databases at room temperature were taken from [HAS 94a, HAS 08, TAL 11];

– the database at 350°C was performed by [TAL 14].

4.7.1. *304SS at room temperature [HAS 08]*

Several studies available in the literature have underlined the complexity of the mechanical responses of this material under uniaxial and biaxial stress-controlled loading histories. It has been demonstrated that ratcheting simulations may fail due to the poor representation of the shape of the hysteresis curve. Moreover, the simulation of the biaxial ratcheting responses is controlled by the evolution of the yield surface, which is determined by the kinematic hardening rule of the model. Consequently, most models failed to simulate ratcheting responses.

Following the work of Aubin *et al.* [AUB 03], who conducted a set of experiments demonstrating the influence of non-proportionality on ratcheting responses of duplex stainless steel, a set of experiments was performed on 304SS under various biaxial stress-controlled non-proportional histories. This new set of data contains original effects that allow us to discriminate between various model formulations.

Biaxial stress-controlled experiments were performed on 304SS tubular specimens not only under tensile and shear loadings, but also with cross and square paths.

The model used is the version proposed by Taleb *et al.* [TAL 06]. The set of constitutive equations is summarized in Table 3.3. It is worth noting that although this model involves only 10 material parameters, it gives reasonable results compared to the Chaboche-type model with about 40 material parameters.

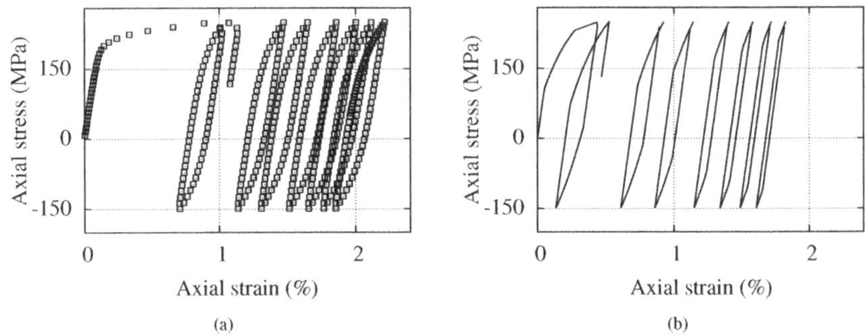

Figure 4.16. *304SS [HAS 08]: One-dimensional ratcheting tests: (a) experiment, (b) simulation*

The stress–strain responses related to the different loading histories are detailed in [HAS 08]. The most significant results are shown below. In Figure 4.16(a), the axial stress–strain hysteresis loops at transient cycles are shown. The viscous effect that produces a smooth unloading at the first cycle decreases gradually in the subsequent cycles. The transient ratcheting rate is a function of cyclic hardening of materials, as noted in [HAS 94a, HAS 94b, JIA 08]. As shown in Figure 4.16, the cyclic softening induces a small increase in the width of the hysteresis loops. This is in agreement with other results obtained in [HAS 94a, HAS 94b]. A comparison between numerical simulations and experiments is shown in Figures 4.16 and 4.17. Table 4.9 shows the values of the material parameters.

4.7.2. *304SS at room temperature [TAL 11]*

The contribution of creep in the cyclic accumulation of the inelastic strain in the stress-controlled tests was first investigated by [TAL 11] for 304SS at room temperature. In this work, the key role of time-dependent phenomena was pointed out thanks to classical creep tests with various hold times, intensities and directions of the applied stress. Furthermore, ratcheting tests have been performed by considering two different stress rates; meanwhile, mean stress (50 MPa) and stress amplitude (100 MPa) remain unchanged. The specimen was subjected to 100 cycles at a stress rate of 40 MPa/s, followed by 100 cycles at a stress rate of 8 MPa/s. The obtained axial strain peaks versus the number of cycles clearly demonstrate the time dependency (Figure 4.18).

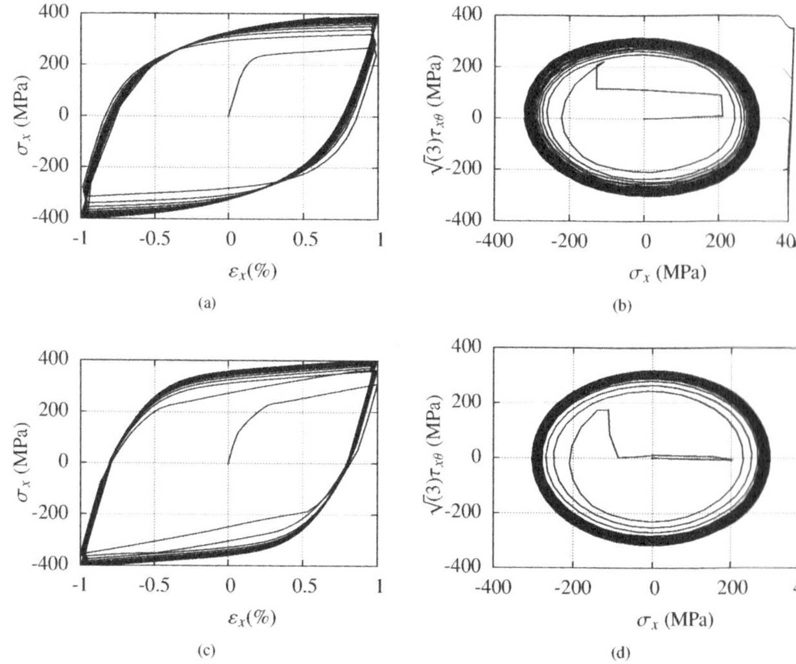

Figure 4.17. *304SS [HAS 08]. Comparison of the mechanical responses for proportional and non-proportional loading: (a), (b) experiment; (c), (d) simulation*

R_0	Q	b	η	C_{11}	C_{22}	C_{12}	D_1	D_2
200	200	20	1	481,056	13,533	5,458	1,513	21

Table 4.9. *304SS at room temperature [HAS 08]. Identified parameters for the 2M1C model (MPa, s)*

In order to discriminate the contributions of creep and ratcheting, a typical creep–ratcheting test composed of two sequences was performed: the first one is a classical creep test where the maximum stress σ_M is applied and kept constant during a given hold time. The second part is a classical ratcheting test where the specimen is submitted to a cyclic stress between σ_M and a selected minimum stress, taken as a parameter of the study. Figure 4.19 shows the stress–strain evolution of the creep–ratcheting test, where the mean stress and the amplitude are, respectively, 50 and 200 MPa. The material exhibits plastic shakedown or a very small ratcheting, which confirms that the cyclic inelastic strain evolution observed is mainly due to creep. The same result was also obtained for 316SS at room temperature [TAL 13a].

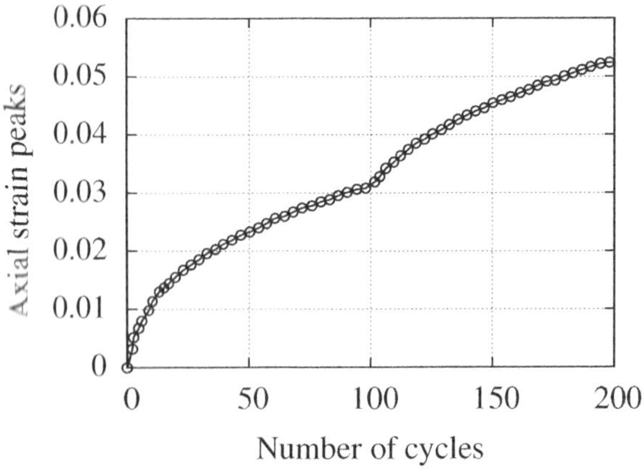

Figure 4.18. *304SS at room temperature [TAL 11]. Stress-controlled test. Axial strain peaks in the ratcheting test with two stress rates: 100 cycles at 40 MPa/s followed by 100 cycles at 8 MPa/s*

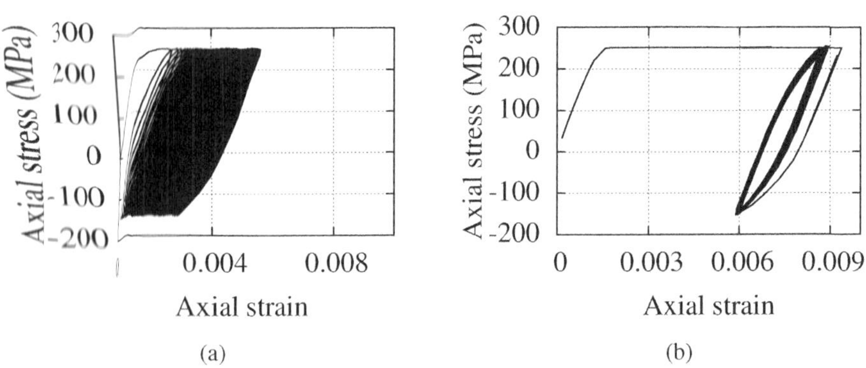

Figure 4.19. *304SS at room temperature [TAL 11]. (a) Axial stress-strain plots illustrating the ratcheting phenomenon; (b) creep-ratcheting test*

4.7.3. *304SS at 350°C [TAL 14]*

The well-documented paradox of austenitic stainless steel is that the behavior is affected by strain rate at room temperature, but this is no longer the case at 350°. This section deals with cyclic behavior at this final temperature [TAL 14], with a simulation by the 2M1C model.

The two local stresses are computed according to a static scheme. The kinematic hardening is described by two linear state variables (Table 3.3). To describe the size of the elastic domain, the isotropic strain hardening variable is the sum of a first variable accounting for the strain memory effect in the plastic strain space and a second variable to represent the non-proportional behavior (see section 2.5.1).

4.7.3.1. *Tests performed under load control*

Two cyclic loadings under stress control were performed. The first one is proportional (tension–compression), while the second follows a circular path in the shear–axial stress space, starting with an initial axial load. In the first cycle of each test, the peak (axial) stress is maintained constant for 15 min. A total of 100 cycles were applied in each test. The two tests considered in tension–compression were performed with (σ_m=35 MPa, σ_a=195 MPa) and (σ_m=70 MPa, σ_a=160 MPa), σ_m being the mean (axial) stress and σ_a the amplitude. The test with the circular path (σ_m=35 MPa, σ_a=195 MPa) starts by a hold time of 15 min at the peak axial stress of the first cycle (230 MPa) followed by 100 cycles along a circular path centered in (35, 0) with an equivalent (von Mises) amplitude (radius) of 170.

4.7.3.2. *Tests performed under strain control*

Strain-controlled experiments were also performed under proportional and non-proportional loading paths. For the proportional loading, each specimen is applied a two-block loading: 20 cycles of tension–compression with an amplitude of 1% followed by 20 cycles of fully reversed cyclic torsion with the same equivalent amplitude (1%). Note that, in the first tension–compression cycle, the peak strain is maintained constant for 15 min in order to quantify the relaxation effect.

The non-proportional tests introduce a "cross" path. The specimen is successively submitted to 20 blocks made of one tension–compression cycle with an amplitude of 1%, followed by one cycle of fully reversed torsion with the same equivalent amplitude.

4.7.3.3. *Numerical results*

Figure 4.20 shows a comparison between the experimental data and the simulated responses. The parameters of the 2M1C model are presented in Table 4.10. In the cross tests, a small difference is observed between the amplitudes of the axial cycles (Exp_axial) and the shear cycles (Exp_shear). The equivalent amplitudes were calculated using the von Mises criterion. The general agreement is satisfactory.

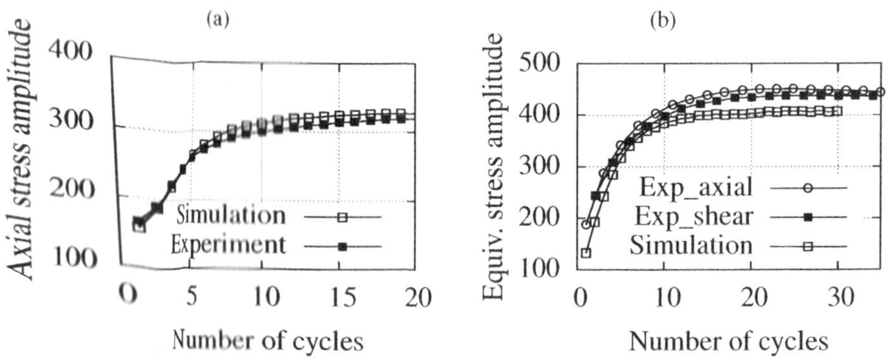

Figure 4.20. *304SS at 350°C [TAL 14]. Simulation of the tests performed under strain control. (a) 20 first cycles of the tension compression test, (b) cross test. "Exp_axial" refers to the amplitude of the axial cycles, meanwhile "Exp_shear" refers to the amplitude of the torsion cycles*

R_0	Q_{10}	Q_1	b_1	η_{mem}	μ	Q_{20}	Q_2
150	0	202	9.75	0.0015	1,498	0	226
b_2	d_2	n	K	C_{22}	C_{12}	D_1	D_2
38.11	228	20	10	4,800	–20,844	0	0

Table 4.10. *Identified parameters for the 304SS at 350°C (MPa, s), $C_{11} = E = 170,000$*

4.7.4. *304SS at room temperature [HAS 94a], 2M1C-3M1C*

In this section, the 2M1C and 3M1C models are compared to the experiments obtained on 304SS by [HAS 94a]. The set of experimental responses considered involves a cyclic uniaxial tension–compression test under strain control, tension–compression ratcheting tests under various mean stresses and amplitudes, and a tension–torsion out-of-phase test.

The specimens were heat-treated at 1070℃ for 40 min before testing. The result is that the first loading induces a sharp elastic–plastic transition, which is no longer present in the subsequent hysteresis loops. Indeed, it was shown by the authors [HAS 08] that the first loading and the mechanical steady state cannot be represented by the same set of parameters in the framework of a single-potential plasticity model. This restriction can be suppressed in the MM models thanks to the different deformation regimes. The tests considered are the following:

– a fully reversed strain-controlled test with a strain amplitude of 1%;

– four uniaxial ratcheting tests with the following (mean stress, stress amplitude) couples (in MPa): (36, 250), (36, 220), (36, 235) and (46, 235);

– a 90-degree out-of-phase strain-controlled test.

The list of the calibrated coefficients for the 2M1C and 3M1C models is given in Tables 4.11 and 4.12, and the corresponding comparisons between simulated responses and experimental data are shown in Figures 4.21–4.23. The material parameters of the 2M1C model are determined according to the following procedure:

– the time-independent plasticity is obtained for large values of the parameters n or $1/K$;

– the parameters of isotropic hardening (R_0, Q_{10}, Q_{1m}, b_1, η_{mem}, μ, b_3 and Q_3) are adjusted in order to fit the first and stable hysteresis loops;

– the degree of additional hardening is managed by the isotropic parameters (Q_{2m}, b_2 and d_2);

– the kinematic parameters are finally estimated to simulate the ratcheting responses as accurately as possible.

The material parameters of the 3M1C model are determined as follows:

– the isotropic parameters R_0, Q_{10}, Q_{1m}, Q_3 and Q_{2m} are deduced by multiplying by $\sqrt{3} / \sqrt{2}$ the values obtained for the 2M1C model;

– the remaining isotropic parameters and the kinematic parameters are then adjusted for the 2M1C models.

Isotropic parameters										
R_0	Q_{10}	Q_{1m}	b_1	η_{mem}	μ	Q_{2m}	b_2	d_2	Q_3	b_3
229	–20.5	209	95	0.15	150	426	5.5	27.5	–8.2	100
Kinematic parameters										
C_{11}	C_{22}	C_{12}	D_1	D_2	η					
219,875	3,544	600	10	20	0.98					
Elasticity, Norton parameters										
E	ν	n	K							
190,000	0.3	11	25							

Table 4.11. *Identified parameters of the 2M1C model for 304SS at room temperature. The experimental database is taken from Hassan and Kyriakides [HAS 94a] (MPa, s)*

Isotropic parameters										
R_0	Q_{10}	Q_{1m}	b_1	η_{mem}	μ	Q_{2m}	b_2	d_2	Q_3	b_3
280	–25	255	150	0.2	180	530	90	40	–10	100
Kinematic parameters										
C_{11}	C_{22}	C_{33}	C_{12}	C_{13}	C_{23}	D_1	D_2	D_3	η	
200,000	1,750	4,810	4,100	2,210	4,550	5	0	5	0.98	
Elasticity, Norton parameters										
E	ν	n	K							
190,000	0.3	11	31							

Table 4.12. *Identified parameters of the 3M1C model for 304SS at room temperature. The experimental database is taken from Hassan and Kyriakides [HAS 94a] (MPa, s)*

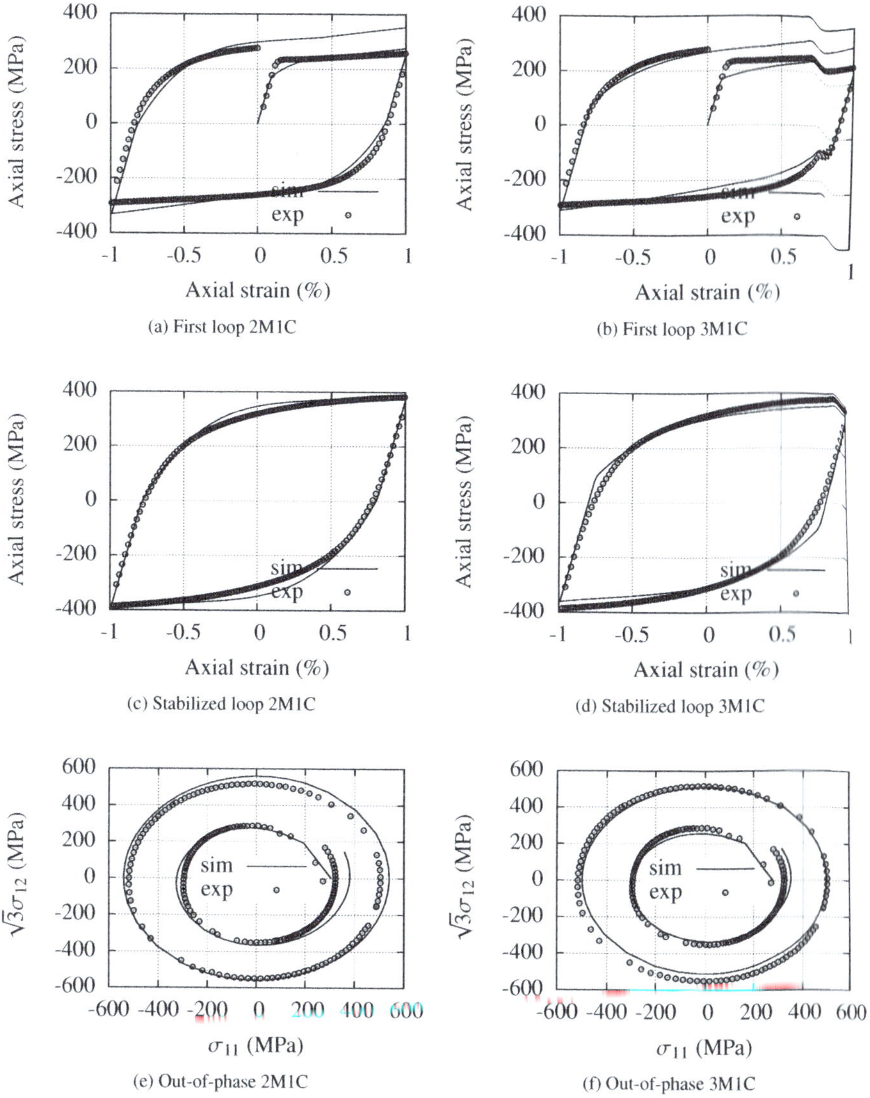

Figure 4.21. *304SS at room temperature [HAS 94a]. Comparison between experiments and simulations for the MM models. (a)–(d) Fully reversed strain control tests $R_\varepsilon = -1$. (e), (f) Out-of-phase test (first and stabilized loops)*

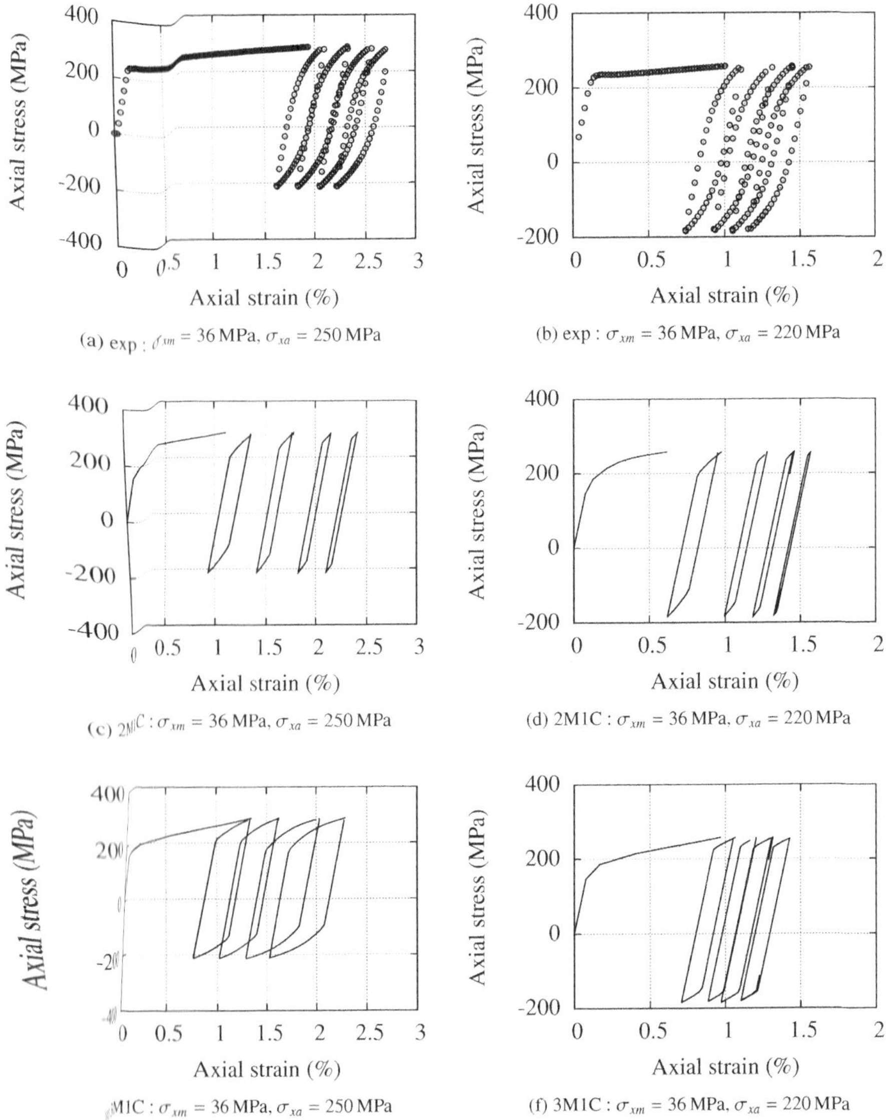

Figure 4.22. *304SS at room temperature [HAS 94a]. Comparison between experiments and simulations for the MM models. Uniaxial ratcheting test (mean stress σ_{xm}, stress amplitude σ_{xa}) 10th, 30th, 60th and 90th cycles*

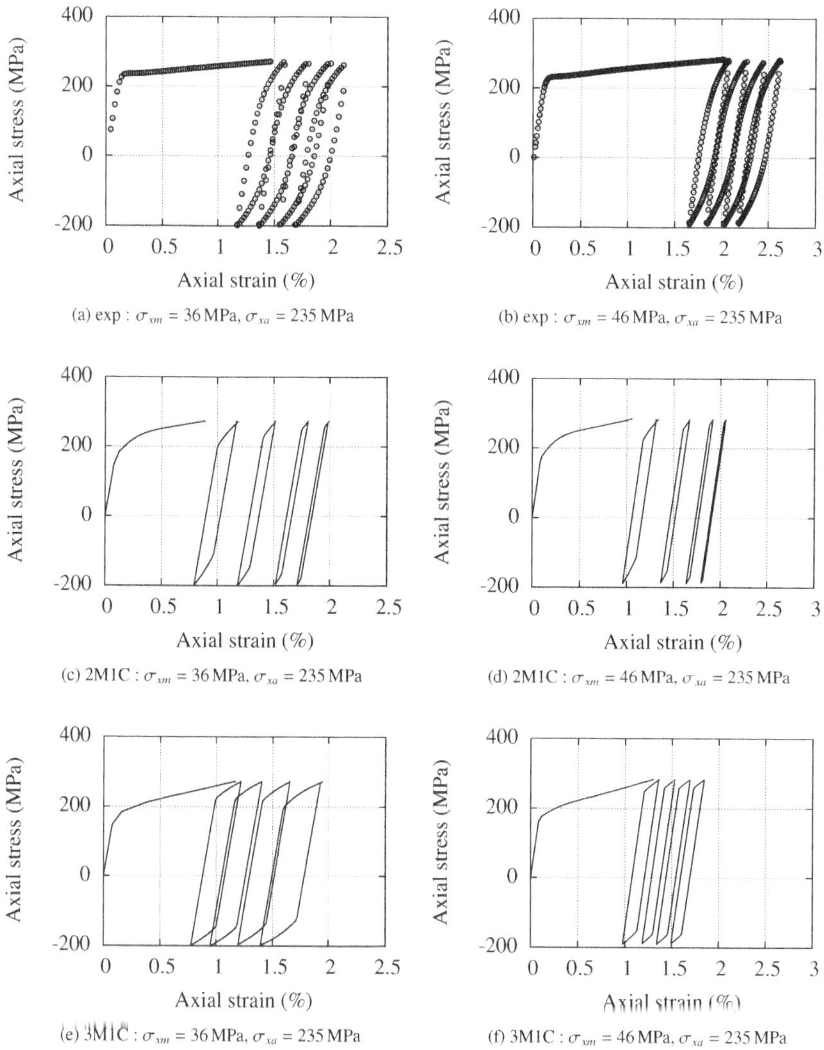

(a) exp : σ_{xm} = 36 MPa, σ_{xa} = 235 MPa

(b) exp : σ_{xm} = 46 MPa, σ_{xa} = 235 MPa

(c) 2M1C : σ_{xm} = 36 MPa, σ_{xa} = 235 MPa

(d) 2M1C : σ_{xm} = 46 MPa, σ_{xa} = 235 MPa

(e) 3M1C : σ_{xm} = 36 MPa, σ_{xa} = 235 MPa

(f) 3M1C : σ_{xm} = 46 MPa, σ_{xa} = 235 MPa

Figure 4.23. *304SS at room temperature [HAS 94a]. Comparison between experiments and simulations for the MM models. Uniaxial ratcheting test (mean stress σ_{xm}, stress amplitude σ_{xa}) 10th, 30th, 60th and 90th cycles*

The agreement between the simulated responses and the experimental results for the set of material parameters is globally satisfactory for both models. They are able to capture the major trends in the strain-controlled tests. Specifically, the out-of-phase test is well described by the two models.

The good simulation of the shape of the stress–strain loops for the four given cycles has been preferred to the evolution of the peak strain for the ratcheting tests; nevertheless, the strain shift between the cycles is well described. Improvements in terms of the first monotonic stage and the shape of the hysteresis loops are obtained with the 3M1C model, due to the larger number of parameters.

4.7.5. *304SS at room temperature [HAS 08, TAL 10], 2M1C–3M1C*

The capabilities of the 2M1C and 3M1C models are tested against a database containing cyclic proportional and non-proportional paths under stress control or combined stress–strain control. The following five experiments are taken from [HAS 08, TAL 10]:

– a fully reversed strain-controlled test with a strain amplitude of 1%;

– "Uni-Cross". A cross stress path [HAS 08] in which 100 tension–compression cycles are followed by 100 blocks of cycles. One block starts with a tension–compression loading of +250/ – 150 MPa followed by one fully reversed cycle in shear at the same equivalent stress. For all cycles, the equivalent amplitude is 200 MPa;

– "Tri-Inv-Tri". A triangular path ratcheting [HAS 08] that consists of a combination of stress- and strain-controlled loading segments connecting linearly the following points in the plan (axial stress, equivalent shear strain): (0, 0), (275, 0), (0, 0.5%), (0, 0), (275, 0), (0, –0.5%), (0, 0) for the so-called "forward triangle" sequence, followed by (0, 0), (–275, 0), (0, 0.5%), (0, 0), (–275, 0), (0, -0.5%), (0, 0) for the "inverse triangle" sequence;

– "Creep-Ax-Ratch-Ax". A creep followed by a ratcheting test [TAL 11] that starts with a tensile stress of 275 MPa maintained 2.5 h, followed by a tension–compression test with a mean stress of 125 MPa and an amplitude of 150 MPa;

– "Creep-Ax-Ratch-Cir". A creep followed by a circular ratcheting test [TAL 11] that starts with a tensile stress of 275 MPa maintained 2.5 h, followed by 100 stress-controlled cycles, then a circular path (center (125, 0); radius 150 MPa) in the axial-equivalent shear stress plane.

A time-dependent framework is chosen to simulate this database [TAL 10]. The identification process uses only the strain-controlled experiments. For the

3M1C model, a first guess of the isotropic parameters can be deduced from those related to the 2M1C model by multiplying them by the ratio $\sqrt{3/2}$. The rest of the parameters of the 3M1C model are determined by best fit to the experiments.

The list of the material parameters is given, respectively, in Tables 4.13 and 4.14 for the 2M1C and 3M1C models, and the comparison between the simulated responses and the experimental data is shown in Figures 4.24, 4.25 and 4.26. Note that the axial strain peak for the first cycle corresponds to the value at the end of the preceding creep sequence (Figure 4.26).

Isotropic parameters								
R_0	Q_{10}	Q_{1m}	b_1	η_{mem}	μ	Q_{2m}	b_2	d_2
160	–60	210	33	0.15	117	204	14	51
Kinematic parameters								
C_{11}	C_{22}	C_{12}	D_1	D_2	η			
190,000	218,304	13,539	0.94	5,025	0.98			
Elasticity, Norton parameters								
E	ν	n	K					
190,000	0.3	3.7	650					

Table 4.13. *304SS at room temperature [TAL 11]: identified parameters of the 2M1C model (MPa, s) (b_3 and Q_3 are not used)*

Isotropic parameters									
R_0	Q_{10}	Q_{1m}	b_1	η_{mem}	μ	Q_{2m}	b_2	d_2	
176.	–100	256	40.6	0.1	200	248	13	70	
Kinematic parameters									
C_{11}	C_{22}	C_{33}	C_{12}	C_{13}	C_{23}	D_1	D_2	D_3	η
575,428	79,896	196,573	97.5	–16,536	3,269	0	175	3,219	0.99
Elasticity, Norton parameters									
E	ν	n	K						
190,000	0.3	3.12	700						

Table 4.14. *Identified parameters of the 3M1C model for 304SS at room temperature (MPa, s) (b_3 and Q_3 are not used)*

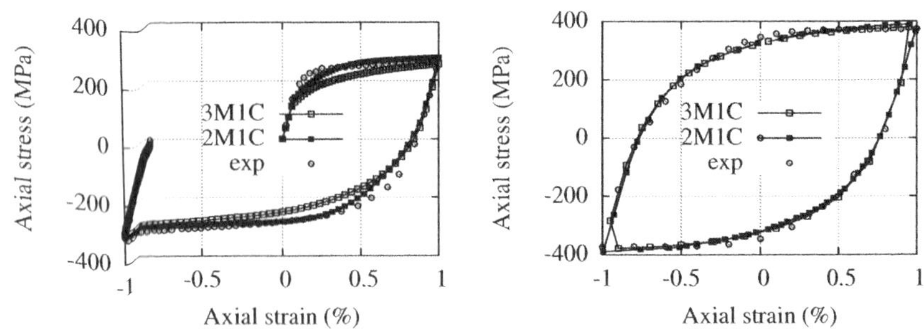

(a) First loop (b) Stabilized loop

Figure 4.24. *Comparison between experiments [TAL 11] (304SS) and simulations for the MM models: fully reversed strain control test R_ε=–1*

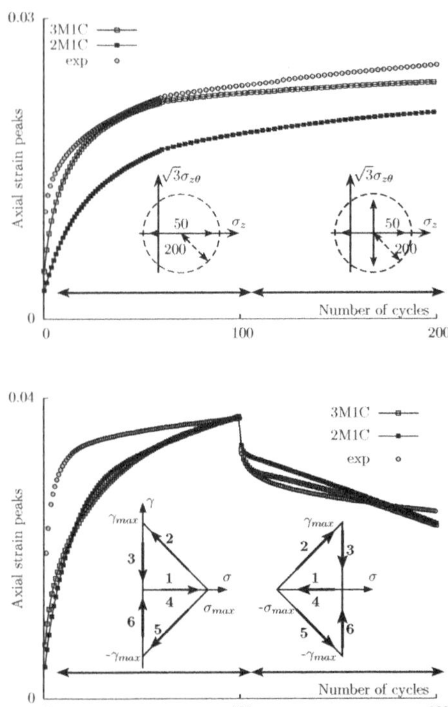

Figure 4.25. *Evolution of the axial strain peaks versus the number of cycles for the test "Uni-Cross"-"Tri-Inv-Tri"*

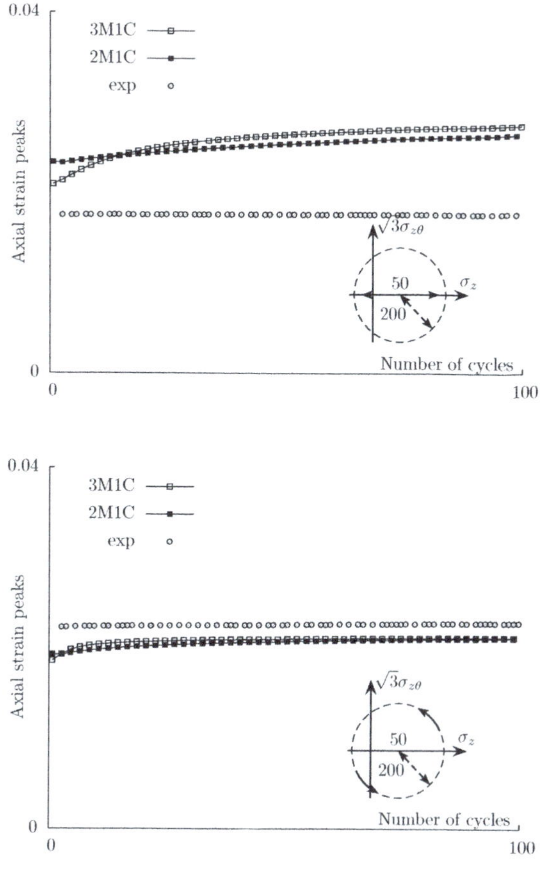

Figure 4.26. *Evolutions of the axial strain peaks versus the number of cycles for the test "Creep-Ax-Ratch-Ax" –"Creep-Ax-Ratch-Cir"*

From Figures 4.25 and 4.26, it can be concluded that the correlation between the simulated responses and experimental results for the optimal set of material parameters is globally satisfactory:

– as expected, the 3M1C model gives a more precise description of non-proportional experimental results;

– the 2M1C model is able to capture the major trends in the tests. Specifically, the creep followed by the ratcheting test is better described by the 2M1C model than by the 3M1C model.

4.8. 316 austenitic stainless steel

Various experimental databases related to 316SS are considered in this section, namely:

– the first database, performed at room temperature, is taken from [POR 00]. It aims to check the capabilities of the models 2M2C, 2M1C and 3M1C;

– the second database [TAL 13a] attempts to discriminate between creep at room temperature and ratcheting [TAL 11]. A particular attention is then paid to the viscous parameters;

– the third database characterizes the behavior at 350℃, where creep is not significant [TAL 14]. The 2M1C model is selected for the numerical simulations of the effect on non-proportional loading.

The main reason for introducing several mechanisms for 316SS is the presence of dislocation patterns in the grains. They form either walls or cells according to the level of the loading and its type. The wall/interwall areas can then induce two deformation modes.

4.8.1. 316SS at room temperature [POR 00]

A numerical simulation of the experimental database of 316SS at room temperature taken from [POR 00] was performed with the 2M1C and 2M2C versions proposed in [SAÏ 07] and recalled in Table 3.4. A special attention is paid to the ratcheting effect and the additional hardening due to out-of-phase loading. In these MM versions, the main features are:

– the inelastic strain is the average of the irreversible deformations of the two mechanisms according to $(1 - z)$ and z, which are the volume fractions attributed to the first and second mechanisms, respectively;

– the use of the β-transition rule of the micro-mechanical models (see, for instance, [CAI 94]) in order to control the kinematic hardening matrix characteristics;

– the modification of the fading memory terms according to the model proposed by [BUR 87] to calibrate multi-axial ratcheting. As underlined in Table, the fading memory terms in the kinematic hardening rule of each mechanism are expressed using the back stresses $\underset{\sim}{\mathbf{X}}^I$ rather than the internal variables $\underset{\sim}{\alpha}^I$.

The experimental database performed by [POR 00] at room temperature (25°C) involves the following loading histories:

– monotonic tensile test, $\dot{\varepsilon} = 3.10^{-4}s^{-1}$;

– cyclic uniaxial tension–compression for three strain ranges, $\Delta\varepsilon_{11} = 0.5$, 0.65 and 0.8%;

– tension–torsion ratcheting tests with two values of the tensile stress (σ_{11} =80 and 100 MPa) and with various shear strain ranges, $\Delta\varepsilon_{12}$=0.1%, 0.2% and 0.5%;

– tension–torsion out-of-phase test at steady state.

The evolution rules of the accommodation variables are taken to be linear for the two models in order to reduce the number of parameters in the identification procedure. For the 2M2C model, one kinematic hardening variable is taken as quasi-linear to reduce ratcheting strain, which is too large with two nonlinear kinematic variables. The list of the calibrated coefficients for the 2M1C and 2M2C models is given in Tables 4.15 and 4.16, and the corresponding comparison between the simulated responses and experimental data is shown in Figure 4.27. In addition to the tension–compression ratcheting test, which is not included in the identification procedure, two tension–torsion ratcheting tests are used to validate *a posteriori* the prediction of the proposed model (Figure 4.28).

R_0 = 163 MPa	Q = 129 MPa	b = =2.8	z=0.04	μ'=20 GPa
C_{11}=115.6 GPa	C_{22}=12.9 GPa	C_{12}=19.5 GPa	D_1=112.5	D_2=1464
δ_1=0.108	δ_2=0.0025	d_1=0	d_2=0	

Table 4.15. *Material parameters of the 2M1C model for the β-scale transition rule. 316SS at 25°C, [POR 00], E=192 GPa, ν = 0.3*

R_{01}=145.6 MPa	Q_1=0 MPa	b_1=0		
R_{02}=220 MPa	Q_2 =200 MPa	b_2=4		
C_{11}=13 GPa	C_{22}=11 GPa	C_{12}=-7.3 GPa	D_1=0	D_2=388
δ_1=0.038	δ_2=0.022	d_1 = 0	d_2 = 0	z=0.373 μ'=19.7 GPa

Table 4.16. *Material parameters of the 2M2C model for the β-scale transition rule. 316SS at 25°C, [POR 00], E=192 GPa, ν = 0.3*

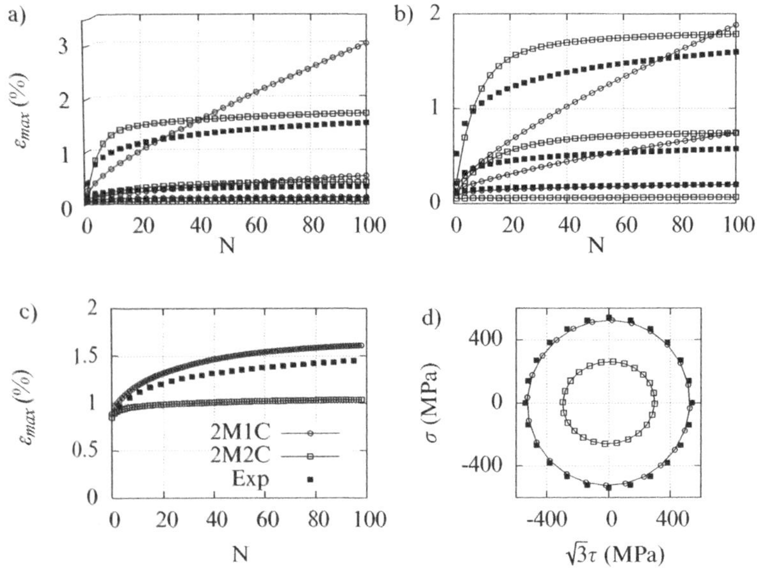

Figure 4.27. *Comparison between experiments [POR 00] performed on 316SS and simulations for the MM model using the β-scale transition rule: (a) 2D ratcheting,* σ_{max} = *80 MPa with various* $\Delta\varepsilon_{12}$, *(b) 2D ratcheting,* σ_{max} = *100 MPa with various* $\Delta\varepsilon_{12}$, *(c) 1D ratcheting and (d) out of phase*

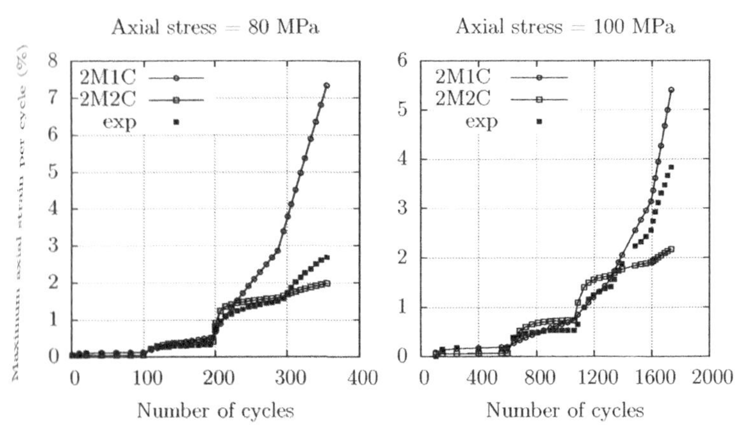

Figure 4.28. *Validation of the MM models using the β-scale transition rule: tension–torsion ratcheting tests [POR 00] with increasing shear strain amplitude*

In these tests, the specimens are submitted to a constant axial strain with an increasing shear strain amplitude. The correlation between the simulated responses and experimental results for the optimal set of material parameters is globally satisfactory. In fact:

– the 2M1C model does not reproduce each tension–torsion ratcheting test exactly, but it is able to capture the major trends in the tests. Specifically, the out-of-phase test is well described by this model;

– the 2M2C model gives a more precise description of the 2D ratcheting tests but underestimates the additional hardening under non-proportional cyclic loading.

4.8.2. *316SS at room temperature [TAL 15]*

The 316L SS behavior is now investigated among the six different metallic materials considered in [TAL 13a]. The numerical simulation was performed thanks to the 2M1C model using a static scheme resumed in Table 3.3, [TAL 15].

In order to identify the material parameters of the model, strain-controlled experiments were performed under proportional and non-proportional loading paths. The proportional loading consists of 20 cycles of tension–compression with an amplitude of 1% (Figure 4.12(a)). For the non-proportional tests, the specimen is subjected to a cross path, namely: one cycle of tension–compression with an amplitude ε_a, followed by one cycle of fully reversed torsion with the same equivalent amplitude ε_a. A total of 20 blocks were applied (Figure 4.12(b)).

Other tests will then be used to evaluate the predictive capability of the model:

– a cyclic tension–compression test under stress control with a non-zero mean stress: σ_{max}=250 MPa, σ_{min}=–150 MPa;

– a triangular path combining a prescribed segment under stress control between 0 and σ_{max}=250 MPa, followed by a segment to the torsional strain of $\gamma_{max} = 0.4\sqrt{3}\%$ and then $-\gamma_{max} = -0.4\sqrt{3}\%$. The next segment comes back to σ_{max}=250 MPa;

– cyclic tension–compression under strain control, 20 cycles: $\varepsilon_a = 1\%$.

Due to the time-dependent behavior, a 15-min hold time is applied at the first peak strain in order to evaluate the relaxation phenomenon. There is no hold time for the subsequent 19 cycles. In the absence of specific tests, and in order to simplify the identification procedure, the following phenomena were assumed to be identical in 316SS and the previously identified 304SS at room temperature [TAL 15]:

– extra-hardening due to the non-proportional loading path;

– memory of the strain hardening;

– time-dependent phenomena.

The successive iterations of the optimization process tend to adopt very small values for D_1 and D_2. It is then decided to take $D_1 = D_2 = 0$. The rest of the parameters are given in Table 4.17, and the corresponding comparisons between the simulated responses and experimental data are shown in Figure 4.29 for the tests used in the identification process and 4.30 for the tests not used in the identification process.

R_0	Q_{10}	Q_{1m}	b_1	η_{mem}	μ	Q_2	Q_{30}
212	0	180	11.1	0.5	117	0	0
Q_{3m}	b_3	d_3	n	K	C_{11}	C_{22}	C_{12}
450	14	51	3.7	650	103,776	5,840	–20,000

Table 4.17. *Material parameters for the 316L steel (MPa, s)*

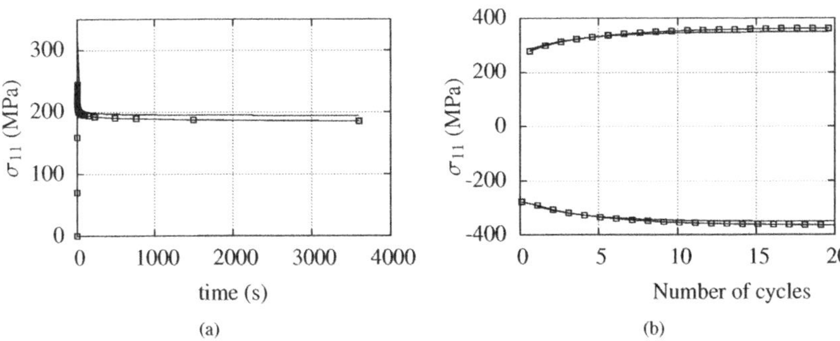

Figure 4.29. *316SS at room temperature. Comparison between experiments [TAL 13a] and simulation of the strain-controlled test: (a) relaxation test and (b) peak and valley of axial stress*

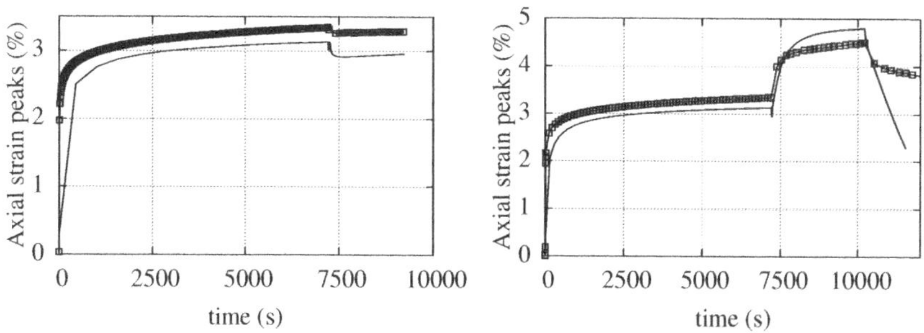

(a) (b)

Figure 4.30. *316SS at room temperature. Comparison between experiments [TAL 13a] and simulation of the stress-controlled test: (a) cyclic tension–compression under stress control and (b) triangle–inverse triangle*

4.8.3. *316SS at 350°C [TAL 13b, TAL 14]*

In the previous section, the cyclic accumulation of the inelastic strain exhibited by 316SS at room temperature under tension–compression stress control is mostly attributed to creep. The cyclic behavior of 316SS at 350°C is investigated in [TAL 14] under stress as well as strain control. For this temperature, it is demonstrated that rate-dependent phenomena are not significant.

4.8.3.1. *Tests performed under load control*

Two cyclic loading paths under stress control were considered. The first one is proportional (tension–compression), while the second one is non-proportional, following a circular path in the axial-shear stress space starting with an axial load. In each test, 100 cycles were applied. In order to estimate the viscous effect, the peak (axial) stress was maintained constant for 15 min during the first cycle. Two tension–compression tests were performed with (σ_m = 30 MPa, σ_a = 170 MPa) and (σ_m = 60 MPa, σ_a = 140 MPa), where σ_m is the mean (axial) stress and σ_a is the amplitude. The test related to the circular path was characterized by (σ_m = 30 MPa, σ_a = 170 MPa), where σ_a is the von Mises equivalent amplitude.

4.8.3.2. *Tests performed under strain control*

The first test is made of two blocks: 20 tension–compression cycles with an amplitude ε_{zz} of 1%, followed by 20 cycles of fully reversed cyclic torsion with the same equivalent amplitude ($2\varepsilon_{z\theta}/\sqrt{3} = \pm 1\%$). In the first cycle, the peak strain is maintained constant for 15 min in order to characterize the relaxation phenomenon.

In the second test, the same load levels are applied, but a torsional cycle directly follows a tension–compression cycle, forming a cross-shape signal. A total of 20 crosses were performed.

4.8.3.3. *Numerical results*

Figure 4.31 shows the comparison between the test results used in the identification process and the simulation by the MM model using the material parameters of Table 4.18. A good agreement was found between them.

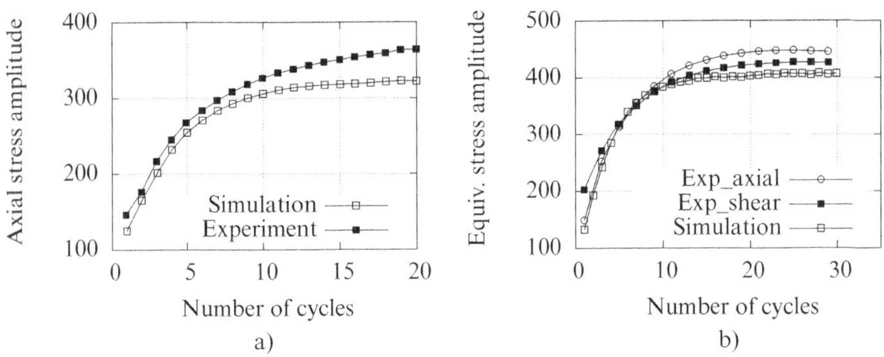

Figure 4.31. *316SS at 350°C. Simulation of the tests performed under strain control: (a) 20 cycles of tension–compression, (b) cross test; "Exp_axial" refers to the amplitude of the axial cycles, meanwhile "Exp_shear" refers to the amplitude of the torsion cycles*

R_0	Q_{10}	Q_{1m}	b_1	η_{mem}	μ	Q_{20}	Q_{2m}
113	0	230	9.75	0.015	1,498	0	175
b_2	d_2	n	K	C_{22}	C_{12}	D_1	D_2
38.11	228	20	10	4,800	-20,844	0	0

Table 4.18. *Identified parameters for 316SS at 350°C (MPa, s),* $C_{11} = E = 170,000$

4.8.4. *316SS at room temperature [POR 00], 3M1C model*

In this section, the 3M1C model is used to simulate an experimental database taken from [POR 00]. The main features of the model are given in section 3.1. In short, the 3M1C model involves, in addition to Young's modulus and Poisson's ratio, 17 independent parameters:

– three parameters devoted to the averaging process, z_I $(I = 1, 3)$ constrained by the condition $z_1 + z_2 + z_3 = 1$ (two free parameters);

– three parameters for isotropic hardening , R_0, Q and b;

– six moduli for the kinematic hardening, C_{IJ};

– six parameters used in the recall terms of the kinematic hardening, D_I and δ_I $(I = 1, 3)$.

The sensitivity of the parameters δ_i is first tested on biaxial ratcheting simulations. The list of the calibrated material parameters is given in Table 4.19, and the corresponding comparisons between the simulated responses and experimental data are shown in Figure 4.32 and Figure 4.33.

R_0=156 MPa	Q=228.8 MPa	b =1.43	z_1=0.45	z_2=0.42	z_3=0.03
C_{11}=86.6 GPa	C_{22}=519 GPa	C_{33}=28.5 GPa	D_1=0	D_2=8	D_3=13
C_{12}=202.6 MPa	C_{13}=461 MPa	C_{23}=455 MPa	N=2	A=2,700 MPa	d=50
δ_1=0.21	δ_2=0.99	δ_3=0.81			

Table 4.19. *316SS at room temperature [POR 00]. Material parameters of the 3M1C model, E=192 GPa, ν = 0.3*

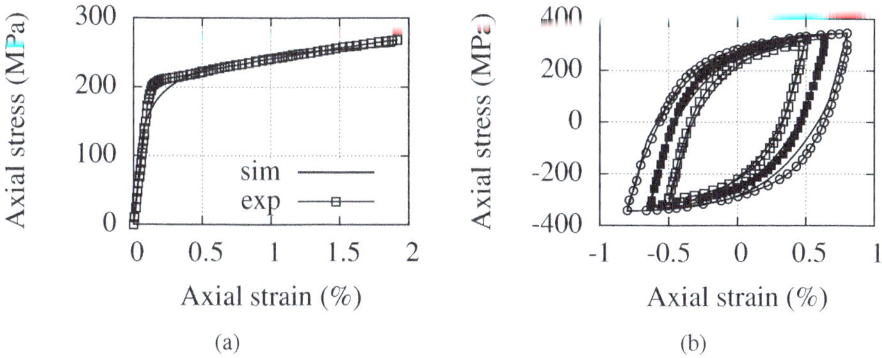

Figure 4.32. *Comparison between experiments from [POR 00] and simulations using the 3M1C model: (a) tensile test and (b) strain-controlled tests $\Delta\varepsilon$=1, 1.3 and 1.6%*

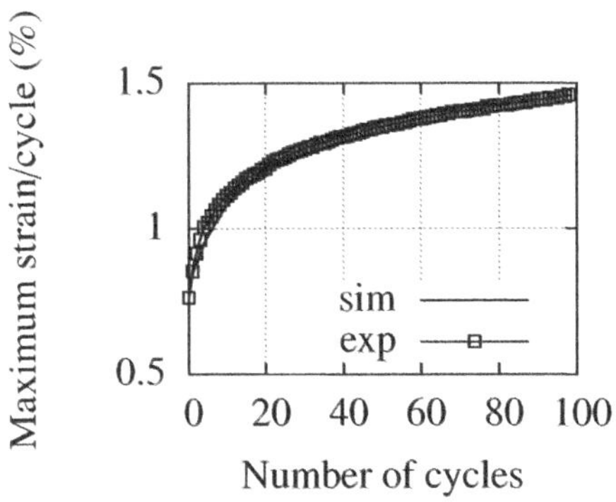

(a)

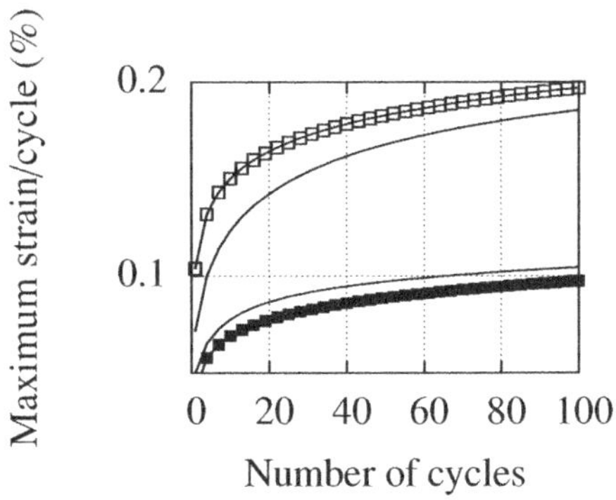

(b)

Figure 4.33. *316SS at room temperature. Comparison between experiments from [POR 00] and simulations using the 3M1C model: (a) 1D ratcheting test (σ_{min}=-40 MPa, σ_{max}=240 MPa) and (b) tension–torsion ratcheting tests $\Delta\varepsilon_{12}$=0.1% (σ_{11}=80 MPa, σ_{11}=100 MPa)*

4.9. Recrystallized Zirconium alloy 4 [PRI 08]

4.9.1. *Context of the case study*

The multi-axial creep behavior at 20°C and 400°C of recrystallized Zircaloy-4 is simulated by means of the MM model [PRI 08]. Zirconium alloys are used in the nuclear industry for cladding fuel elements. These alloys are then subjected to severe thermo-mechanical loadings and neutronic irradiation. In the operating temperature range, and in small strains, prismatic slip is the predominant plasticity mechanism. Indeed, this mechanism controls the mechanical behavior in one-dimensional loading conditions. However, under combined loadings (e.g. internal pressure and axial loading), secondary slip systems such as pyramidal systems are also activated. Following the statistical analysis of [GEY 99], an MM model was developed by [PRI 08]. Two experimental databases obtained on (i) Zircaloy-4 at 20°C taken from the work of [ROB 95] and (ii) Zircaloy-4 at 400°C performed by [GRO 09] were selected.

4.9.2. *Particular model features*

In the authors' notation, the subscript P refers to prismatic slip and NP refers to non-prismatic systems. The flow rules are generated by the following potentials:

$$F^P = f^P(\underset{\sim}{\sigma}) - \left(\sigma_y^P + R_P\right) \qquad F^{NP} = f^{NP}(\underset{\sim}{\sigma}) - \left(\sigma_y^{NP} + R_{NP}\right) \qquad [4.12]$$

So that

$$\dot{\underset{\sim}{\varepsilon}}^{vp} = \dot{p}_P \frac{\partial F^P}{\partial \underset{\sim}{\sigma}} + \dot{p}_{NP} \frac{\partial F^{NP}}{\partial \underset{\sim}{\sigma}} \qquad [4.13]$$

where $\dot{p}_P$ and $\dot{p}_{NP}$ are the equivalent viscoplastic strain rates involved in the evolution laws. They are written using hyperbolic sine rules to account for the strong nonlinearity of the creep rate as a function of stress:

$$\dot{p}_P = \dot{\underset{\sim}{\varepsilon}}_0^P \sinh\left(\frac{F^P V_0^P}{k_B T}\right) \qquad \dot{p}_{NP} = \dot{\underset{\sim}{\varepsilon}}_0^{NP} \sinh\left(\frac{F^{NP} V_0^{NP}}{k_B T}\right) \qquad [4.14]$$

The evolution of the isotropic hardening variables is deduced from the average density of mobile dislocations:

$$R_P = \alpha_P \mu b \sqrt{\rho_P + H\rho_{NP}} \qquad R_{NP} = \alpha_{NP} \mu b \sqrt{\rho_{NP} + H\rho_P} \qquad [4.15]$$

where μ is the shear modulus, b is the magnitude of Burgers vector and H is a coupling parameter between the two mechanisms. Following the dislocation evolution used in the crystal plasticity models, the flow rules can be written as

$$\dot{\rho}_P = \frac{\dot{p}_P}{b}\left(\frac{1}{L_P} - G_P\,\rho_P\right) \qquad \dot{\rho}_{NP} = \frac{\dot{p}_{NP}}{b}\left(\frac{1}{L_{NP}} - G_{NP}\,\rho_{NP}\right) \qquad [4.16]$$

where L_P, G_P, L_{NP} and G_{NP} are phenomenological material parameters.

4.9.3. *Numerical results*

The previous model is first tested on a large set of experiments taken from [ROB 95]. The experimental database involves:

– uniaxial tensile tests for two strain rates (6.6×10^{-7} s^{-1} and 6.6×10^{-4} s^{-1});

– tension–torsion loading for different proportional loading paths (ε_{zz},$\varepsilon_{z\theta}$). Three tests are performed at the same equivalent strain rate 6.6×10^{-4} s^{-1}: $\dot{\varepsilon}_{z\theta}/\dot{\varepsilon}_{zz} = +\infty$; $\sqrt{3}$; $\sqrt{3}/2$; $\sqrt{3}/4$;

– combined internal pressure and axial loading for different proportional loading paths ($\varepsilon_{\theta\theta}$,$\varepsilon_{zz}$). Three tests are performed at the same equivalent strain rate of 6.6×10^{-4} s^{-1}: $\dot{\varepsilon}_{zz}/\dot{\varepsilon}_{\theta\theta} = +\infty$; 1; -2.

For the sake of simplicity, only isotropic hardening is considered since there are no tests available in cyclic loading. The list of the parameters is given in Table 4.20, and the corresponding predicted responses are shown in Figure 4.34.

Parameter	E (GPa)	ν	b (nm)	ρ_0 (m/m^3)	H	
Value	95	0.37	0.323	6.7 10^{+10}	0.1	
Mechanism	σ_y (MPa)	V_0 (b^3)	p_0 (s^{-1})	α	L (b)	G (b)
P	70.1	41.7	8.93 10^{-35}	0.26	49	25
NP	78.6	39.9	1.40 10^{-34}	0.35	25	56

Table 4.20. *Material parameters of the MM models: recrystallized Zircaloy-4 at 20°C according to [PRI 08]*

4.10. Semi-crystalline polymers [REG 09b]

4.10.1. *Context of the case study*

The microstructure of semi-crystalline polymers (SCP) can be seen as a two-phase composite, consisting of flat crystalline lamellae separated by amorphous layers. Each phase is then assigned a deformation mechanism. Indeed, molecular chains are oriented uniformly in the crystalline phase, while they are randomly distributed in the amorphous phase. The evolution of the inelastic strain in the crystalline phase is mainly governed by slip along the molecular chain. Although the amorphous regions are hardly isolated from the SCP, two deformation mechanisms of amorphous layers have been identified: interlamellar shear and interlamellar separation.

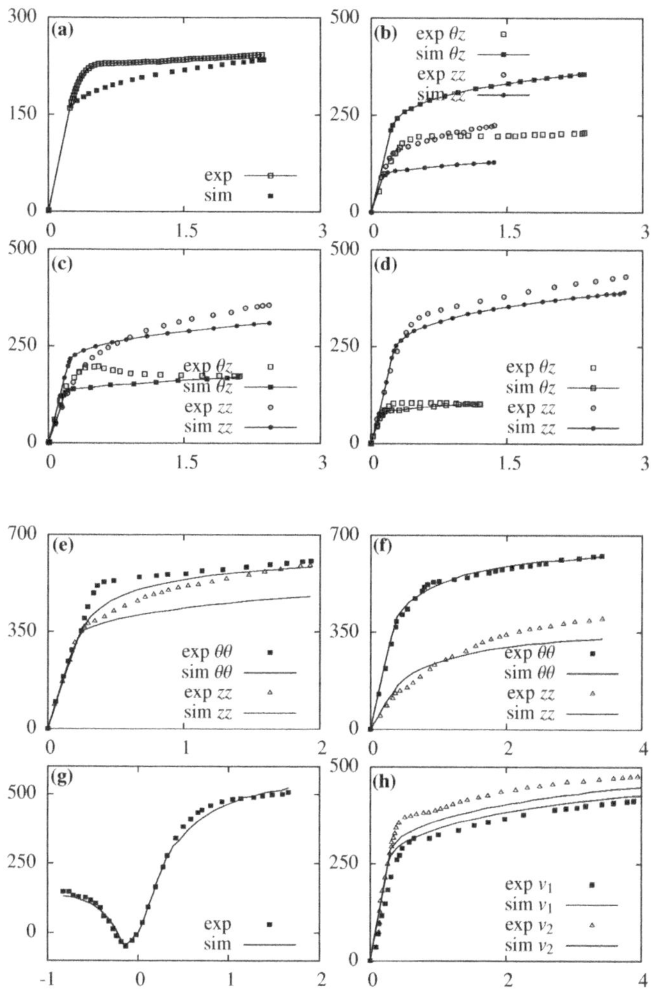

Figure 4.34. *Zircaloy-4 [PRI 08]: comparison of the simulated and experimental stress–strain curves for a tensile–torsion load [ROB 95]. (a)* $\dot{\varepsilon}_{z\theta}/\dot{\varepsilon}_{zz} = +\infty$, $\sigma_{z\theta}$ *versus* $\varepsilon_{z\theta}$, *(b)* $\dot{\varepsilon}_{z\theta}/\dot{\varepsilon}_{zz} = \sqrt{3}$, $\sigma_{z\theta}$ *versus* $\varepsilon_{z\theta}$ *and* σ_{zz} *versus* ε_{zz}, *(c)* $\dot{\varepsilon}_{z\theta}/\dot{\varepsilon}_{zz} = \sqrt{3}/2$, $\sigma_{z\theta}$ *versus* $\varepsilon_{z\theta}$ *and* σ_{zz} *versus* ε_{zz}, *(d)* σ_{zz} *versus* ε_{zz}. *Comparison of the simulated and experimental stress–strain curves for a combined loading under internal pressure and axial tension [ROB 95]. (e)* $\dot{\varepsilon}_{zz}/\dot{\varepsilon}_{\theta\theta} = 1$, σ_{zz} *versus* ε_{zz} *and* $\sigma_{\theta\theta}$ *versus* $\varepsilon_{\theta\theta}$, *(f)* $\dot{\varepsilon}_{zz}/\dot{\varepsilon}_{\theta\theta} = +\infty$, σ_{zz} *versus* ε_{zz} *and* $\sigma_{\theta\theta}$ *versus* $\varepsilon_{\theta\theta}$, *(g)* $\dot{\varepsilon}_{zz}/\dot{\varepsilon}_{\theta\theta} = -2$, σ_{zz} *versus* ε_{zz} *and* $\sigma_{\theta\theta}$ *versus* $\varepsilon_{\theta\theta}$ *and (h)* σ_{zz} *versus* ε_{zz} *with low (v1) and high (v2) strain rates*

4.10.2. *Particular model features*

The modeling approach introduces the inelastic behavior of each phase, weighted by the crystallinity ratio z. A finite strain (updated Lagragian) formalism is used (see section 1.4 for a detailed presentation). The constitutive relations are expressed in the corotational frame, so that the treatment is equivalent to a hypoelastic formulation and introduces a Green–Naghdi stress rate. An additive strain partition is allowed, with one elastic and two inelastic components. The first inelastic strain (I = 1) is associated with the amorphous phase, whereas the second one (I = 2) refers to the crystalline phase. The total inelastic strain $\underset{\sim}{\varepsilon}^{in}$ is the weighted sum ((1 – z) and z, respectively) of the irreversible deformation of each phase. The experimental results are taken from Regrain [REG 09a]. Two modeling frameworks have been tested, namely 2M2C and 2M1C.

4.10.3. *Numerical results*

The tests used for the identification were performed on a smooth specimen with a cylindrical part, which was simulated by means of an axisymmetric finite element, in order to account for finite strains. A comparison between the experimental and simulated data for creep and tensile tests is shown in Figure 4.35 for both models. The material parameters are given in Table 4.21.

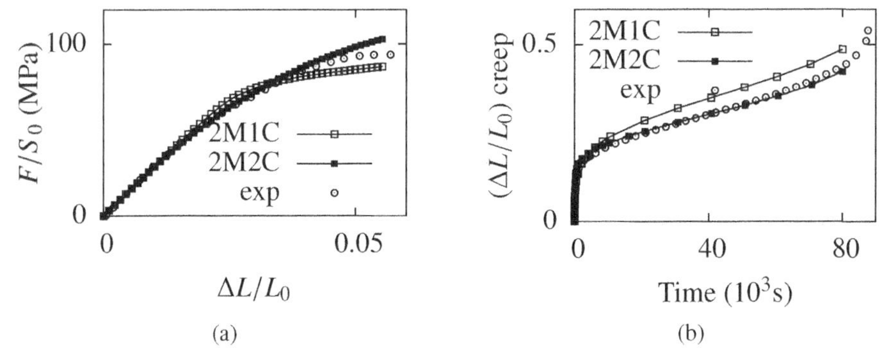

Figure 4.35. *Polyamide 6: comparison between experimental and simulated data for (a) stress versus strain for a tensile monotonic test at 0.026 s^{-1} and (b) creep strain history for a smooth specimen with σ=76 MPa*

Parameters		2M2C	Units	Parameters		2M1C	Units
	n_1	2.5					
	K_1	244	MPa.s$^{1/n_1}$				
Criterion 1	R_{01}	3.9	MPa				
	Q_1	27.4	MPa		n	11.2	
	b_1	120			K	143	MPa.s$^{1/n}$
				Criterion	R_0	0.4	MPa
	n_2	5.3			Q	9.1	MPa
	K_2	859	MPa.s$^{1/n_2}$		b	203	
Criterion 2	R_{02}	0.9	MPa				
	Q_2	56.3	MPa				
	b_2	6.2					
Mechanism 1	C_{11}	317	MPa	Mechanism 1	C_{11}	1,187	MPa
	D_1	3.2			D_1	8.5	
Mechanism 2	C_{22}	166	MPa	Mechanism 2	C_{22}	260	MPa
	D_2	34.9			D_2	1.8	
	A	5.4	MPa		A	17.1	MPa
Interaction	C_{12}	-6.9	MPa	Interaction	C_{12}	-83.3	MPa
	d_1	16.1			d_1	20.3	
	d_2	17.3			d_2	20.3	

Table 4.21. *Material parameters of the 2M1C and 2M2C models, polyamide 6.* $E = 2,800$ *MPa,* $\nu = 0.4$, $z = 0.43$

Even if the local stress and strain in each phase cannot be obtained from experiments, the values provided by the simulation are inspected, in order to evaluate the relevance of the models. As shown in Figure 4.36, the plastic strain is larger in the amorphous (soft) phase than in the crystalline (hard) phase, especially for the 2M2C model, which seems physically consistent.

This MM devoted to the SCP is then generalized to account for the mechanical damage behavior. The constitutive equations will be shown in section 5.1. To assess model reliability, finite-element simulations of notched specimens subjected to creep loadings were performed and compared in terms of creep notch opening displacement (δ_c). Two notched round bars with $r_0 = 4$ mm and $r_0 = 0.45$ mm are considered. The material parameters of the undamaged model are those used to simulate the smooth specimens (Figure 4.35). The optimization procedure used to calibrate the damage parameters is detailed in [SAÏ 11b]. Figures 4.38(a) and 4.38(b), respectively, illustrate the comparison between experimental and simulated results for r_0 = 0.45 mm and r_0 = 4 mm. A good agreement is observed in Figure 4.37(a) between the two curves in terms of δ_c rate and time to creep failure. With the same set of material coefficients, it is observed in Figure 4.37(b) that the simulation overestimates δ_c rate and accordingly underestimates the creep time to failure. However, it was shown in [SAÏ 11b] that this set of coefficients is satisfactory in terms of local parameters. The location of the damage must correspond to the surface where the ductile fracture is observed, which is centered or ring-shaped, as shown in the following section.

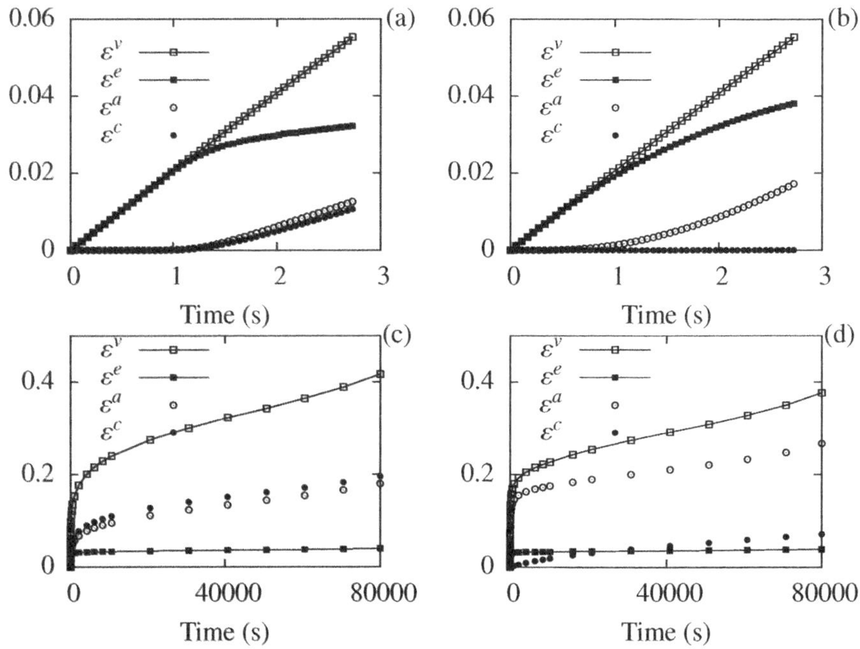

Figure 4.36. *Polyamide 6: comparison between viscoplastic strains of amorphous and crystalline phases. Tensile monotonic test at 0.026 s^{-1} using (a) the 2M1C model and (b) the 2M2C model. Creep simulation at 76 MPa using (c) the 2M1C model and (d) the 2M2C model. The equations are in section 5.1.1 for 2M1C model and 5.1.2 for 2M2C model. ε^a for the amorphous phase is denoted by ε^1. ε^c for the crystalline phase is denoted by ε^2. $\varepsilon^v = \varepsilon^a + \varepsilon^c$. ε^e is the elastic strain*

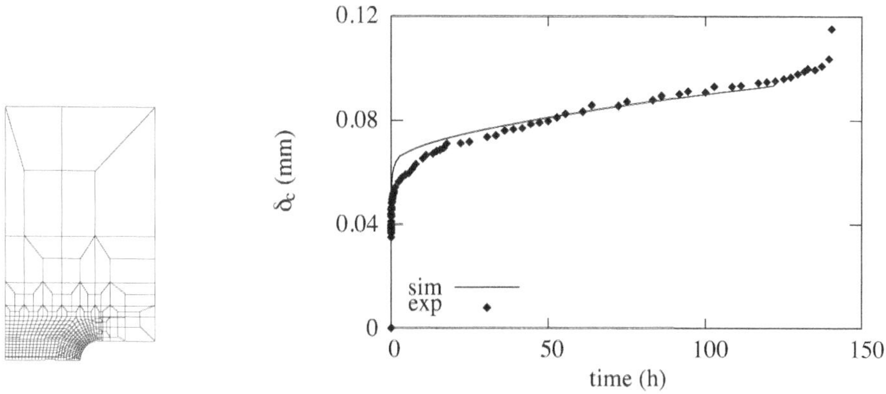

(a) $r_0 = 0.45$ mm and $\sigma_{net} = 73$ MPa

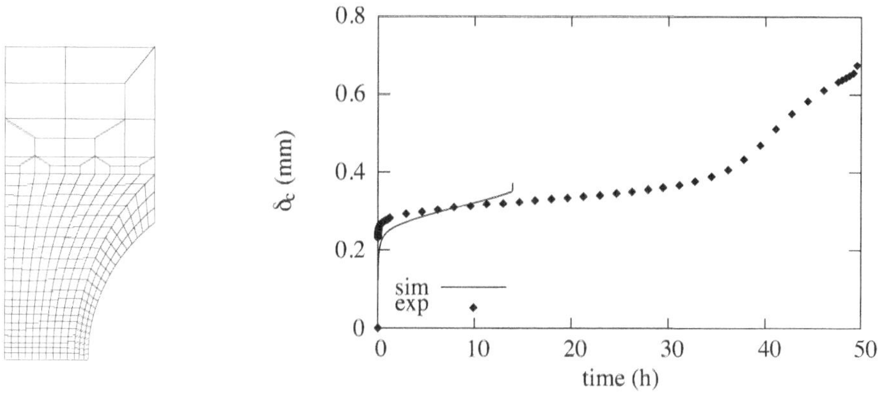

(b) $r_0 = 4$ mm and $\sigma_{net} = 74$ MPa

Figure 4.37. *Mesh and graph showing the comparison of experimental and simulated creep test data on notched round bars [SAÏ 11b]*

4.11. Glassy polymers [JER 14]

4.11.1. *Context of the case study*

The stress–strain curves of most of the glassy polymers present three successive regimes: (i) a nonlinear evolution with growing values until a maximum stress, the yield stress, is reached, (ii) a stress decrease due to softening and (iii) hardening for larger strains due to molecular orientation.

4.11.2. *Particular model features*

A 3M3C (three mechanisms and three criteria) model was applied. Finite strains were modeled by a Green–Naghdi formulation in an updated Lagrangian formalism, already described in section 1.4. For the sake of simplicity, a static scale transition is assumed, so that the local stress for each mechanism is nothing but the macroscopic stress tensor $\underset{\sim}{\sigma}$: $\underset{\sim}{\sigma} = \underset{\sim}{\sigma}_1 = \underset{\sim}{\sigma}_2 = \underset{\sim}{\sigma}_3$. The stress tensors ($\underset{\sim}{\sigma}_1, \underset{\sim}{\sigma}_2$ and $\underset{\sim}{\sigma}_3$) are involved in yield functions (f_1, f_2 and f_3) to define the elastic domain related to each mechanism. Because the present study is only dedicated to polymer behavior under monotonic loading, the kinematic hardening variables were not considered:

$$f^1 = J(\underset{\sim}{\sigma}_1) - R^1 - R_{01} \quad f^2 = J(\underset{\sim}{\sigma}_2) - R^2 - R_{02} \quad f^3 = J(\underset{\sim}{\sigma}_3) - R^3 - R_{03} \qquad [4.17]$$

where R_{01}, R_{02} and R_{03} are the initial sizes of the elastic domain for mechanisms 1, 2 and 3, respectively. The isotropic hardening variables (R^1, R^2 and R^3) that characterize the expansion of the elastic domain related to each yield surface depend on three state variables (r^1, r^2 and r^3), respectively. An isotropic–isotropic coupling has been introduced between the hardening variables to account for the interaction between the different mechanisms:

$$\begin{pmatrix} R^1 \\ R^2 \\ R^3 \end{pmatrix} = \begin{pmatrix} Q_{11} & Q_{12} & Q_{13} \\ Q_{21} & Q_{22} & Q_{23} \\ Q_{31} & Q_{32} & Q_{33} \end{pmatrix} \begin{pmatrix} b_1 r^1 \\ b_2 r^2 \\ b_3 r^3 \end{pmatrix} \qquad [4.18]$$

where Q_{11}, Q_{22} and Q_{33} are isotropic hardening moduli; and ($Q_{IJ}, I \neq J$) are coupling parameters that play an essential role in the polymer behavior. For example, a positive value of Q_{12} allows mechanism 2 (if active) to refrain mechanism 1. Increasing Q_{12} may deactivate mechanism 1. On the contrary, a negative value of Q_{21} allows mechanism 1 to produce an early activation of mechanism 2. This is the reason why the third mechanism is active after the softening stage only. According to the normality flow rule, the viscoplastic strain rates may be expressed as

$$\dot{\underset{\sim}{\varepsilon}}^{v1} = \dot{v}^1 \underset{\sim}{\mathbf{n}}^1 \qquad \dot{\underset{\sim}{\varepsilon}}^{v2} = \dot{v}^2 \underset{\sim}{\mathbf{n}}^2 \qquad \dot{\underset{\sim}{\varepsilon}}^{v3} = \dot{v}^3 \underset{\sim}{\mathbf{n}}^3 \qquad [4.19]$$

where $\dot{v}^1$, $\dot{v}^2$ and $\dot{v}^3$ are three equivalent viscoplastic strain rates. In the present case, a `sinh` is used instead of a power function to account for the strong nonlinearity of the creep rate as a function of equivalent stress:

$$\dot{v}^I = \dot{v}_{0I} \sinh\left(\left\langle \frac{f^I}{K_I} \right\rangle^{n_I}\right) \qquad I = 1,3 \tag{4.20}$$

where K_1, n_1, K_2, n_2, K_3 and n_3 characterize the viscosity of the different mechanisms. The evolution of the isotropic hardening variable is defined by

$$\dot{r}^1 = \dot{v}^1(1 - b_1 r^1) \qquad \dot{r}^2 = \dot{v}^2(1 - b_2 r^2) \qquad \dot{r}^3 = \dot{v}^3(1 - b_3 r^3) \tag{4.21}$$

It is worth noting that the evolution rules of the isotropic internal variables were intentionally left more simple, compared to those suggested by the thermodynamic framework, where "flow" coupling is considered in addition to "state" coupling (equation [4.18]):

$$\dot{r}^1 = \dot{v}^1(1 - \frac{R^1}{Q_{11}}) \qquad \dot{r}^2 = \dot{v}^2(1 - \frac{R^2}{Q_{22}}) \qquad \dot{r}^3 = \dot{v}^3(1 - \frac{R^3}{Q_{33}}) \tag{4.22}$$

4.11.3. *Numerical results*

The model was compared to tensile and relaxation stress–strain curves for a PC at different strain rates at room temperature taken from the work of Zaïri *et al.* [ZAÏ 05]. The authors describe an apparent Young's modulus E, determined in a previous study, that depends on strain rate. In our approach, the strain rate effect is given by the second and the third mechanisms. The material parameters are determined in three steps as follows:

– Material parameters of mechanism 1: the first step of the optimization procedure is devoted to the determination of the material parameters of mechanism 1 only. It must be activated at the beginning of the test, so that $R_{01} = 0$. The other mechanisms are imposed to remain inactive during this step. The isotropic hardening modulus Q_{11} and the material parameter b_1 are set to describe the first branch of the tensile curve. The strain rate sensitivity is accounted for thanks to the viscosity parameters K_1 and n_1. These parameters will be kept unchanged for the rest of the identification. Figure 4.38(a) shows the superimposition of the experimental results and the simulation when only the first mechanism is active.

– Material parameters of mechanism 2: mechanism 2 is devoted to the description of the stress drop with increasing deformation. It is made active by setting R_{02} to 65 MPa, the maximum of the tensile curve. The softening is modeled through a negative value of the asymptotic value Q_{22}. The softening rate is well described by a value of 25 for b_2. The strain rate sensitivity is defined by K_2 and n_2. Finally, the coupling between Q_{12} and Q_{21} is neglected, so that $R^1 = b_1 Q_{11} r^1 + b_1 Q_{13} r^3$ and $R^2 = b_2 Q_{22} r^2 + b_2 Q_{23} r^3$. The simulated curves after the activation of mechanisms 1 and 2 (Figure 4.38(b)) show that the latter is predominant.

– Material parameters of mechanism 3: the most difficult task is to activate the third mechanism while decreasing stress for a high value of R_{03}. To solve this problem, negative values are assigned to the coupling term between Q_{31} and Q_{32}. The evolution of the third mechanism is, according to equation [4.18], $R^3 = b_3 Q_{31} r^1 + b_3 Q_{32} r^2 + b_3 Q_{33} r^3$. R^3 will be attenuated by the negative terms $b_3 Q_{31} r^1$ and $b_3 Q_{32} r^2$. On the other hand, the softening of mechanism 2 is attenuated by activating the material parameter Q_{23}; meanwhile; Q_{13} is kept to be zero. Material parameters are then numerically optimized with a sequential quadratic programming (SQP) method [STO 85]. The results are shown in Figure 4.38.

The list of the calibrated coefficients for the MM models applied to polycarbonate at room temperature is given in Table 4.22. Moreover, two relaxation tests are used to validate *a posteriori* the prediction of the proposed model. In these tests, the specimens are submitted to a true strain rate of 10^{-3} s^{-1}, followed by stress relaxation with different strain levels. Figure 4.38(d) shows a comparison between experiments and a prediction for the two models for the relaxation tests.

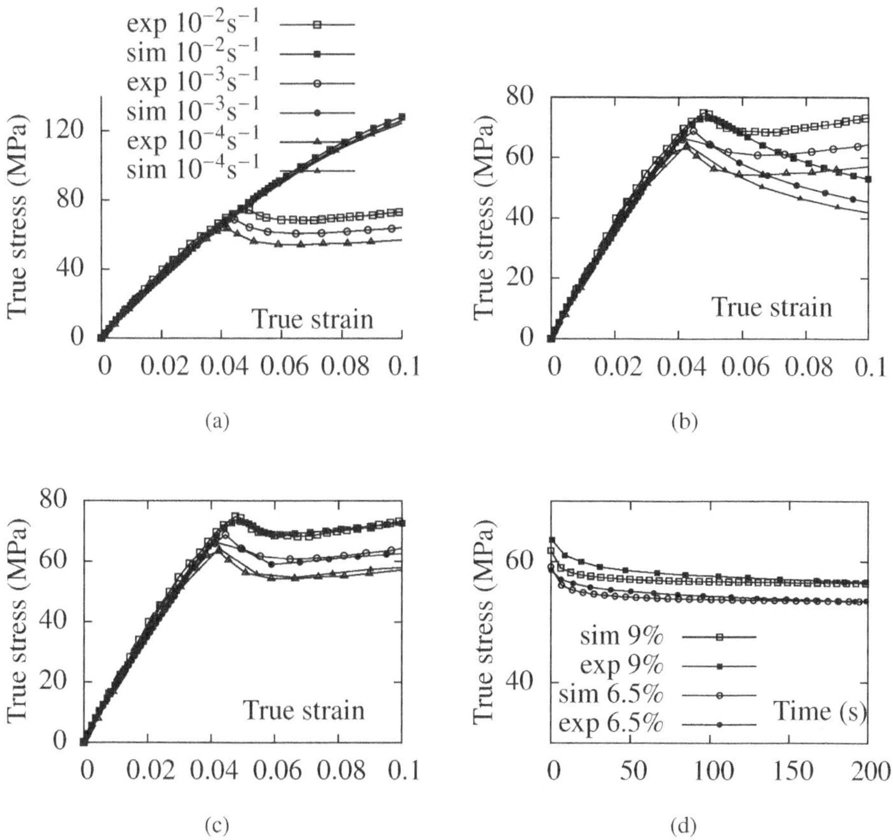

Figure 4.38. *Comparison between the MM model and tensile tests taken from [ZAÏ 05]. Strain rates are 10^{-4}, 10^{-3} and 10^{-2} s^{-1}: (a) only the first mechanism is activated ($R_{02} \rightarrow \infty$, $R_{03} \rightarrow \infty$), (b) mechanisms 1 and 2 are activated ($R_{03} \rightarrow \infty$), (c) three mechanisms are activated, (d) validation of the proposed MM model and comparison with two relaxation tests performed by Zaïri et al. [ZAÏ 05]*

Mechanism	R_{0i}	K_i	n_i	$\dot{v}_{0i}$	b_i	Q_{ii}
1	0	50	5	10	50	8,000
2	65	200	3	25	25	-688.25
3	113	10	3	25	5	3,287.3
Coupling	$Q_{23} = 1,500$	$Q_{31} = -2,600$	$Q_{32} = -2,200$			
Unused	$Q_{12} = 0$	$Q_{13} = 0$	$Q_{21} = 0$			

Table 4.22. *Identified material parameters for polycarbonate (MPa, s)*

4.12. Copper-zinc alloy CuZn27 [TAL 15]

4.12.1. *Context of the case study*

The behavior of CuZn27 alloy was investigated in [TAL 15]. This material is used in automotive industry. The specimens were obtained by the casting process performed by FAVI Company (http://www.favi.com/ang/index.php). The numerical simulation was performed thanks to the 2M1C model summarized in Table 3.3.

4.12.2. *Numerical results*

The identification process follows three main steps:

– Step 1. Initial guess: the asymptotic values of the isotropic variables are estimated. For this purpose, strain-controlled experiments under proportional (identification of Q_{10}, Q_{1S}, Q_2) and non-proportional (identification of Q_{30} and Q_{3inf}) loading paths were considered. The identified values represent generally the minimum (for continuous cyclic hardening) or the maximum (for continuous cyclic softening) of the parameters in question (Figure 4.39).

– Step 2. Optimization process: using the preceding values of the parameters, the optimization process allows us to find the other parameters using strain-controlled experiments in addition to creep and/or relaxation tests. If the optimization process selects very small values for some parameters, they are set to zero in order to have the simplest version of the model.

– Step 3. Manual improvement of the parameters: the cost function used for the automatic optimization may produce a result that is not satisfactory for the user. In such a case, some of the parameters may be adjusted to improve the fit according to the user's wishes.

The identification is made on the strain-controlled experiments of Figure 4.12. Additional stress- and strain-controlled experiments were also performed in order to study the ratcheting phenomenon (Figure 4.13). Specifically, the tests considered here are the following:

– CuZn27_50-250: $\sigma_{max} = 300$ MPa, $\sigma_{min} = -200$ MPa;

– CuZn27_300-0.4: $\sigma_{max} = 300$ MPa, $\frac{\gamma_{max}}{\sqrt{3}} = 0.4\%$;

– CuZn27_Ax_1%: 20 cycles: $\varepsilon_{zz} = \pm 1\%$;

– CuZn27_Cross_Ax-Sh_1% 20 blocks: $\varepsilon_{zz} = \pm 1\% \rightarrow \gamma / \sqrt{3} = \pm 1\%$.

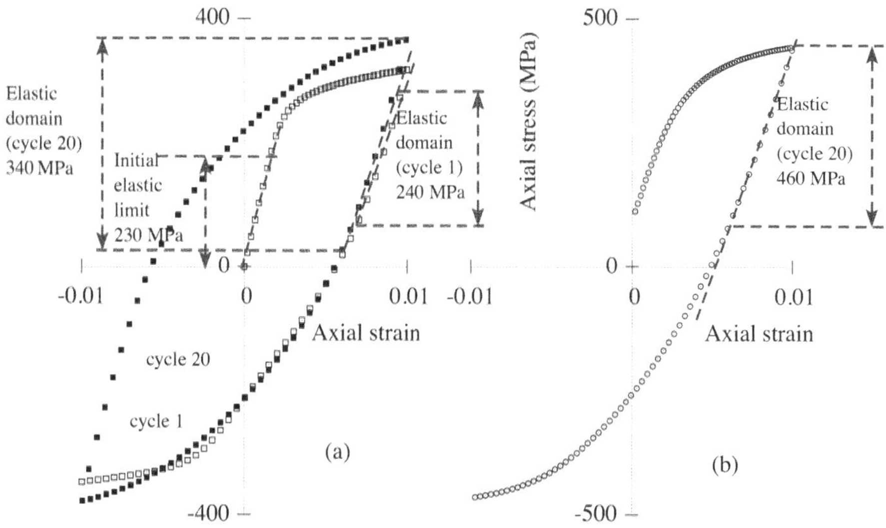

Figure 4.39. *First guess of the saturation values of the isotropic hardening variables under proportional and non-proportional loading paths: (a) estimation of the elastic domain during the unloading part of the first cycle (evaluation of Q_{10}) and the 20th cycle (evaluation of Q_{1m}); (b) estimation of the elastic domain after 20 blocks in the test "CuZn27_Cross_Ax-Sh_1%" (evaluation of Q_{3m}). For a color version of the figure, see www.iste.co.uk/cailletaud/multi-mechanism.zip*

The following assumptions were made for this material during the identification process:

– no significant memory of the strain hardening;

– time-dependent phenomena were assumed small, so that $K = 10$ and $n = 20$.

As the optimization process suggests very small values for D_1 and D_2, we choose $D_1 = D_2 = 0$. The other parameters are given in Table 4.23, and the corresponding comparisons between the simulated responses and experimental data are shown in Figures 4.40 and 4.41.

R_0	Q_{10}	Q_{1m}	b_1	Q_2	b_2	Q_{30}	Q_{3m}	b_3	d_3	C_{11}	C_{22}	C_{12}
255	-40	0	301	120	21	0	150	300	202	186,233	7,710	82

Table 4.23. *Identified parameters for the CuZn27 alloy (MPa, s).*

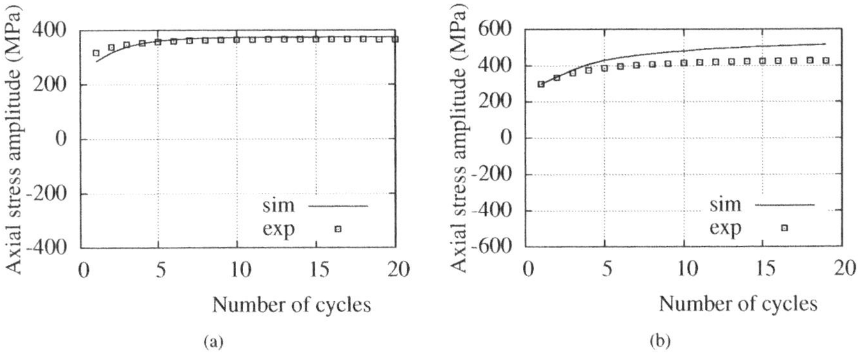

Figure 4.40. *CuZn27: comparison of the experimental results and the numerical simulations for the strain-controlled tests. (a) Test CuZn27_Ax_1%, (b) test CuZn27_Cross_Ax-Sh_1%*

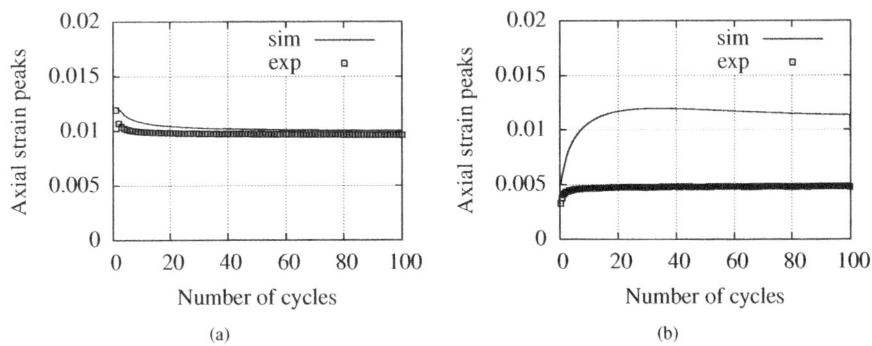

Figure 4.41. *CuZn27: comparison of the experiments and simulations of the stress-controlled tests. (a) Test CuZn27_50-250, (b) test CuZn27_250-0.4*

4.13. Ferritic steel 35NiCrMo16 [TAL 15]

4.13.1. *Context of the case study*

The behavior of the 35NiCrMo16 alloy was investigated in [TAL 15]. This material is also known as 35NCD16. The numerical simulation was performed thanks to the 2M1C model summarized in Table 3.3.

4.13.2. *Numerical results*

The material parameters were obtained from strain-controlled tests under proportional and non-proportional loadings. For the case of proportional loading, the specimen was subjected to 20 tension–compression cycles with an amplitude of 1% (Figure 4.12(a)). For the non-proportional tests, a cross path was chosen, where the specimen was successively subjected to 20 blocks made of one tension–compression cycle with an amplitude of ε_a, followed by one torsion cycle at the same equivalent amplitude ε_a (Figure 4.12(b)).

Specifically, the two tests performed under strain control are

– 35NCD16_Ax_1 that corresponds to 20 cycles at $\varepsilon_{zz} = \pm 1\%$;

– 35NCD16_Cross_Ax-Sh_1 that corresponds to 20 blocks composed of $\varepsilon_{zz} = \pm 1\% \rightarrow \gamma / \sqrt{3} = \pm 1\%$.

The following assumptions were made for the 35NCD16 steel during the identification process:

– there is no significant extra-hardening under non-proportional loading: $Q_3 = 0$;

– there is no significant memory effect of the maximum strain;

– time-dependent phenomena are limited: $K = 10$ MPa.s$^{1/n}$ and $n = 20$.

As the optimization leads to very small values of D_1 and D_2, it is assumed that $D_1 = D_2 = 0$. The other parameters are given in Table 4.24, and the comparisons between the simulated responses and experimental data are shown in Figures 4.42 and 4.43.

R_0	Q_{10}	Q_{1m}	b_1	Q_2	b_2	C_{11}	C_{22}	C_{12}
849	700	0	100	-211	5.025	1,419,347	77	51

Table 4.24. *Identified parameters for the 35NCD16 steel (MPa, s)*

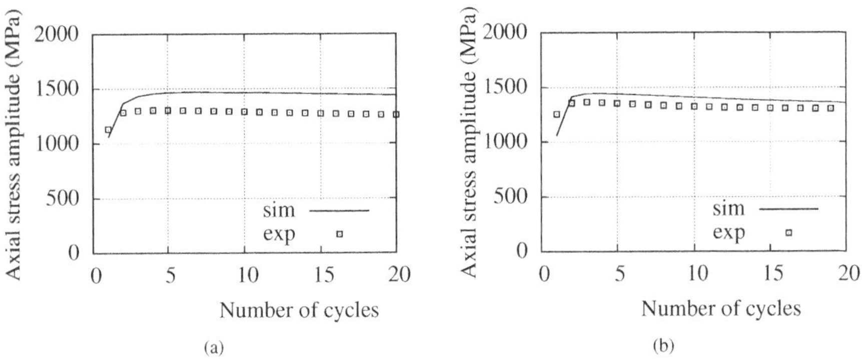

Figure 4.42. *35NCD16: comparison between the strain-controlled experiments and their simulations. Axial stress amplitude versus the number of cycles for the tests: (a) 35NCD16_Ax_1 and (b) 35NCD16_Cross_Ax-Sh_1*

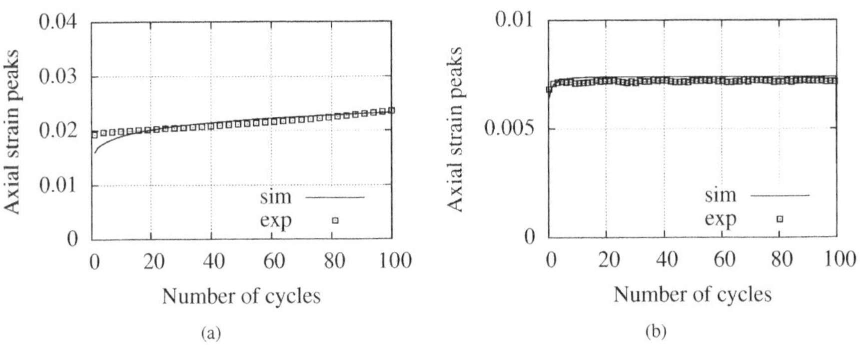

Figure 4.43. *35NCD16: comparison between the stress-controlled experiments and their simulations. Axial strain peaks versus the number of cycles for the tests: (a) 35NCD16_200-1050 and (b) 35NCD16_900-0.4*

The validation of the model was performed through two cyclic loading paths: one proportional and the other non-proportional. The proportional loading path is a non-symmetric tension–compression under load control, while the non-proportional path is a combination of axial force and torsion following a triangular shape (Figure 4.13). The loadings applied in these tests are the following:

– 35NCD16_200-1050, for which $\sigma_{max} = 1,250$ MPa, $\sigma_{min} = -850$ MPa;

– 35NCD16_900-0.4, for which $\sigma_{max} = 900$ MPa, $\frac{\gamma_{max}}{\sqrt{3}} = 0.4\%$.

4.14. Ferritic steel XC18 [TAL 13a]

4.14.1. *Context of the case study*

The behavior of XC18 ferritic steel was investigated in [TAL 15]. The numerical simulation was performed thanks to the 2M1C model summarized in Table 3.3.

4.14.2. *Numerical results*

The material parameters of the model were identified from the strain-controlled experiments of Figure 4.12. The tests performed under stress control (Figure 4.13) were then simulated, in order to evaluate the predictive capabilities of the model. Specifically, the tests are:

– XC18_50-320: $\sigma_{max} = 370$ MPa, $\sigma_{min} = -270$ MPa;

– XC18_250-0.4: $\sigma_{max} = 250$ MPa, $\frac{\gamma_{max}}{\sqrt{3}} = 0.4\%$;

– XC18_Ax_1.5: 20 cycles: $\varepsilon_{zz} = \pm 1.5\%$;

– XC18_Cross_Ax-Sh_1.5: 20 blocks composed of $\varepsilon_{zz} = \pm\%1.5 \rightarrow \gamma/\sqrt{3} = \pm 1.5\%$.

The following assumptions were made during the identification process:

– no significant memory of the strain hardening;

– time-dependent phenomena are negligible: $K = 10$ MPa.s$^{1/n}$ and $n = 20$. This fact is verified in creep tests, where no significant strain evolution is observed under constant stress.

As the optimization leads to a very small value of C_{12}, the coupling was neglected ($C_{12} = 0$). The other parameters are given in Table 4.25, and the comparisons between the simulated responses and experimental data are shown in Figures 4.44 and 4.45.

R_0	Q_{10}	b_1	Q_2	b_2	Q_{3m}	b_3	d_3	η	C_{11}	C_{22}	D_1
424	–110	312	80	12.1	200	10,000	100	0.9	53,395	3,358	143

Table 4.25. *Identified parameters for the XC18 steel (MPa, s), $Q_{1m}=Q_{30}=D_2=0$*

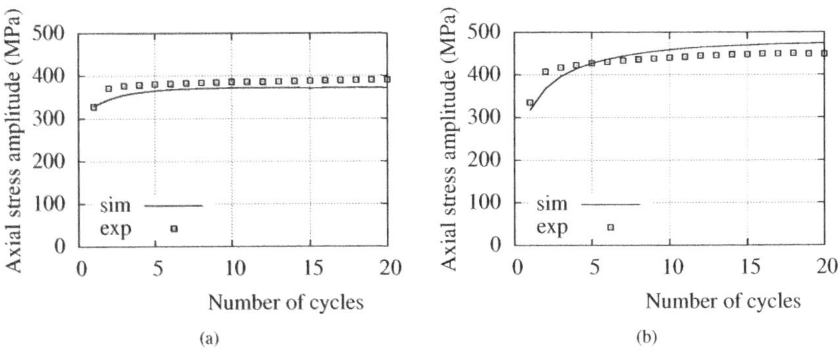

Figure 4.44. *XC18: comparison between the strain-controlled experiments and their simulations. Axial stress amplitude versus the number of cycles for the tests: (a) XC18_Ax_1.5 and (b) XC18_Cross_Ax-Sh_1.5*

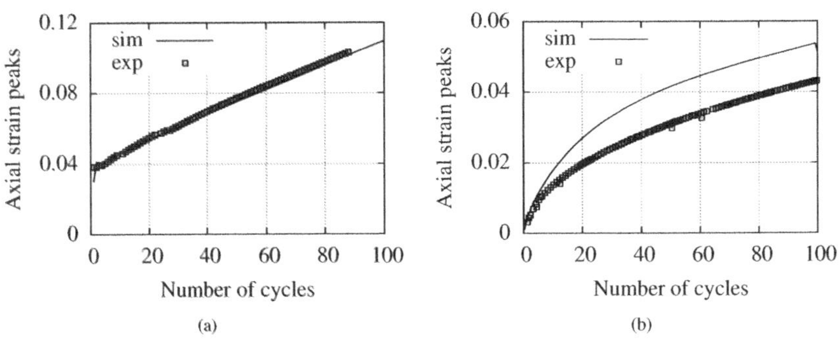

Figure 4.45. *XC18: comparison between the stress-controlled experiments and their simulations. Axial strain peaks versus the number of cycles for the tests: (a) XC18_200-1050 and (b) XC18_900-0.4*

4.15. Phase transformation in titanium alloys Ti6Al4V [LON 09]

4.15.1. *Context of the case study*

Ti-6Al-4V (TA6V) titanium alloy is widely used in industrial applications such as aeronautic and aerospace due to its good mechanical properties at high temperature. Its microstructure strongly depends on thermo-mechanical treatments. The microstructure of the present study is also called $\alpha + \beta$ type, which is composed of HCP grains (phase α) with some stabilized BCC β phase at room temperature. When heated over about 980℃ (beta transus), α is totally transformed into β, which is stable up to melting temperature, at about 1660℃. The effect of phase transformations is taken into account in the simulation, which is performed by the multi-mechanism model proposed in [LON 09]. The model is then applied to two laser-assisted processes: direct laser fabrication and laser welding.

4.15.2. *Particular model features*

Following the approach already mentioned in [SAÏ 07], the scale transition is made by means of a β-rule [CAI 94], with the assumption of uniform elasticity. For each phase Φ ($\Phi = \alpha, \beta, \alpha'$), the residual stress is defined by an intergranular accommodation variable $\underset{\sim}{\beta}_\Phi$. The local stress $\underset{\sim}{\sigma}_\Phi$ is then proportional to the difference between $\underset{\sim}{\beta}_\Phi$ and the average of these variables in all phases:

$$\underset{\sim}{\sigma}_\Phi = \underset{\sim}{\sigma} + \mu'\left(\underset{\sim}{\beta} - \underset{\sim}{\beta}_\Phi\right) \qquad \underset{\sim}{\beta} = \sum_\Phi z_\Phi \underset{\sim}{\beta}_\Phi \qquad [4.23]$$

where z_Φ is the volume fraction of the phase Φ. The evolution of $\underset{\sim}{\beta}_\Phi$ has a driving term and a dynamic recovery term:

$$\underset{\sim}{\beta}_\Phi = \underset{\sim}{\varepsilon}_\Phi - D_\Phi \underset{\sim}{\beta}_\Phi \|\dot{p}_\Phi\| \qquad [4.24]$$

Each phase is described by a viscoplastic potential, which is a power function of the von Mises criterion defining its yield surface. A kinematic hardening variable is introduced to represent the Bauschinger effect involving a dynamic recovery term and a static recovery term:

$$\underset{\sim}{\mathbf{X}}_{\Phi} = \frac{2}{3} C_{\Phi} \underset{\sim}{\alpha}_{\Phi} \quad [4.25]$$

$$\dot{\underset{\sim}{\alpha}}_{\Phi} = \dot{\underset{\sim}{\varepsilon}}_{\Phi} - d_{\Phi} \underset{\sim}{\alpha}_{\Phi} \dot{p}_{\Phi} - \left(\frac{J(\underset{\sim}{\alpha}_{\Phi})}{M_{\Phi}} \right)^{m}_{\Phi} \frac{\underset{\sim}{\alpha}_{\Phi}}{J(\underset{\sim}{\alpha}_{\Phi})} \quad [4.26]$$

The total strain $\underset{\sim}{\varepsilon}$ is decomposed into several components: the elastic strain $\underset{\sim}{\varepsilon}^{e}$, the viscoplastic strain $\underset{\sim}{\varepsilon}^{vp}$, the thermal dilatation $\underset{\sim}{\varepsilon}^{th}$ and the intrinsic volume change due to phase transformation $\underset{\sim}{\varepsilon}^{pc}$:

$$\underset{\sim}{\varepsilon} = \underset{\sim}{\varepsilon}^{e} + \underset{\sim}{\varepsilon}^{vp} + \underset{\sim}{\varepsilon}^{th} + \underset{\sim}{\varepsilon}^{pc} \quad [4.27]$$

A secant isotropic thermal expansion coefficient is used, so that the thermal dilatation term can be written as

$$\underset{\sim}{\varepsilon}^{th} = (\alpha(T)(T - T_0) - \alpha(T_{ini})(T_{ini} - T_0)) \underset{\sim}{\mathbf{I}} \quad [4.28]$$

where T is the current temperature, T_{ini} is the initial temperature and T_0 is the reference temperature. The HCP phase, being most compact, is taken as reference, so that $\underset{\sim}{\varepsilon}^{pc}$ is proportional to the volume fraction z_{β}:

$$\underset{\sim}{\varepsilon}^{pc} = \delta_{\alpha \to \beta} z_{\beta} \underset{\sim}{\mathbf{I}} \quad [4.29]$$

4.15.3. *Numerical results*

Two types of tests are available for the identification:

– strain-controlled cyclic tests under symmetric loading (±1%);

– cyclic relaxation tests (60 s dwell time).

The temperature-dependent parameters are represented by a function, which allows us to obtain a best fit for non-anisothermal loadings. A typical result is shown in Figure 4.46(a), which shows the comparison between the experiments and simulation at 800℃ for different strain rates. For high temperatures (more than 1,000℃), the stress decreases to zero for the selected strain rates. There is a good agreement between the experimental stress in Figure 4.46(b) and the simulated results in Figure 4.46(c).

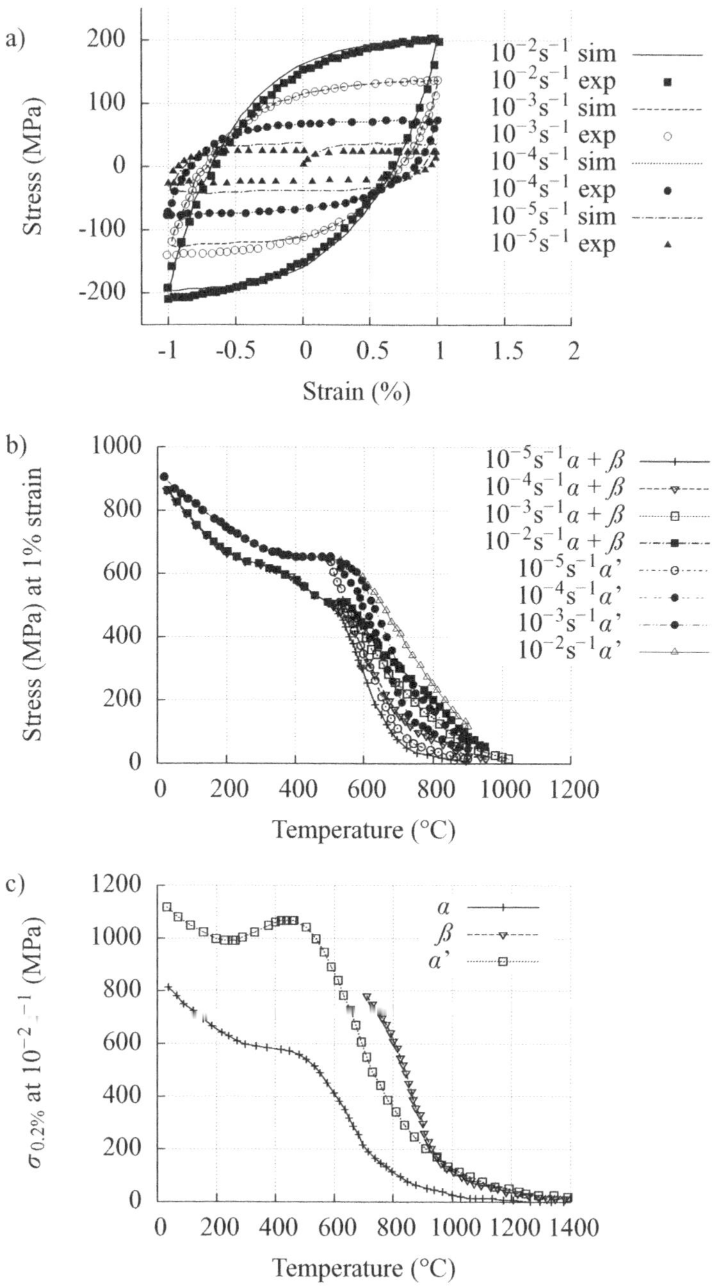

Figure 4.46. *TA6V [LON 09]. (a) Comparison between the experiment and the simulation at 800°C for different strain rates for the $\alpha + \beta$ structure. (b) Stress at 1% strain with respect to temperature for different strain rates. (c) Stress at 0.2% strain and a strain rate of 10^{-2} s^{-1} for each phase*

5

MM Damage–Plasticity Models

The MM models previously investigated are extended in this chapter to account for material degradation. Damage is an additional internal variable that is coupled with the pre-existing equations. Two classical approaches have emerged in the literature: continuum damage mechanics (CDM) and porous plasticity models. The first class of model is known to provide good predictions under shear tensile loading conditions where typically low stress triaxialities are encountered, whereas at high stress triaxialities, a porous plasticity model type is more appropriate.

– In the CDM approach, damage is represented at the macroscopic level by means of tensorial or scalar variables that produce a drop in the elastic moduli and may also affect plastic behavior, according to the formulation. This approach can be extended in order to include volumetric plastic strains that play an important role in ductile fracture at high triaxialities;

– Porous plasticity models are inspired by the solutions representing micro void nucleation, growth and coalescence of micro-cracks. Void growth is linked to plastic strain and the stress triaxiality level. Improvements have been proposed in the literature in order to take into account the effects of low/null triaxiality during shear-dominated loading paths.

This chapter is restricted to small deformation and uses an assumption of uniform elasticity. The total strain is then decomposed into an elastic part and a plastic part:

$$\underset{\sim}{\varepsilon} = \underset{\sim}{\varepsilon}^{e} + \underset{\sim}{\varepsilon}^{in} \qquad [5.1]$$

Plastic strain is the average of the irreversible deformation of each mechanism throughout this chapter:

$$\underset{\sim}{\varepsilon}^{in} = (1 - z)\underset{\sim}{\varepsilon}^1 + z\underset{\sim}{\varepsilon}^2 \qquad [5.2]$$

5.1. MM models based on the GTN approach

The most popular porous plasticity model, proposed originally by Gurson [GUR 77], is based on the concept of a plastic yield surface which is a function of the void volume fraction (or porosity). The evolution of porosity results from the mass conservation principle. Two MM sub-models can be distinguished as follows:

– the first version attributes a fraction of the porosity to each mechanism, so that the total porosity f is obtained as a weighted sum of the values f^1 and f^2 for each mechanism $f = (1 - z)f^1 + zf^2$;

– damage is classically described by a unique overall variable. Accordingly, an overall damage variable f may be considered for the MM models.

The Gurson model is based on the definition of effective quantities $a_\star$ for any stress tensor $\underset{\sim}{\mathbf{a}}$ and a value of the porosity:

$$\underset{\sim}{\mathbf{a}} : \frac{\partial a_\star}{\partial \underset{\sim}{\mathbf{a}}} = a_\star \qquad [5.3]$$

According to Besson *et al.* [BES 03], the effective stresses related to the two mechanisms $\sigma_\star^1$ and $\sigma_\star^2$ are implicitly defined by the following conditions:

$$\begin{cases} G^1(\underset{\sim}{\sigma}^1, f_*, \sigma_{1\star}) = \left(\dfrac{J(\underset{\sim}{\sigma}^1)}{\sigma_{1\star}}\right)^2 + 2q_1 f_* \cosh\left(\dfrac{q_2}{2}\dfrac{I(\underset{\sim}{\sigma}^1)}{\sigma_{1\star}}\right) - 1 - (q_1 f_*)^2 = 0 \\ G^2(\underset{\sim}{\sigma}^2, f_*, \sigma_{2\star}) = \left(\dfrac{J(\underset{\sim}{\sigma}^2)}{\sigma_{2\star}}\right)^2 + 2q_1 f_* \cosh\left(\dfrac{q_2}{2}\dfrac{I(\underset{\sim}{\sigma}^2)}{\sigma_{2\star}}\right) - 1 - (q_1 f_*)^2 = 0 \end{cases} \qquad [5.4]$$

The material parameters q_1 and q_2 account for interactions [FAL 98]. The function f_* depends on the porosity f, in order to account for void coalescence:

$$f_* = \begin{cases} f & \text{if} \quad f < f_c \\ f_c + \delta(f - f_c) & \text{if} \quad f \geq f_c \end{cases} \qquad [5.5]$$

where δ and f_c are material parameters. The value of the porosity at the onset of coalescence is f_c. In the next two sections, damage is assumed to be described by a unique variable f. We only consider the "void growth" phase. Consequently, void nucleation is neglected and the evolution of the porosity due to the void growth is simply obtained from mass conservation:

$$\dot{f} = (1 - f)\mathrm{Tr}\,(\dot{\underset{\sim}{\varepsilon}}^{in}) \qquad [5.6]$$

Using the decomposition of plastic strain rate (equation [5.2]), the contribution of each mechanism to damage is then:

$$\dot{f}^1 = (1 - z)(1 - f)\, tr\,(\dot{\underset{\sim}{\varepsilon}}^1) \quad \text{and} \quad \dot{f}^2 = z(1 - f)\, tr\,(\dot{\underset{\sim}{\varepsilon}}^2) \qquad [5.7]$$

For the sake of brevity, the MM models coupled with GTN damage are restricted to nonlinear isotropic hardening. Applications with kinematic hardening can be found elsewhere [SAÏ 11b].

5.1.1. *Damage in the 2M1C model based on the GTN approach*

As for the original 2M1C model for "undamaged" materials, the potential used introduces a quadratic combination of the two mechanisms:

$$\phi = \sqrt{(\sigma_{1\star})^2 + \left(\sigma^1_\star\right)^2} - \frac{R}{(1 - f)} - R_0 \qquad [5.8]$$

The effective stresses $\sigma_{1\star}$ and $\sigma_{2\star}$ are defined according to equation [5.4]. The effective size change of the elastic domain is affected by the damage level f and linked to the internal variables r according to

$$R = (1 - f)bQr \qquad [5.9]$$

The inelastic strain rates are given by

$$\begin{cases} \dot{\underset{\sim}{\varepsilon}}^1 = (1 - f)\dot{p}\underset{\sim}{\mathbf{n}}^1 & \text{with} \quad \underset{\sim}{\mathbf{n}}^1 = \dfrac{\partial \phi}{\partial \underset{\sim}{\sigma}^1} = \dfrac{\sigma_{1\star}}{\sqrt{\sigma^2_{1\star} + \sigma^2_{2\star}}}\left(P^1_d\,\underset{\sim}{\mathbf{s}}^1 + P^1_s\,\underset{\sim}{\mathbf{I}}_1\right) \\[2ex] \dot{\underset{\sim}{\varepsilon}}^2 = (1 - f)\dot{p}\underset{\sim}{\mathbf{n}}^2 & \text{with} \quad \underset{\sim}{\mathbf{n}}^2 = \dfrac{\partial \phi}{\partial \underset{\sim}{\sigma}^2} = \dfrac{\sigma_{2\star}}{\sqrt{\sigma^2_{1\star} + \sigma^2_{2\star}}}\left(P^2_d\,\underset{\sim}{\mathbf{s}}^2 + P^2_s\,\underset{\sim}{\mathbf{I}}_2\right) \end{cases} \qquad [5.10]$$

In the viscoplasticity framework, a power law can be used by introducing the material parameters K and n to express the plastic multiplier:

$$\dot{p} = \left\langle \frac{\phi}{K} \right\rangle^n \qquad [5.11]$$

It can be seen from equation [5.10] that both normals $\underset{\sim}{\mathbf{n}}^1$ and $\underset{\sim}{\mathbf{n}}^2$ are decomposed into the sum of a deviatoric part and a spherical part. The non-null spherical terms account for the volume change at the mechanism level and contribute to overall volume change:

$$\begin{cases} P_d^1 = \dfrac{3\sigma_{1\star}}{2J^2(\underset{\sim}{\sigma}_1) + f_* q_1 q_2 I(\underset{\sim}{\sigma}_1)\sigma_{1\star} \sinh\left(\dfrac{q_2}{2}\dfrac{I(\underset{\sim}{\sigma}_1)}{\sigma_{1\star}}\right)} \\ P_d^2 = \dfrac{3\sigma_{2\star}}{2J^2(\underset{\sim}{\sigma}_2) + f_* q_1 q_2 I(\underset{\sim}{\sigma}_2)\sigma_{2\star} \sinh\left(\dfrac{q_2}{2}\dfrac{I(\underset{\sim}{\sigma}_2)}{\sigma_{2\star}}\right)} \\ P_s^1 = \dfrac{q_1 q_2 f_*(\sigma_{1\star})^2 \sinh\left(\dfrac{q_2}{2}\dfrac{I(\underset{\sim}{\sigma}_1)}{\sigma_{1\star}}\right)}{2J^2(\underset{\sim}{\sigma}_1) + f_* q_1 q_2 I(\underset{\sim}{\sigma}_1)\sigma_{1\star} \sinh\left(\dfrac{q_2}{2}\dfrac{I(\underset{\sim}{\sigma}_1)}{\sigma_{1\star}}\right)} \\ P_s^2 = \dfrac{q_1 q_2 f_*(\sigma_{2\star})^2 \sinh\left(\dfrac{q_2}{2}\dfrac{I(\underset{\sim}{\sigma}_2)}{\sigma_{2\star}}\right)}{2J^2(\underset{\sim}{\sigma}_2) + f_* q_1 q_2 I(\underset{\sim}{\sigma}_2)\sigma_{2\star} \sinh\left(\dfrac{q_2}{2}\dfrac{I(\underset{\sim}{\sigma}_2)}{\sigma_{2\star}}\right)} \end{cases} \qquad [5.12]$$

The evolution law of the isotropic hardening variable is written as

$$\dot{r} = \dot{p}\left(\frac{1}{(1-f)} - \frac{R}{Q(1-f)^2}\right) \qquad [5.13]$$

5.1.2. *Damage in the 2M2C model based on the GTN approach*

The flow potentials for the two mechanisms are respectively:

$$\begin{cases} \phi^1 = \sigma_{1\star} - \dfrac{R^1}{(1-f)} - R_{01} \\ \phi^2 = \sigma_{2\star} - \dfrac{R^2}{(1-f)} - R_{02} \end{cases} \qquad [5.14]$$

The size of the elastic domain of each mechanism is affected by the damage level f. They are related to two internal variables r^1 and r^2, such as

$$\begin{cases} R^1 = (1-f)b_1 Q_1 r^1 \\ R^2 = (1-f)b_2 Q_2 r^2 \end{cases} \qquad [5.15]$$

The viscoplastic strain rates are given by

$$\begin{cases} \dot{\underset{\sim}{\varepsilon}}^1 = (1-f)\dot{p}^1 \underset{\sim}{\mathbf{n}}^1 \quad \text{with} \quad \underset{\sim}{\mathbf{n}}^1 = \dfrac{\partial \phi^1}{\partial \underset{\sim}{\sigma}^1} = P_d^1 \underset{\sim}{\mathbf{s}}^1 + P_s^1 \underset{\sim}{\mathbf{I}}^1 \\ \dot{\underset{\sim}{\varepsilon}}^2 = (1-f)\dot{p}^2 \underset{\sim}{\mathbf{n}}^2 \quad \text{with} \quad \underset{\sim}{\mathbf{n}}^2 = \dfrac{\partial \phi^2}{\partial \underset{\sim}{\sigma}^2} = P_d^2 \underset{\sim}{\mathbf{s}}^2 + P_s^2 \underset{\sim}{\mathbf{I}}^2 \end{cases} \qquad [5.16]$$

A Norton law is used to compute the plastic multipliers of the two mechanisms:

$$\begin{cases} \dot{p}^1 = \left\langle \dfrac{\phi^1}{K_1} \right\rangle^{n_1} \\ \dot{p}^2 = \left\langle \dfrac{\phi^2}{K_2} \right\rangle^{n_2} \end{cases} \qquad [5.17]$$

P_d^1, P_s^1, P_d^2 and P_s^2 are defined in the same way as in the 2M1C model (equation [5.12]).

The evolution laws of the isotropic hardening variable are written as

$$\begin{cases} \dot{r}^1 = \dot{p}^1 \left(\dfrac{1}{1-f} - \dfrac{R^1}{(1-f)^2 Q_1} \right) \\ \dot{r}^2 = \dot{p}^2 \left(\dfrac{1}{1-f} - \dfrac{R^2}{(1-f)^2 Q_2} \right) \end{cases} \qquad [5.18]$$

5.2. MM models coupled with CDM theory

In the next sections, the use of the MM models in the framework of CDM is demonstrated. It is then assumed that crack initiation is preceded by a progressive internal deterioration of the material (i.e. micro-cracks, micro-defects) which induces a loss of elastic properties, represented through the effective stress concept. Damage is supposed to be isotropic. Similarly to the case of the MM model based on the GTN approach, two sub-models are to be considered:

– damage is classically accounted for by means of a unique variable D, varying between 0 and 1 [CHA 87]. This case will be considered thereafter;

– each of the two mechanisms is characterized by its own damage variable, respectively D^1 and D^2. The overall damage D is the average of the local damage of each phase $D = (1 - z)D^1 + zD^2$. The use of two damage variables was also introduced by Boudifa *et al.* [BOU 09] and investigated in the case of MM models by Jeridi *et al.* [JER 15].

The CDM theory is based on a thermodynamic framework with internal variables. The free energy $\rho\Psi$ defines the macroscopic stress $\underset{\sim}{\sigma}$ and the hardening variables A^I, knowing the elastic strain $\underset{\sim}{\varepsilon}^e$ and the internal variables α^I:

$$\underset{\sim}{\sigma} = \rho\frac{\partial\Psi}{\partial\underset{\sim}{\varepsilon}^e} \quad \text{and} \quad A^I = \rho\frac{\partial\Psi}{\partial\alpha^I} \qquad [5.19]$$

where ρ is the density of the material. Assuming uncoupling between elastic and plastic parts, Ψ can be considered as the sum of two contributions: an elastic part (Ψ_e) and a plastic part (Ψ_p). Following Saanouni [SAA 88], a state coupling is considered between damage and hardening:

$$\Psi = \Psi_e\left(\underset{\sim}{\varepsilon}^e, D\right) + \Psi_p\left(\alpha^I, D\right) \qquad [5.20]$$

The inelastic part of the free energy will be specified according to the considered MM model. The elastic potential is written in all cases:

$$\rho\Psi_e = \frac{1}{2}(1 - D)\,\underset{\sim}{\varepsilon}^e : \underset{\approx}{\mathbf{\Lambda}} : \underset{\sim}{\varepsilon}^e \qquad [5.21]$$

The overall stress tensor is then deduced:

$$\underset{\sim}{\sigma} = (1 - D)\underset{\approx}{\mathbf{\Lambda}} : \underset{\sim}{\varepsilon}^e \qquad [5.22]$$

In opposition to the GTN approach, for which only the effect of the damage variable on plastic behavior is taken into account, the influence of damage is introduced into plastic behavior, and also into elastic behavior. The thermodynamic force Y associated with D is the sum of two contributions, Y^e and Y^p. The first comes from the elastic part of the model:

$$Y^e = \frac{1}{2}(\underset{\sim}{\varepsilon}^e : \underset{\approx}{\mathbf{\Lambda}} : \underset{\sim}{\varepsilon}^e) \qquad [5.23]$$

The assumption of “Strain Equivalence” is adopted in sections 5.2.1 and 5.2.2, whereas the “Energy Equivalence” principle is the subject of sections 5.2.3 and 5.2.4.

5.2.1. *2M1C model “Strain Equivalence”*

In this section, damage is used to define an effective tensor $\overline{\underset{\sim}{\mathbf{T}}}$ from a given tensor $\underset{\sim}{\mathbf{T}}$ according to

$$\overline{\underset{\sim}{\mathbf{T}}} = \frac{\underset{\sim}{\mathbf{T}}}{1-D} \qquad [5.24]$$

Consequently, damage influences plastic flow by affecting either the elastic domain or the evolution of the hardening variables. Assuming that effective stress, effective kinematic and effective isotropic variables are used, the elastic limit criterion can be expressed as

$$\begin{cases} \phi = \sqrt{J(\overline{\underset{\sim}{\sigma}}^1 - \overline{\underset{\sim}{\mathbf{X}}}^1)^2 + J(\overline{\underset{\sim}{\sigma}}^2 - \overline{\underset{\sim}{\mathbf{X}}}^2)^2} - \overline{R} - R_0 = 0 \\ \left(J(\overline{\underset{\sim}{\sigma}}^1 - \overline{\underset{\sim}{\mathbf{X}}}^1)\right)^2 = \frac{3}{2}(\overline{\underset{\sim}{\mathbf{s}}}^1 - \overline{\underset{\sim}{\mathbf{X}}}^1) : (\overline{\underset{\sim}{\mathbf{s}}}^1 - \overline{\underset{\sim}{\mathbf{X}}}^1) \\ \left(J(\overline{\underset{\sim}{\sigma}}^2 - \overline{\underset{\sim}{\mathbf{X}}}^2)\right)^2 = \frac{3}{2}(\overline{\underset{\sim}{\mathbf{s}}}^2 - \overline{\underset{\sim}{\mathbf{X}}}^2) : (\overline{\underset{\sim}{\mathbf{s}}}^2 - \overline{\underset{\sim}{\mathbf{X}}}^2) \\ \overline{J}^2 = \left(\left(J(\overline{\underset{\sim}{\sigma}}^1 - \overline{\underset{\sim}{\mathbf{X}}}^1)\right)^2 + \left(J(\overline{\underset{\sim}{\sigma}}^2 - \overline{\underset{\sim}{\mathbf{X}}}^2)\right)^2\right) \end{cases} \qquad [5.25]$$

The inelastic part of the free energy function is then a function of the kinematic internal variables $\underset{\sim}{\alpha}^1$, $\underset{\sim}{\alpha}^2$ and of the isotropic internal variable r:

$$\rho\Psi_p = \frac{1-D}{3}\left(C_{11}\underset{\sim}{\alpha}^1 : \underset{\sim}{\alpha}^1 + C_{12}\underset{\sim}{\alpha}^1 : \underset{\sim}{\alpha}^2 + C_{12}\underset{\sim}{\alpha}^2 : \underset{\sim}{\alpha}^1 + C_{22}\underset{\sim}{\alpha}^2 : \underset{\sim}{\alpha}^2\right) + \frac{1-D}{2}bQr^2 \qquad [5.26]$$

The hardening variables are then

$$\begin{cases} \overline{\underset{\sim}{\mathbf{X}}}^1 = \dfrac{\underset{\sim}{\mathbf{X}}^1}{1-D} = \dfrac{1}{1-D}\rho\dfrac{\partial \Psi_p}{\partial \underset{\sim}{\alpha}^1} = \frac{2}{3}\left(C_{11}\underset{\sim}{\alpha}^1 + C_{12}\underset{\sim}{\alpha}^2\right) \\ \overline{\underset{\sim}{\mathbf{X}}}^2 = \dfrac{\underset{\sim}{\mathbf{X}}^2}{1-D} = \dfrac{1}{1-D}\rho\dfrac{\partial \Psi_p}{\partial \underset{\sim}{\alpha}^2} = \frac{2}{3}\left(C_{12}\underset{\sim}{\alpha}^1 + C_{22}\underset{\sim}{\alpha}^2\right) \\ \overline{R} = \dfrac{R}{1-D} = \dfrac{1}{1-D}\rho\dfrac{\partial \Psi_p}{\partial r} = bQr \end{cases} \quad [5.27]$$

The evolution laws of these variables are generated by a potential that introduces a quadratic combination of the effective hardening variables in addition to the damage term:

$$F = \phi + \frac{3D_1}{4C_{11}}\overline{\underset{\sim}{\mathbf{X}}}^1 : \overline{\underset{\sim}{\mathbf{X}}}^1 + \frac{3D_2}{4C_{22}}\overline{\underset{\sim}{\mathbf{X}}}^2 : \overline{\underset{\sim}{\mathbf{X}}}^2 + \frac{1}{2Q}\overline{R}^2 + \frac{1}{s+1}\left(\frac{Y}{S}\right)^{s+1} \quad [5.28]$$

It leads to

$$\begin{cases} (1-D)\dot{\underset{\sim}{\varepsilon}}^1 = \underset{\sim}{\mathbf{n}}^1\dot{p} & (1-D)\dot{\underset{\sim}{\varepsilon}}^2 = \underset{\sim}{\mathbf{n}}^2\dot{p} \\ (1-D)\dot{\underset{\sim}{\alpha}}^1 = (\underset{\sim}{\mathbf{n}}^1 - \dfrac{3D_1}{2C_{11}}\overline{\underset{\sim}{\mathbf{X}}}^1)\dot{p} & (1-D)\dot{\underset{\sim}{\alpha}}^2 = (\underset{\sim}{\mathbf{n}}^2 - \dfrac{3D_2}{2C_{22}}\overline{\underset{\sim}{\mathbf{X}}}^2)\dot{p} \\ (1-D)\dot{r} = (1-br)\dot{p} & \\ \dot{D} = \left(\dfrac{Y}{S}\right)^s \dot{p} & \\ \dot{p} = \left\langle \dfrac{\phi}{K} \right\rangle^n & \end{cases} \quad [5.29]$$

In the previous equations, $\underset{\sim}{\mathbf{n}}^1$ and $\underset{\sim}{\mathbf{n}}^2$ are defined as

$$\underset{\sim}{\mathbf{n}}^1 = \frac{3}{2}\frac{\left(\underset{\sim}{\mathbf{s}}^1 - \underset{\sim}{\mathbf{X}}^1\right)}{J(\underset{\sim}{\mathbf{s}}^1 - \underset{\sim}{\mathbf{X}}^1)} \qquad \underset{\sim}{\mathbf{n}}^2 = \frac{3}{2}\frac{\left(\underset{\sim}{\mathbf{s}}^2 - \underset{\sim}{\mathbf{X}}^2\right)}{J(\underset{\sim}{\mathbf{s}}^2 - \underset{\sim}{\mathbf{X}}^2)} \quad [5.30]$$

5.2.2. *2M2C model "Strain Equivalence"*

Instead of affecting all the variables, as in the previous example with the 2M1C model, the present version of the 2M2C model introduces only a partial coupling, with the first mechanism only. The second yield criterion is the same as for the standard version, meanwhile the first depends on the effective stress and the effective hardening variables:

$$\begin{cases} \phi^1 = J(\overline{\underset{\sim}{\sigma}}^1 - \overline{\underset{\sim}{\mathbf{X}}}^1) - \overline{R}^1 - R_{01} \\ \phi^2 = J(\underset{\sim}{\sigma}^2 - \underset{\sim}{\mathbf{X}}^2) - R^2 - R_{02} \end{cases} \qquad [5.31]$$

The inelastic part of the free energy function is expressed as a function of the kinematic internal variables $\underset{\sim}{\alpha}^1$, $\underset{\sim}{\alpha}^2$ and the isotropic internal variables r^1 and r^2:

$$\begin{cases} \rho\Psi_p = \frac{1}{3}\left((1-D)\, C_{11}\underset{\sim}{\alpha}^1 : \underset{\sim}{\alpha}^1 + C_{12}\underset{\sim}{\alpha}^1 : \underset{\sim}{\alpha}^2 + C_{12}\underset{\sim}{\alpha}^2 : \underset{\sim}{\alpha}^1 + C_{22}\underset{\sim}{\alpha}^2 : \underset{\sim}{\alpha}^2\right) \\ \qquad + \frac{1-D}{2} b_1 Q_1 (r^1)^2 + \frac{1}{2} b_2 Q_2 (r^2)^2 \end{cases} \qquad [5.32]$$

The hardening variables are written as

$$\begin{cases} \overline{\underset{\sim}{\mathbf{X}}}^1 = \dfrac{\underset{\sim}{\mathbf{X}}^1}{1-D} = \dfrac{1}{1-D}\rho\dfrac{\partial \Psi_p}{\partial \underset{\sim}{\alpha}^1} = \frac{2}{3}\left(C_{11}\underset{\sim}{\alpha}^1 + C_{12}\underset{\sim}{\alpha}^2\right) \\ \underset{\sim}{\mathbf{X}}^2 = \rho\dfrac{\partial \Psi_p}{\partial \underset{\sim}{\alpha}^2} = \frac{2}{3}\left(C_{12}\underset{\sim}{\alpha}^1 + C_{22}\underset{\sim}{\alpha}^2\right) \\ \overline{R}^1 = \dfrac{R^1}{1-D} = \dfrac{1}{1-D}\rho\dfrac{\partial \Psi_p}{\partial R} = b^1 Q^1 r^1 \\ R^2 = \rho\dfrac{\partial \Psi_p}{\partial R} = b_2 Q_2 r^2 \end{cases} \qquad [5.33]$$

The evolution of these variables is defined by two potentials, which introduce quadratic combinations of the effective hardening variables

$$\begin{cases} F^1 = \phi^1 + \dfrac{3D_1}{4C_{11}}\overline{\underset{\sim}{\mathbf{X}}}^1 : \overline{\underset{\sim}{\mathbf{X}}}^1 + \dfrac{1}{2Q_1}(\overline{R}^1)^2 + \dfrac{1}{s+1}\left(\dfrac{Y}{S}\right)^{s+1} \\ F^2 = \phi^2 + \dfrac{3D_2}{4C_{22}}\underset{\sim}{\mathbf{X}}^2 : \underset{\sim}{\mathbf{X}}^2 + \dfrac{1}{2Q_2}(R^2)^2 \end{cases} \qquad [5.34]$$

It leads to

$$\begin{cases} (1-D)\dot{\underset{\sim}{\varepsilon}}^1 = \underset{\sim}{\mathbf{n}}^1 \dot{p}^1 & \dot{\underset{\sim}{\varepsilon}}^2 = \underset{\sim}{\mathbf{n}}^2 \dot{p}^2 \\ (1-D)\dot{\underset{\sim}{\alpha}}^1 = (\underset{\sim}{\mathbf{n}}^1 - \dfrac{3D_1}{2C_{11}}\overline{\underset{\sim}{\mathbf{X}}}^1)\dot{p}^1 & \dot{\underset{\sim}{\alpha}}^2 = (\underset{\sim}{\mathbf{n}}^2 - \dfrac{3D_2}{2C_{22}}\underset{\sim}{\mathbf{X}}^2)\dot{p}^2 \\ (1-D)\dot{r}^1 = (1 - b_1 r^1)\dot{p}^1 & \dot{r}^2 = (1 - b_2 r^2)\dot{p}^2 \\ \dot{D} = \left(\dfrac{Y}{S}\right)^s \dot{p}^1 & \\ \dot{p}^1 = \left\langle \dfrac{\phi^1}{K_1} \right\rangle^{n_1} & \dot{p}^2 = \left\langle \dfrac{\phi^2}{K_2} \right\rangle^{n_2} \end{cases} \qquad [5.35]$$

In the previous equations, $\underset{\sim}{\mathbf{n}}^1$ and $\underset{\sim}{\mathbf{n}}^2$ are defined as

$$\underset{\sim}{\mathbf{n}}^1 = \frac{3}{2}\frac{\left(\underset{\sim}{\sigma}^1 - \underset{\sim}{\mathbf{X}}^1\right)}{J(\underset{\sim}{\overline{\sigma}}^1 - \underset{\sim}{\overline{\mathbf{X}}}^1)} \qquad \underset{\sim}{\mathbf{n}}^2 = \frac{3}{2}\frac{\left(\underset{\sim}{\mathbf{s}}^2 - \underset{\sim}{\mathbf{X}}^2\right)}{J(\underset{\sim}{\sigma}^2 - \underset{\sim}{\mathbf{X}}^2)}$$

5.2.3. *2M1C model "Energy Equivalence"*

In this section, an energy equivalence principle coming from CDM is introduced in a 2M1C model. The concept of effective tensors will not be used here. The yield function is defined according to

$$\begin{cases} \phi = \dfrac{J - R}{\sqrt{(1-D)}} - R_0 = 0 \\ J = \sqrt{\left(J(\underset{\sim}{\sigma}^1 - \underset{\sim}{\mathbf{X}}^1)\right)^2 + \left(J(\underset{\sim}{\sigma}^2 - \underset{\sim}{\mathbf{X}}^2)\right)^2} \end{cases} \tag{5.36}$$

The inelastic part of the free energy is taken from (equation [5.26]), so that the kinematic and isotropic variables are

$$\begin{cases} \underset{\sim}{\mathbf{X}}^1 = (1-D)\frac{2}{3}\left(C_{11}\underset{\sim}{\alpha}^1 + C_{12}\underset{\sim}{\alpha}^2\right) \\ \underset{\sim}{\mathbf{X}}^2 = (1-D)\frac{2}{3}\left(C_{12}\underset{\sim}{\alpha}^1 + C_{22}\underset{\sim}{\alpha}^2\right) \\ R = (1-D)bQr \end{cases} \tag{5.37}$$

For each mechanism, the expression of the normal to the yield surface is written as

$$\begin{cases} \underset{\sim}{\mathbf{n}}^1 = \dfrac{3}{2}\dfrac{1}{\sqrt{(1-D)}}\dfrac{\underset{\sim}{\mathbf{s}}^1 - \underset{\sim}{\mathbf{X}}^1}{J} \\ \underset{\sim}{\mathbf{n}}^2 = \dfrac{3}{2}\dfrac{1}{\sqrt{(1-D)}}\dfrac{\underset{\sim}{\mathbf{s}}^2 - \underset{\sim}{\mathbf{X}}^2}{J} \end{cases} \tag{5.38}$$

The potential used to generate the evolution laws differs slightly from the potential used for the strain energy principle:

$$F = f + \frac{3D_1}{4(1-D)C_{11}}\underset{\sim}{\mathbf{X}}^1 : \underset{\sim}{\mathbf{X}}^1 + \frac{3D_2}{4(1-D)C_{22}}\underset{\sim}{\mathbf{X}}^2 : \underset{\sim}{\mathbf{X}}^2 + \frac{R^2}{2(1-D)Q} + \frac{1}{s+1}\left(\frac{Y}{S}\right)^{s+1} \tag{5.39}$$

The viscoplastic flow, the hardening rate and the damage rate are respectively defined by equations [5.40]:

$$\dot{\underset{\sim}{\varepsilon}}^1 = \dot{p}\underset{\sim}{\mathbf{n}}^1 \quad \dot{\underset{\sim}{\varepsilon}}^2 = \dot{p}\underset{\sim}{\mathbf{n}}^2 \tag{5.40a}$$

$$\dot{p} = \left\langle \frac{\phi}{K} \right\rangle^n \tag{5.40b}$$

$$\dot{r} = \dot{p}\left(\frac{1}{\sqrt{(1-D)}} - \frac{R}{(1-D)Q}\right) \tag{5.40c}$$

$$\dot{\underset{\sim}{\alpha}}^1 = \dot{p}\left(\underset{\sim}{\mathbf{n}}^1 - \frac{3}{2}\frac{D_1}{(1-D)C_{11}}\underset{\sim}{\mathbf{X}}^1\right), \qquad \dot{\underset{\sim}{\alpha}}^2 = \dot{p}\left(\underset{\sim}{\mathbf{n}}^2 - \frac{3}{2}\frac{D_2}{(1-D)C_{22}}\underset{\sim}{\mathbf{X}}^2\right) \tag{5.40d}$$

$$\dot{D} = \dot{p}\left(\frac{Y}{S}\right)^s \tag{5.40e}$$

5.2.4. *2M2C model "Energy Equivalence"*

The damage-induced plastic volume change is controlled by the shape of the yield functions [CHA 06]:

$$\begin{cases} \phi^1 = \dfrac{1}{\sqrt{(1-D)}}\left(J(\underset{\sim}{\sigma}^1 - \underset{\sim}{\mathbf{X}}^1) - R^1\right) - R_{01} \\ \phi^2 = \dfrac{1}{\sqrt{(1-D)}}\left(J(\underset{\sim}{\sigma}^2 - \underset{\sim}{\mathbf{X}}^2) - R^2\right) - R_{02} \end{cases} \tag{5.41}$$

In the free energy, the two phases are supposed to be affected by damage:

$$\begin{cases} \rho\Psi_p = \frac{1}{3}(1-D)\left(C_{11}\underset{\sim}{\alpha}^1 : \underset{\sim}{\alpha}^1 + C_{12}\underset{\sim}{\alpha}^1 : \underset{\sim}{\alpha}^2 + C_{12}\underset{\sim}{\alpha}^2 : \underset{\sim}{\alpha}^1 + C_{22}\underset{\sim}{\alpha}^2 : \underset{\sim}{\alpha}^2\right) \\ \qquad + \frac{1-D}{2}b_1Q_1(r^1)^2 + \frac{1-D}{2}b_2Q_2(r^2)^2 \end{cases} \tag{5.42}$$

The hardening thermodynamic forces are

$$\begin{cases} \underset{\sim}{\mathbf{X}}^1 = (1-D)\,\frac{2}{3}\left(C_{11}\underset{\sim}{\alpha}^1 + C_{12}\underset{\sim}{\alpha}^2\right) \\ \underset{\sim}{\mathbf{X}}^2 = (1-D)\,\frac{2}{3}\left(C_{12}\underset{\sim}{\alpha}^1 + C_{22}\underset{\sim}{\alpha}^2\right) \\ R^1 = (1-D)\,b_1Q_1r^1 \\ R^2 = (1-D)\,b_2Q_2r^2 \end{cases} \tag{5.43}$$

Damage is described by a unique scalar variable D associated with a thermodynamic force Y. The viscoplastic thermodynamic force Y^p is written as

$$Y^p = \frac{1}{2}b_1 Q_1 (r^1)^2 + \frac{1}{2}b_2 Q_2 (r^2)^2 \quad [5.44]$$

The overall thermodynamic force Y associated with D is the sum of the two previous thermodynamic forces:

$$Y = Y^e + Y^p = \frac{1}{2}(1-D)(\underset{\sim}{\varepsilon}^e : \underset{\approx}{\Lambda} : \underset{\sim}{\varepsilon}^e) + \frac{1}{2}b_1 Q_1 (r^1)^2 + \frac{1}{2}b_2 Q_2 (r^2)^2 \quad [5.45]$$

The normals to the yield surfaces are calculated as

$$\underset{\sim}{\mathbf{n}}^1 = \frac{3}{2}\frac{1}{\sqrt{(1-D)}}\frac{\underset{\sim}{\mathbf{s}}^1 - \underset{\sim}{\mathbf{X}}^1}{J(\underset{\sim}{\sigma}^2 - \underset{\sim}{\mathbf{X}}^2)} \qquad \underset{\sim}{\mathbf{n}}^2 = \frac{3}{2}\frac{1}{\sqrt{(1-D)}}\frac{\underset{\sim}{\mathbf{s}}^2 - \underset{\sim}{\mathbf{X}}^2}{J(\underset{\sim}{\sigma}^2 - \underset{\sim}{\mathbf{X}}^2)} \quad [5.46]$$

Damage is described by two material parameters s and S. Finally, the viscoplastic flow, the hardening rate and the damage rate are respectively defined by equations [5.47]:

$$\dot{\underset{\sim}{\varepsilon}}^1 = \dot{p}^1 \underset{\sim}{\mathbf{n}}^1 \quad \dot{\underset{\sim}{\varepsilon}}^2 = \dot{p}^2 \underset{\sim}{\mathbf{n}}^2 \quad [5.47a]$$

$$\dot{p}^1 = \left\langle \frac{\phi^1}{K_1} \right\rangle^{n_1} \quad \dot{p}^2 = \left\langle \frac{\phi^2}{K_2} \right\rangle^{n_2} \quad [5.47b]$$

$$\dot{r}^1 = \dot{p}^1 \left(\frac{1}{\sqrt{(1-D)}} - \frac{R^1}{(1-D)Q_1} \right) \quad [5.47c]$$

$$\dot{r}^2 = \dot{p}^2 \left(\frac{1}{\sqrt{(1-D)}} - \frac{R^2}{(1-D)Q_2} \right) \quad [5.47d]$$

$$\dot{\underset{\sim}{\alpha}}^1 = \dot{p}^1 \left(\underset{\sim}{\mathbf{n}}^1 - \frac{3}{2}\frac{D_1}{(1-D)C_{11}} \underset{\sim}{\mathbf{X}}^1 \right) \quad [5.47e]$$

$$\dot{\underset{\sim}{\alpha}}^2 = \dot{p}^2 \left(\underset{\sim}{\mathbf{n}}^2 - \frac{3}{2}\frac{D_2}{(1-D)C_{22}} \underset{\sim}{\mathbf{X}}^2 \right) \quad [5.47f]$$

$$\dot{D} = ((1-z)\dot{p}^1 + z\dot{p}^2)\left(\frac{Y}{S}\right)^s \quad [5.47g]$$

5.3. Two plastic mechanisms combined with a damage mechanism

The present version of the MM-damage coupling involves a 2M1C model and a damage variable that is controlled by means of a distinct evolution rule. We also assume that $\underset{\sim}{\sigma}^1 = \underset{\sim}{\sigma}^2 = \underset{\sim}{\sigma}$:

$$\begin{cases} f = \sqrt{J(\overline{\underset{\sim}{\sigma}} - \overline{\underset{\sim}{\mathbf{X}}}^1)^2 + J(\overline{\underset{\sim}{\sigma}} - \overline{\underset{\sim}{\mathbf{X}}}^2)^2} - \overline{R} - R_0 = 0 \\ f_d = Y - \omega(D) - \omega_0 = 0 \qquad \text{with} \qquad Y = \frac{1}{2}\underset{\sim}{\varepsilon}^e : \underset{\approx}{\mathbf{\Lambda}} : \underset{\sim}{\varepsilon}^e \end{cases} \qquad [5.48]$$

As in the 2M1C model based on the CDM theory (section 5.2.1), the following effective quantities are used:

$$\overline{\underset{\sim}{\sigma}} = \underset{\approx}{\mathbf{\Lambda}} : \underset{\sim}{\varepsilon}^e \qquad \overline{R} = bQr \qquad [5.49a]$$

$$\overline{\underset{\sim}{\mathbf{X}}}^1 = \frac{2}{3}\left(C_{11}\underset{\sim}{\alpha}^1 + C_{12}\underset{\sim}{\alpha}^2\right) \qquad \overline{\underset{\sim}{\mathbf{X}}}^2 = \frac{2}{3}\left(C_{12}\underset{\sim}{\alpha}^1 + C_{22}\underset{\sim}{\alpha}^2\right) \qquad [5.49b]$$

The evolution laws of the internal variables are also the same:

$$(1 - D)\dot{\underset{\sim}{\varepsilon}}^1 = \underset{\sim}{\mathbf{n}}^1\dot{p} \quad (1 - D)\dot{\underset{\sim}{\varepsilon}}^2 = \underset{\sim}{\mathbf{n}}^2\dot{p} \qquad [5.49c]$$

$$(1 - D)\dot{r} = (1 - br)\dot{p} \qquad [5.49d]$$

$$(1 - D)\dot{\underset{\sim}{\alpha}}^1 = \left(\underset{\sim}{\mathbf{n}}^1 - \frac{3D_1}{2C_{11}}\overline{\underset{\sim}{\mathbf{X}}}^1\right)\dot{p} \qquad [5.49e]$$

$$(1 - D)\dot{\underset{\sim}{\alpha}}^2 = \left(\underset{\sim}{\mathbf{n}}^2 - \frac{3D_2}{2C_{22}}\overline{\underset{\sim}{\mathbf{X}}}^2\right)\dot{p} \qquad [5.49f]$$

The damage evolution rule is generated by a damage potential f_d:

$$\dot{D} = \dot{p}_d\frac{\partial f_d}{\partial Y} = \dot{p}_d \qquad [5.50]$$

Two cases can be distinguished:

– For time-dependent plasticity, the equivalent viscoplastic strain rate is determined, for example, by a power of the yield function. The damage rate is also chosen as a power function:

$$\dot{p} = \left\langle \frac{f}{K} \right\rangle^{n} \qquad \dot{p}_d = \left\langle \frac{f_d}{K_d} \right\rangle^{n_d} \tag{5.51}$$

– In the case of time-independent plasticity, two independent multipliers, $\dot{p}$ for plasticity and $\dot{p}_d$ for damage are to be determined by the consistency conditions:

$$\dot{p} = (1-D)\frac{\left(\underset{\sim}{\mathbf{n}}^1 + \underset{\sim}{\mathbf{n}}^2\right) : \dot{\underset{\sim}{\overline{\sigma}}}}{H_{X1} + H_{X2} + H_R} \tag{5.52a}$$

$$H_{X1} = (2/3)\left(\underset{\sim}{\mathbf{n}}^1 C_{11} + \underset{\sim}{\mathbf{n}}^2 C_{12}\right)\left(\underset{\sim}{\mathbf{n}}^1 - \frac{D_1}{C_{11}}\left(C_{11}\underset{\sim}{\alpha}^1 + C_{12}\underset{\sim}{\alpha}^2\right)\right) \tag{5.52b}$$

$$H_{X2} = (2/3)\left(\underset{\sim}{\mathbf{n}}^1 C_{12} + \underset{\sim}{\mathbf{n}}^2 C_{22}\right)\left(\underset{\sim}{\mathbf{n}}^2 - \frac{D_2}{C_{22}}\left(C_{12}\underset{\sim}{\alpha}^1 + C_{22}\underset{\sim}{\alpha}^2\right)\right) \tag{5.52c}$$

$$H_R = bQ(1 - br) \tag{5.52d}$$

It is worth noting that the derivative of the effective stress with respect to time is given by

$$\dot{\underset{\sim}{\overline{\sigma}}} = \frac{\dot{\underset{\sim}{\sigma}}}{1-D} + \frac{\underset{\sim}{\overline{\sigma}}}{1-D}\dot{p}_d \tag{5.53}$$

The damage multiplier is obtained from the condition $\dot{f}_d = f_d = 0$:

$$\left[(1-D)\omega'(D) - \underset{\sim}{\varepsilon}^e : \underset{\sim}{\overline{\sigma}}\right]\dot{p}_d = \underset{\sim}{\varepsilon}^e : \dot{\underset{\sim}{\sigma}} \tag{5.54}$$

The two multipliers are the solution of two linear equations. In a stress controlled test, the system of equations is written as

$$\begin{cases} \left(H_{X1} + H_{X2} + H_R\right)\dot{p} - \left(\underset{\sim}{\mathbf{n}}^1 + \underset{\sim}{\mathbf{n}}^2\right)\underset{\sim}{\overline{\sigma}}\dot{p}_d = \left(\underset{\sim}{\mathbf{n}}^1 + \underset{\sim}{\mathbf{n}}^2\right)\dot{\underset{\sim}{\sigma}} \\ \left[(1-D)\omega'(D) - \underset{\sim}{\varepsilon}^e : \underset{\sim}{\overline{\sigma}}\right]\dot{p}_d = \underset{\sim}{\varepsilon}^e : \dot{\underset{\sim}{\sigma}} \end{cases} \tag{5.55}$$

whereas in a strain controlled test, the system of equations is as follows:

$$\begin{cases} \left(H_{X1} + H_{X2} + H_R + \left(\underset{\sim}{\mathbf{n}}^1 + \underset{\sim}{\mathbf{n}}^2\right) : \underset{\approx}{\mathbf{\Lambda}} : \left(\underset{\sim}{\mathbf{n}}^1 + \underset{\sim}{\mathbf{n}}^2\right)\right)\dot{p} \\ = (1 - D)\left(\underset{\sim}{\mathbf{n}}^1 + \underset{\sim}{\mathbf{n}}^2\right) : \underset{\approx}{\mathbf{\Lambda}} : \dot{\underset{\sim}{\varepsilon}} \\ \overline{\underset{\sim}{\sigma}} \frac{\underset{\sim}{\mathbf{n}}^1 + \underset{\sim}{\mathbf{n}}^2}{1-D}\dot{p} + \omega'(D)\dot{p}_d = \overline{\underset{\sim}{\sigma}} : \dot{\underset{\sim}{\varepsilon}} \end{cases} \quad [5.56]$$

For a one-dimensional strain-controlled loading, the two multipliers are given by

$$\begin{aligned} \dot{p}_d &= \frac{(H_{X1} + H_{X2} + H_R)\, E \left\langle \varepsilon^e \dot{\varepsilon} \right\rangle}{E\left(n^1 + n^2\right)^2 + H_{X1} + H_{X2} + H_R} \\ \dot{p} &= \frac{(1 - D)\left(n^1 + n^2\right) E\dot{\varepsilon}}{E\left(n^1 + n^2\right)^2 + H_{X1} + H_{X2} + H_R} \end{aligned} \quad [5.57]$$

For a one-dimensional stress-controlled loading, it is given by

$$\begin{aligned} \dot{p}_d &= \frac{\left\langle \varepsilon^e \dot{\sigma} \right\rangle}{(1 - D)\omega'(D) - E(\varepsilon^e)^2} \\ \dot{p} &= \frac{\left(n^1 + n^2\right) E\varepsilon^e \dot{p}_d + \left(n^1 + n^2\right)\dot{\sigma}}{H_{X1} + H_{X2} + H_R} \end{aligned} \quad [5.58]$$

If we consider that ω is proportional to damage ($\omega(D) = KD$), the damage multiplier is simply written as

$$\dot{p}_d = \frac{\varepsilon^e \dot{\sigma}}{(1 - D)K - E(\varepsilon^e)^2} \quad [5.59]$$

Combining $f_d = 0$ and $\dot{p}_d > 0$ leads to the condition

$$D < 1/3 - 2\omega_0/(3K) \quad [5.60]$$

The asymptotic value of damage is then 1/3. The transient evolution has a closed form. For example, for $\omega_0 = 0$,

$$D = \frac{2}{3} - \left(- \left(\frac{\sigma^2}{4EK} \right) + \frac{1}{27} + \frac{1}{2} \sqrt{ \left(\frac{\sigma^2}{2EK} \right)^2 - \left(\frac{2\sigma^2}{27EK} \right) } \right)^{1/3}$$

$$+ \left(- \left(\frac{\sigma^2}{4EK} \right) + \frac{1}{27} - \frac{1}{2} \sqrt{ \left(\frac{\sigma^2}{2EK} \right)^2 - \left(\frac{2\sigma^2}{27EK} \right) } \right)^{1/3} \quad [5.61]$$

5.4. MM models taking into account volume change (CDM theory)

The classical CDM theories and GTN approach differ mainly by the effect of triaxiality on the mechanical response. Volume change is naturally accounted for in GTN models since the hydrostatic pressure is involved in the expression of the criterion. In the following, we consider pressure sensitive criteria in the framework of CDM. For the sake of brevity, we restrict ourselves to isotropic hardening.

The volume change can be introduced in a 2M1C-type model in several ways:

– assuming that the model involves the hydrostatic pressure in the first mechanism and the second invariant of the stress tensor in the second mechanism:

$$f = \sqrt{(I_1(\underset{\sim}{\sigma}^1))^2 + (J_2(\underset{\sim}{\sigma}^2))^2} - R - R_0 \quad [5.62]$$

– using a combination of the two local stresses in the trace as well as in the von Mises equivalent term:

$$\begin{cases} f = \sqrt{I^2 + J^2} - R - R_0 \\ I = \sqrt{(I_1(\underset{\sim}{\sigma}^1))^2 + (I_1(\underset{\sim}{\sigma}^2))^2} \\ J = \sqrt{(J_2(\underset{\sim}{\sigma}^1))^2 + (J_2(\underset{\sim}{\sigma}^2))^2} \end{cases} \quad [5.63]$$

– choosing the two equivalent stresses as combinations of the second and the first invariants of the stress tensors as in [CHA 06]:

$$\begin{cases} f = \sqrt{(\sigma^1_{eq})^2 + (\sigma^2_{eq})^2} - R - R_0 \\ \sigma^1_{eq} = \left((1-\theta)(J_2(\underset{\sim}{\sigma}^1))^\xi + \theta(I_1(\underset{\sim}{\sigma}^1))^\xi\right)^{\frac{1}{\xi}} \\ \sigma^2_{eq} = \left((1-\theta)(J_2(\underset{\sim}{\sigma}^2))^\xi + \theta(I_1(\underset{\sim}{\sigma}^2))^\xi\right)^{\frac{1}{\xi}} \end{cases} \quad [5.64]$$

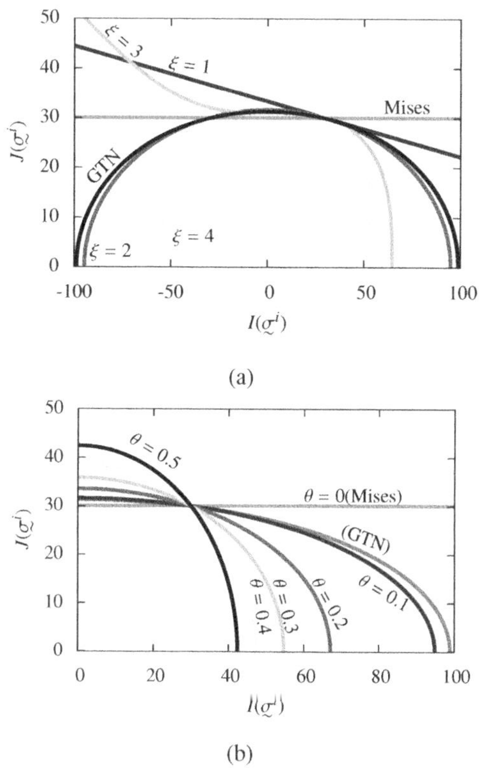

Figure 5.1. *(I, J) functions given by equations 5.64 (a)* $\theta = 0.1$ *(b)* $\xi = 2$. *For a color version of the figure, see www.iste.co.uk/cailletaud/multi-mechanism.zip*

In the following, the 2M1C type criterion of equation [5.64] used in the framework of CDM theory provides a description of a volume variation linked to damage during plastic flow. A parametric study is shown in Figure 5.1 to illustrate typical yield criteria given by these equations for various values of the parameters ξ and θ. It can be seen that the GTN model can be (almost) recovered by ξ=2 and θ=0.1. The results given by the MM coupled with GTN and the MM coupled with CDM theory are then expected to be substantially close. Note also that a classical version of the Drucker-Prager model is recovered by choosing $\xi = 1$.

Using the energy equivalence principle for the 2M1C model, the yield criterion can be expressed as

$$\phi = \frac{1}{\sqrt{(1-D)}}\left(\sqrt{(\sigma_{eq}^1)^2 + (\sigma_{eq}^2)^2} - R\right) - R_0 \qquad [5.65]$$

σ_{eq}^1 and σ_{eq}^2 are defined by equation [5.64]. The isotropic hardening is given by

$$R = (1-D)bQr \qquad [5.66]$$

The normals to the yield surface of the two mechanisms are

$$\begin{cases} \underset{\sim}{\mathbf{n}}^1 = \dfrac{\sigma_{eq}^{(\frac{1}{\xi}-1)}}{\sqrt{(1-D)}}\left(\dfrac{3}{2}(1-\theta)(J_2(\underset{\sim}{\sigma}^1))^{(\xi-2)}\underset{\sim}{\mathbf{s}}^1 + \theta(I(\underset{\sim}{\sigma}^1))^{(\xi-1)}\underset{\sim}{\mathbf{I}}\right) \\ \underset{\sim}{\mathbf{n}}^2 = \dfrac{\sigma_{eq}^{(\frac{1}{\xi}-1)}}{\sqrt{(1-D)}}\left(\dfrac{3}{2}(1-\theta)(J_2(\underset{\sim}{\sigma}^2))^{(\xi-2)}\underset{\sim}{\mathbf{s}}^2 + \theta(I(\underset{\sim}{\sigma}^2))^{(\xi-1)}\underset{\sim}{\mathbf{I}}\right) \end{cases} \qquad [5.67]$$

The isotropic flow, the isotropic hardening rate and the damage rate are respectively defined by equations [5.68]:

$$\dot{\underset{\sim}{\varepsilon}}^1 = \dot{p}\underset{\sim}{\mathbf{n}}^1 \dot{\underset{\sim}{\varepsilon}}^2 = \dot{p}\underset{\sim}{\mathbf{n}}^2 \qquad [5.68a]$$

$$\dot{p} = \left\langle\frac{\phi}{K}\right\rangle^n \qquad [5.68b]$$

$$\dot{r} = \dot{p}\left(\frac{1}{\sqrt{(1-D)}} - \frac{R}{(1-D)Q}\right) \qquad [5.68c]$$

$$\dot{D} = \dot{p}\left(\frac{Y}{S}\right)^s \qquad [5.68d]$$

5.4.1. *2M2C model for compressible materials, CDM theory*

The volume change can be taken into account by means of a Drucker–Prager criterion:

$$\begin{cases} f(\underset{\sim}{\sigma}^1) = (1-\alpha_1)J(\underset{\sim}{\sigma}^1) + \alpha_1 I_1(\underset{\sim}{\sigma}^1) - R_1 - R_{01} \\ f(\underset{\sim}{\sigma}^2) = (1-\alpha_2)J(\underset{\sim}{\sigma}^2) + \alpha_2 I_1(\underset{\sim}{\sigma}^2) - R_2 - R_{02} \end{cases} \qquad [5.69]$$

or using yield criteria as in [CHA 06]:

$$\begin{cases} f(\underset{\sim}{\sigma}^1) = \sigma_{eq}^1 - R_1 - R_{01} \\ f(\underset{\sim}{\sigma}^2) = \sigma_{eq}^2 - R_2 - R_{02} \\ \sigma_{eq}^1 = \left((1-\theta)(J_2(\underset{\sim}{\sigma}^1))^{\xi} + \theta(I_1(\underset{\sim}{\sigma}^1))^{\xi}\right)^{\frac{1}{\xi}} \\ \sigma_{eq}^2 = \left((1-\theta)(J_2(\underset{\sim}{\sigma}^2))^{\xi} + \theta(I_1(\underset{\sim}{\sigma}^2))^{\xi}\right)^{\frac{1}{\xi}} \end{cases} \qquad [5.70]$$

Using the energy equivalence principle and the yield criteria of equation [5.70] for the 2M2C model leads to

$$\begin{cases} \phi^1 = \dfrac{1}{\sqrt{(1-D)}}\left(\sigma_{eq}^1 - R^1\right) - R_{01} \\ \phi^2 = \dfrac{1}{\sqrt{(1-D)}}\left(\sigma_{eq}^2 - R^2\right) - R_{02} \end{cases} \qquad [5.71]$$

If the two phases are supposed to be affected by damage, the isotropic hardening is defined by

$$\begin{cases} R^1 = (1-D)b_1 Q_1 r^1 \\ R^2 = (1-D)b_2 Q_2 r^2 \end{cases} \qquad [5.72]$$

The normals to the yield surface of the two phases can be calculated as

$$\begin{cases} \underset{\sim}{\mathbf{n}}^1 = \dfrac{J(\underset{\sim}{\sigma}^1)^{(\frac{1}{\xi}-1)}}{\sqrt{(1-D)}}\left(\dfrac{3}{2}(1-\theta)(J(\underset{\sim}{\sigma}^1))^{(\xi-2)}\underset{\sim}{\mathbf{s}} + \theta(I(\underset{\sim}{\sigma}^1))^{(\xi-1)}\underset{\sim}{\mathbf{I}}\right) \\ \underset{\sim}{\mathbf{n}}^2 = \dfrac{J(\underset{\sim}{\sigma}^2)^{(\frac{1}{\xi}-1)}}{\sqrt{(1-D)}}\left(\dfrac{3}{2}(1-\theta)(J(\underset{\sim}{\sigma}^2))^{(\xi-2)}\underset{\sim}{\mathbf{s}} + \theta(I(\underset{\sim}{\sigma}^2))^{(\xi-1)}\underset{\sim}{\mathbf{I}}\right) \end{cases} \qquad [5.73]$$

The isotropic flow, the isotropic hardening rate and the damage rate are respectively defined by equations [5.74]:

$$\dot{\underset{\sim}{\varepsilon}}^1 = \dot{p}^1\underset{\sim}{\mathbf{n}}^1 \quad \dot{\underset{\sim}{\varepsilon}}^2 = \dot{p}^2\underset{\sim}{\mathbf{n}}^2 \qquad [5.74a]$$

$$\dot{p}^1 = \left\langle \frac{\phi^1}{K_1} \right\rangle^{n_1} \quad \dot{p}^2 = \left\langle \frac{\phi^2}{K_2} \right\rangle^{n_2} \qquad [5.74b]$$

$$\dot{r}^1 = \dot{p}^1 \left(\frac{1}{\sqrt{(1-D)}} - \frac{R^1}{(1-D)Q_1} \right) \qquad [5.74c]$$

$$\dot{r}^2 = \dot{p}^2 \left(\frac{1}{\sqrt{(1-D)}} - \frac{R^2}{(1-D)Q_2} \right) \qquad [5.74d]$$

$$\dot{D} = \left((1-z)\dot{p}^1 + z\dot{p}^2 \right) \left(\frac{Y}{S} \right)^s \qquad [5.74e]$$

Each mechanism has its own damage variable	Only one overall damage D is considered
$\phi^1 = \frac{1}{\sqrt{(1-D^1)}} \left(\sigma_{eq} - R^1 \right) - R_{01}$	$\phi^1 = \frac{1}{\sqrt{(1-D)}} \left(\sigma_{eq} - R_1 \right) - R_{01}$
$\phi^2 = \frac{1}{\sqrt{(1-D^2)}} \left(\sigma_{eq} - R_2 \right) - R_{02}$	$\phi^2 = \frac{1}{\sqrt{(1-D)}} \left(\sigma_{eq} - R_2 \right) - R_{02}$
$\sigma_{eq} = \left((1-\theta_1)(J(\underset{\sim}{\sigma}))^{\xi} + \theta_1 (I(\underset{\sim}{\sigma}))^{\xi} \right)^{\frac{1}{\xi}}$	$\sigma_{eq} = \left((1-\theta)(J(\underset{\sim}{\sigma}))^{\xi} + \theta (I(\underset{\sim}{\sigma}))^{\xi} \right)^{\frac{1}{\xi}}$
$\underset{\sim}{\sigma} = (1 - ((1-z)D_1 + zD_2))\underset{\approx}{\mathbf{\Lambda}} : \underset{\sim}{\varepsilon}^e$	$\underset{\sim}{\sigma} = (1-D)\underset{\approx}{\mathbf{\Lambda}} : \underset{\sim}{\varepsilon}^e$
$R^1 = (1-D_1)b_1 Q_1 r^1 \quad R^2 = (1-D_2)b_2 Q_2 r^2$	$R_1 = (1-D)b_1 Q_1 r^1 \quad R_2 = (1-D)b_2 Q_2 r^2$
$Y_1^e = \frac{1}{2}(1-z)(\underset{\sim}{\varepsilon}^e : \underset{\approx}{\mathbf{\Lambda}} : \underset{\sim}{\varepsilon}^e) \quad Y_2^e = \frac{1}{2}z(\underset{\sim}{\varepsilon}^e : \underset{\approx}{\mathbf{\Lambda}} : \underset{\sim}{\varepsilon}^e)$	$Y^e = \frac{1}{2}(\underset{\sim}{\varepsilon}^e : \underset{\approx}{\mathbf{\Lambda}} : \underset{\sim}{\varepsilon}^e)$
$Y_1^{vp} = \frac{1}{2}b_1 Q_1 (r^1)^2 \quad Y_2^{vp} = \frac{1}{2}b_2 Q_2 (r^2)^2$	$Y^{vp} = \frac{1}{2}b_1 Q_1 (r^1)^2 + \frac{1}{2}b_2 Q_2 (r^2)^2$
$Y^1 = Y_1^e + Y_1^{vp} \quad Y^2 = Y_2^e + Y_2^{vp}$	$Y = Y^e + Y^{vp}$
$\dot{\underset{\sim}{\varepsilon}}^1 = \dot{p}^1 \frac{1}{\sqrt{(1-D_1)}} \underset{\sim}{\mathbf{n}}^1 \quad \dot{\underset{\sim}{\varepsilon}}^2 = \dot{p}^2 \frac{1}{\sqrt{(1-D_2)}} \underset{\sim}{\mathbf{n}}^2$	$\dot{\underset{\sim}{\varepsilon}}^1 = \dot{p}^1 \frac{1}{\sqrt{(1-D)}} \underset{\sim}{\mathbf{n}} \quad \dot{\underset{\sim}{\varepsilon}}^2 = \dot{p}_2 \frac{1}{\sqrt{(1-D)}} \underset{\sim}{\mathbf{n}}$
$\dot{p}^1 = \left\langle \frac{\phi^1}{K_1} \right\rangle^{n^1} \quad \dot{p}^2 = \left\langle \frac{\phi^2}{K_2} \right\rangle^{n^2}$	$\dot{p}^1 = \left\langle \frac{\phi^1}{K_1} \right\rangle^{n^1} \quad \dot{p}^2 = \left\langle \frac{\phi^2}{K_2} \right\rangle^{n^2}$
$\underset{\sim}{\mathbf{n}}^1 = \frac{(\underset{\sim}{\sigma}_{eq}^1)^{(\frac{1}{\xi}-1)}}{\sqrt{(1-D^1)}} \left(\frac{3}{2}(1-\theta_1)(J(\underset{\sim}{\sigma}))^{(\xi-2)} \underset{\sim}{\mathbf{s}} + \theta_1 (I(\underset{\sim}{\sigma}))^{(\xi-1)} \underset{\sim}{\mathbf{I}} \right)$	$\underset{\sim}{\mathbf{n}} = \frac{(\underset{\sim}{\sigma}_{eq})^{(\frac{1}{\xi}-1)}}{\sqrt{(1-D)}} (\frac{3}{2}(1-\theta)(J(\underset{\sim}{\sigma}))^{(\xi-2)} \underset{\sim}{\mathbf{s}}$ $+\theta(I(\underset{\sim}{\sigma}))^{(\xi-1)} \underset{\sim}{\mathbf{I}})$
$\underset{\sim}{\mathbf{n}}^2 = \frac{(\underset{\sim}{\sigma}_{eq}^2)^{(\frac{1}{\xi}-1)}}{\sqrt{(1-D_2)}} \left(\frac{3}{2}(1-\theta_2)(J(\underset{\sim}{\sigma}))^{(\xi-2)} \underset{\sim}{\mathbf{s}} + \theta_2 (I(\underset{\sim}{\sigma}))^{(\xi-1)} \underset{\sim}{\mathbf{I}} \right)$	
$\dot{r}^1 = \dot{p}^1 \left(\frac{1}{\sqrt{(1-D^1)}} - \frac{R^1}{(1-D^1)Q_1} \right)$	$\dot{r}^1 = \dot{p}^1 \left(\frac{1}{\sqrt{(1-D)}} - \frac{R^1}{(1-D)Q_1} \right)$
$\dot{r}^2 = \dot{p}^2 \left(\frac{1}{\sqrt{(1-D^2)}} - \frac{R^2}{(1-D^2)Q_2} \right)$	$\dot{r}^2 = \dot{p}^2 \left(\frac{1}{\sqrt{(1-D)}} - \frac{R^2}{(1-D)Q_2} \right)$
$\dot{D}^1 = \dot{p}^1 \left(\frac{Y^1}{S_1} \right)^{s_1} \quad \dot{D}^2 = \dot{p}^2 \left(\frac{Y^2}{S_2} \right)^{s_2}$	$\dot{D} = ((1-z)\dot{p}^1 + z\dot{p}^2) \left(\frac{Y}{S} \right)^s$

Table 5.1. *CDM-based MM models*

5.4.2. *MM models for compressible materials, CDM theory, two damage variables*

Instead of having only one damage variable D, as shown in section 5.4.1, each mechanism can be affected by its own damage variable (D^1 and D^2, respectively). The two model variants are then summarized in Table 5.1, for the particular case $\underset{\sim}{\sigma} = \underset{\sim}{\sigma}^1 = \underset{\sim}{\sigma}^2$.

5.5. Damage behavior of mortar-rubber aggregate mixtures

The reuse of rubber wastes of worn tires in an aggregate form, to serve as a building material, is appreciated to preserve the environment. The mechanical behavior of a mortar-rubber aggregate is investigated here by means of the 2M2C model. This model must include the following aspects:

- the respective volume fraction of mortar and rubber;
- the damage in the mortar phase;
- the plastic strains and stresses at the level of the phases.

$\underset{\sim}{\varepsilon}^p = z\underset{\sim}{\varepsilon}^1 + (1-z)\underset{\sim}{\varepsilon}^2$	$\underset{\sim}{\sigma} = (1-D)\underset{\approx}{\mathbf{\Lambda}} : (\underset{\sim}{\varepsilon} - \underset{\sim}{\varepsilon}^p)$
$\underset{\sim}{\sigma}^1 = \underset{\sim}{\sigma} + A\left(\underset{\sim}{\beta} - \underset{\sim}{\beta}^1\right)$	$\underset{\sim}{\sigma}^2 = \underset{\sim}{\sigma} + A\left(\underset{\sim}{\beta} - \underset{\sim}{\beta}^2\right)$
$\underset{\sim}{\beta} = z\underset{\sim}{\beta}^1 + (1-z)\underset{\sim}{\beta}^2$	
$\underset{\sim}{\mathbf{X}}^1 = \frac{2}{3}\left(C_{11}(1-D)\underset{\sim}{\alpha}^1 + C_{12}\underset{\sim}{\alpha}^2\right)$	$\underset{\sim}{\mathbf{X}}^2 = \frac{2}{3}\left(C_{12}\underset{\sim}{\alpha}^1 + C_{22}\underset{\sim}{\alpha}^2\right)$
$R^1 = Q_1(1-D)r^1$	$R^2 = Q_2 r^2$
$f^1 = \frac{J(\underset{\sim}{\sigma}^1 - \underset{\sim}{\mathbf{X}}^1) - (1-D)R^1}{\sqrt{(1-D)}} - R_{01}$	$f^2 = J(\underset{\sim}{\sigma}^2 - \underset{\sim}{\mathbf{X}}^2) - R^2 - R_{02}$
$\dot{\underset{\sim}{\varepsilon}}^1 = \dot{p}^1\underset{\sim}{n}^1$	$\dot{\underset{\sim}{\varepsilon}}^2 = \dot{p}^2\underset{\sim}{n}^2$
$\dot{\underset{\sim}{\alpha}}^1 = \dot{\underset{\sim}{\varepsilon}}^1 - \frac{3D_1}{2(1-D)C_{11}}\underset{\sim}{\mathbf{X}}^1\dot{p}^1$	$\dot{\underset{\sim}{\alpha}}^2 = \dot{\underset{\sim}{\varepsilon}}^2 - \frac{3D_2}{2C_{22}}\underset{\sim}{\mathbf{X}}^2\dot{p}^2$
$\dot{\underset{\sim}{\beta}}^1 = \dot{\underset{\sim}{\varepsilon}}^1 - d_1\underset{\sim}{\beta}^1\dot{p}^1$	$\dot{\underset{\sim}{\beta}}^2 = \dot{\underset{\sim}{\varepsilon}}^2 - d_2\underset{\sim}{\beta}^2\dot{p}^2$
$\dot{r}^1 = \dot{p}^1\left(\frac{1}{\sqrt{1-D}} - b_1 Q_1\right)$	$\dot{r}^2 = \dot{p}^2(1 - b_2 Q_2)$
$Y = \frac{1}{2}(\underset{\sim}{\varepsilon} - \underset{\sim}{\varepsilon}^p) : \underset{\approx}{\mathbf{\Lambda}} : (\underset{\sim}{\varepsilon} - \underset{\sim}{\varepsilon}^p) + \frac{1}{3}C_{11}\underset{\sim}{\alpha}^1 : \underset{\sim}{\alpha}^1 + \frac{1}{2}Q_1(r^1)^2$	$\dot{D} = \dot{p}^2\left(\frac{Y}{A}\right)^{\gamma}\frac{1}{(1-D)^{\beta}}$

Table 5.2. *Summary of the constitutive equations of the proposed model. 1: mortar, 2: rubber*

The model incorporates an isotropic damage variable to describe the loss of compressive strength. Rubber particles were introduced in the mortar mixture by partial substitution of sand. Uniaxial compression tests with rubber volume fractions of 10%, 30% and 50% are considered. The rubber-free material has been taken as a reference material. The behavior of the reference material is elastic-brittle. For increasing volume fractions of rubber, the material behavior becomes more ductile.

	Parameters	Values				Units
Elasticity	z	0	0.1	.3	.5	
	E	22,500	9,000	4,500	2,100	MPa
	ν	0.3				
Mortar phase	k_1	90				$MPa.s^{1/n_1}$
	n_1	2				
	R_{01}	11	3.5	2.4	1.1	MPa
	Q_1	31.4				MPa
	b_1	20.3				
	C_{11}	1,800				MPa
	D_1	37				
Rubber phase	k_2	90				$MPa.s^{1/n_2}$
	n_2	2				
	R_{02}	14	5.5	3.1	1.3	MPa
	Q_2	21.4				MPa
	b_2	16.9				
	C_{22}	1,400				MPa
	D_2	51				
Interaction	A	25				MPa
	C_{12}	800				MPa
	d_1	5				
	d_2	7				
Damage	B	0.0045				
	γ	5.8				
	β	0.005				

Table 5.3. *Material parameters of the proposed model*

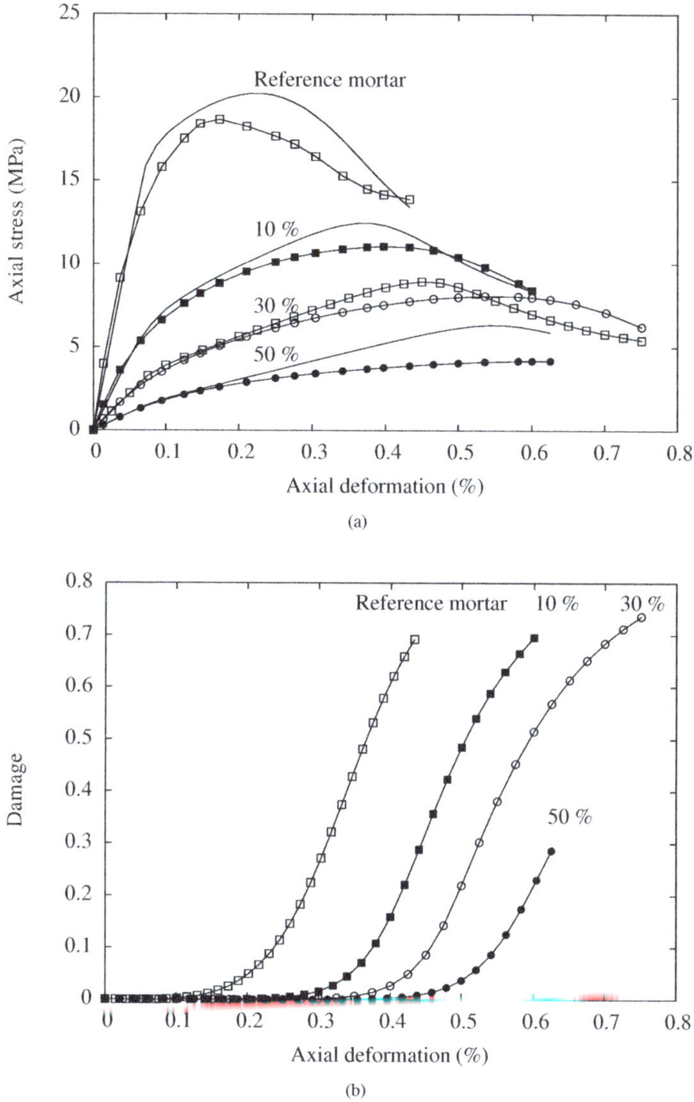

Figure 5.2. *Mechanical and damage behavior of mortar-rubber aggregate mixtures: (a) experimental and modeling results of compressive tests for different rubber aggregate substitutions; (b) evolution of the damage in the compressive tests*

6

Finite Element Implementation

This chapter deals with the numerical implementation of MM equations in a finite element code. The expressions that are given are ready for practical use, so that the reader can directly introduce the model in his own code. The implicit version of the most simple MM models is first presented (section 6.1) and then several extensions are shown according to the following outline:

– Creep–plasticity interaction for the case of creep tests on notched specimens, simulated with a 2M2C model (section 6.2);

– FE simulation of plane-strain forging experiments performed on polycarbonate specimens, based on a 3M3C model (section 6.3);

– FE simulations of the bulging phenomenon based on a MM model describing the phase transformation in 304SS (section 6.4);

– Damage-plasticity interaction applied to the creep behavior of PA6 notched specimens (section 6.5).

6.1. Implementations of particular models

In a classical displacement-based approach, the role of the function called by the FE code to implement the constitutive equations is to propose an update of the stress and the integrated variables for a prescribed strain increment. The so-called consistent tangent operator may be calculated by a perturbation technique, or directly obtained as a sub-product of the integration method. A detailed description of the related procedures is not within the scope of the present book. The reader can refer to classical [SIM 97] or more recent

[BES 09b] books on the subject. In the following, we first show the classical implicit formulation that uses a θ-method solved by a Newton algorithm. An explicit Runge–Kutta scheme may also be convenient in the model development phase, or to solve some very critical cases, provided an efficient automatic time step procedure is available [BES 09b].

6.1.1. *Basic version of the 2M1C model*

Here, the explicit and implicit integration schemes for the initial version of the 2M1C model [CAI 95] are shown. The full set of equations was already given in Table 3.2. The time-dependent case is recalled below:

$$\dot{\underset{\sim}{\varepsilon}} = \dot{\underset{\sim}{\varepsilon}}^e + A_1 \dot{\underset{\sim}{\varepsilon}}^1 + A_2 \dot{\underset{\sim}{\varepsilon}}^2$$

$$f = \left(J(\underset{\sim}{\sigma}^1 - \underset{\sim}{\mathbf{X}}^1)^2 + J(\underset{\sim}{\sigma}^2 - \underset{\sim}{\mathbf{X}}^2)^2 \right)^{1/2} - R - R_0$$

$$\underset{\sim}{\sigma}^1 = A_1 \underset{\sim}{\sigma} \quad \underset{\sim}{\sigma}^2 = A_2 \underset{\sim}{\sigma}$$

$$\begin{pmatrix} \underset{\sim}{\mathbf{X}}^1 \\ \underset{\sim}{\mathbf{X}}^2 \end{pmatrix} = (2/3) \begin{pmatrix} C_{11} & C_{12} \\ C_{12} & C_{22} \end{pmatrix} \begin{pmatrix} \underset{\sim}{\alpha}^1 \\ \underset{\sim}{\alpha}^2 \end{pmatrix}$$

$$R = Q(1 - \exp(-bp))$$

$$\dot{\underset{\sim}{\varepsilon}}^1 = \dot{p} \underset{\sim}{\mathbf{n}}^1 \quad \dot{\underset{\sim}{\varepsilon}}^2 = \dot{p} \underset{\sim}{\mathbf{n}}^2$$

$$\underset{\sim}{\mathbf{n}}^1 = \frac{3}{2} \frac{\left(\underset{\sim}{\mathbf{s}}^1 - \underset{\sim}{\mathbf{X}}^1 \right)}{\left(J(\underset{\sim}{\sigma}^1 - \underset{\sim}{\mathbf{X}}^1)^2 + J(\underset{\sim}{\sigma}^2 - \underset{\sim}{\mathbf{X}}^2)^2 \right)^{1/2}} \quad \underset{\sim}{\mathbf{n}}^2 = \frac{3}{2} \frac{\left(\underset{\sim}{\mathbf{s}}^2 - \underset{\sim}{\mathbf{X}}^2 \right)}{\left(J(\underset{\sim}{\sigma}^1 - \underset{\sim}{\mathbf{X}}^1)^2 + J(\underset{\sim}{\sigma}^2 - \underset{\sim}{\mathbf{X}}^2)^2 \right)^{1/2}}$$

$$\dot{\underset{\sim}{\alpha}}_1 = \dot{p} \left(\underset{\sim}{\mathbf{n}}^1 - (3D_1/2C_{11}) \underset{\sim}{\mathbf{X}}^1 \right) \quad \dot{\underset{\sim}{\alpha}}^2 = \dot{p} \left(\underset{\sim}{\mathbf{n}}^2 - (3D_2/2C_{22}) \underset{\sim}{\mathbf{X}}^2 \right)$$

$$\dot{p} = \left\langle \frac{f}{K} \right\rangle^n$$

The behavior is fully defined by four internal variables: the elastic strain tensor $\underset{\sim}{\varepsilon}^e$, the two local tensors $\underset{\sim}{\alpha}^1$ and $\underset{\sim}{\alpha}^2$, and the accumulated plastic strain p.

6.1.1.1. *Explicit integration*

For the case of an explicit integration, the only information needed by the code is the expression at time t of the derivative of $\underset{\sim}{\varepsilon}^e$, and, in the viscoplastic regime, of derivatives of the integrated variables, $\underset{\sim}{\alpha}^1$, $\underset{\sim}{\alpha}^2$ and p. The actual value of the stress tensor $\underset{\sim}{\sigma}$ is obtained from the elastic strain: $\underset{\sim}{\sigma} = \underset{\approx}{\mathbf{\Lambda}} : \underset{\sim}{\varepsilon}^e$. Similarly, the actual values of the hardening variables $\underset{\sim}{\mathbf{X}}^1$ and $\underset{\sim}{\mathbf{X}}^2$ are calculated after $\underset{\sim}{\alpha}^1$ and $\underset{\sim}{\alpha}^2$ at time t, and R is computed as a function of p. The value of the yield function f can then be calculated. In the elastic regime ($f < 0$),

$$\dot{\underset{\sim}{\varepsilon}}^e = \dot{\underset{\sim}{\alpha}}^1 = \dot{\underset{\sim}{\alpha}}^2 = 0 \qquad \dot{p} = 0$$

If the yield condition is reached ($f > 0$), $\dot{p}$ is first calculated by means of a power law and then applied to the relevant normal tensors $\underset{\sim}{\mathbf{n}}^1$ and $\underset{\sim}{\mathbf{n}}^2$ to determine the two elementary plastic strain rate tensors. The elastic strain rate tensor is then given by

$$\dot{\underset{\sim}{\varepsilon}}^e = \dot{\underset{\sim}{\varepsilon}} - A_1\dot{\underset{\sim}{\varepsilon}}^1 - A_2\dot{\underset{\sim}{\varepsilon}}^2$$

and the expression of the rates of the internal variables $\dot{\underset{\sim}{\alpha}}^1$ and $\dot{\underset{\sim}{\alpha}}^2$ follows easily.

The 2M1C model with the described explicit integration scheme is implemented in the FE code Z-set and in its material library, which can be used with the major commercial codes [TRA 17]. The corresponding user material functions are given in section 6.6.1. Specific subroutines are also shown for Abaqus (section 6.6.2) and Ansys (section 6.6.3).

6.1.1.2. *Implicit integration*

The implicit θ method requires to define a set of residuals. The unknowns of the resulting system are now the increments of the integrated variables $\Delta\underset{\sim}{\varepsilon}^e$, $\Delta\underset{\sim}{\alpha}^1$, $\Delta\underset{\sim}{\alpha}^2$ and Δp. The loading is imposed by a prescribed strain increment $\Delta\underset{\sim}{\varepsilon}$. In the elastic regime, the resolution is immediate. Otherwise, the system to solve is

$$\begin{aligned}
\underset{\sim}{\mathbf{R}}_e &= \Delta\underset{\sim}{\varepsilon}^e + \left(A_1\underset{\sim}{\mathbf{n}}^1 + A_2\underset{\sim}{\mathbf{n}}^2\right)\Delta p - \Delta\underset{\sim}{\varepsilon} = 0\\
\underset{\sim}{\mathbf{R}}_{\alpha 1} &= \Delta\underset{\sim}{\alpha}^1 - \left(\underset{\sim}{\mathbf{n}}^1 - \frac{3D_1}{2C_{11}}\underset{\sim}{\mathbf{X}}^1\right)\Delta p = 0\\
\underset{\sim}{\mathbf{R}}_{\alpha 2} &= \Delta\underset{\sim}{\alpha}^2 - \left(\underset{\sim}{\mathbf{n}}^2 - \frac{3D_2}{2C_{22}}\underset{\sim}{\mathbf{X}}^2\right)\Delta p = 0\\
R_p &= \Delta p - \left(\frac{f}{K}\right)^n \Delta t = 0
\end{aligned}$$

It involves three tensorial and one scalar equations, which are expressed at time $t + \theta\Delta t$. The Jacobian matrix is formed by the partial derivatives of each of them with respect to the variables. It is then defined by several blocks.

The notations $\underset{\approx}{\mathbf{N}}^{ij} = \partial \underset{\sim}{\mathbf{n}}^i / \partial \underset{\sim}{\sigma}^j$ (i,j=1,2) are introduced to define the partial derivatives of the normals with respect to the stress tensors:

$$\underset{\approx}{\mathbf{N}}^{11} = \frac{\partial \underset{\sim}{\mathbf{n}}^1}{\partial \underset{\sim}{\sigma}^1} = \frac{\underset{\approx}{\mathbf{J}} - \underset{\sim}{\mathbf{n}}^1 \otimes \underset{\sim}{\mathbf{n}}^1}{\sqrt{\left(J(\underset{\sim}{\sigma}^1 - \underset{\sim}{\mathbf{X}}^1)\right)^2 + \left(J(\underset{\sim}{\sigma}^2 - \underset{\sim}{\mathbf{X}}^2)\right)^2}}$$

$$\underset{\approx}{\mathbf{N}}^{22} = \frac{\partial \underset{\sim}{\mathbf{n}}^2}{\partial \underset{\sim}{\sigma}^2} = \frac{\underset{\approx}{\mathbf{J}} - \underset{\sim}{\mathbf{n}}^2 \otimes \underset{\sim}{\mathbf{n}}^2}{\sqrt{\left(J(\underset{\sim}{\sigma}^1 - \underset{\sim}{\mathbf{X}}^1)\right)^2 + \left(J(\underset{\sim}{\sigma}^2 - \underset{\sim}{\mathbf{X}}^2)\right)^2}}$$

$$\underset{\approx}{\mathbf{N}}^{12} = \frac{\partial \underset{\sim}{\mathbf{n}}^1}{\partial \underset{\sim}{\sigma}^2} = \frac{\underset{\sim}{\mathbf{n}}^1 \otimes \underset{\sim}{\mathbf{n}}^2}{\sqrt{\left(J(\underset{\sim}{\sigma}^1 - \underset{\sim}{\mathbf{X}}^1)\right)^2 + \left(J(\underset{\sim}{\sigma}^2 - \underset{\sim}{\mathbf{X}}^2)\right)^2}}$$

$$\underset{\approx}{\mathbf{N}}^{21} = \frac{\partial \underset{\sim}{\mathbf{n}}^2}{\partial \underset{\sim}{\sigma}^1} = \frac{\underset{\sim}{\mathbf{n}}^2 \otimes \underset{\sim}{\mathbf{n}}^1}{\sqrt{\left(J(\underset{\sim}{\sigma}^1 - \underset{\sim}{\mathbf{X}}^1)\right)^2 + \left(J(\underset{\sim}{\sigma}^2 - \underset{\sim}{\mathbf{X}}^2)\right)^2}}$$

$\underset{\approx}{\mathbf{J}}$ is the fourth-rank tensor that associates a second-rank tensor with its deviator $\underset{\sim}{\mathbf{s}} = \underset{\approx}{\mathbf{J}} : \underset{\sim}{\sigma}$.

The blocks of the Jacobian matrix are defined below.

Function R_e

$$\begin{aligned}
&\partial \underset{\sim}{\mathbf{R}}_e / \partial \Delta \underset{\sim}{\varepsilon}^e = \underset{\approx}{\mathbf{I}} + \left(A_1(\underset{\approx}{\mathbf{N}}^{11} + \underset{\approx}{\mathbf{N}}^{12}) + A_2(\underset{\approx}{\mathbf{N}}^{21} + \underset{\approx}{\mathbf{N}}^{22})\right) : \underset{\approx}{\mathbf{\Lambda}} \theta \Delta p \\
&\partial \underset{\sim}{\mathbf{R}}_e / \partial \Delta \underset{\sim}{\alpha}^1 = -(2/3)\theta \Delta p \left[A_1(C_{11}\underset{\approx}{\mathbf{N}}^{11} + C_{12}\underset{\approx}{\mathbf{N}}^{12}) + A_2(C_{11}\underset{\approx}{\mathbf{N}}^{21} + C_{12}\underset{\approx}{\mathbf{N}}^{22})\right] \\
&\partial \underset{\sim}{\mathbf{R}}_e / \partial \Delta \underset{\sim}{\alpha}^2 = -(2/3)\theta \Delta p \left[A_1(C_{12}\underset{\approx}{\mathbf{N}}^{11} + C_{22}\underset{\approx}{\mathbf{N}}^{12}) + A_2(C_{12}\underset{\approx}{\mathbf{N}}^{21} + C_{22}\underset{\approx}{\mathbf{N}}^{22})\right] \\
&\partial \underset{\sim}{\mathbf{R}}_e / \Delta p = A_1 \underset{\sim}{\mathbf{n}}^1 + A_2 \underset{\sim}{\mathbf{n}}^2
\end{aligned}$$

Function $\underset{\sim}{\mathbf{R}}_{\alpha 1}$

$$\begin{aligned}
&\partial \underset{\sim}{\mathbf{R}}_{\alpha_1} / \partial \Delta \underset{\sim}{\varepsilon}^e = -\theta \Delta p (\underset{\approx}{\mathbf{N}}^{11} + \underset{\approx}{\mathbf{N}}^{12}) \underset{\approx}{\mathbf{\Lambda}} \\
&\partial \underset{\sim}{\mathbf{R}}_{\alpha_1} / \partial \Delta \underset{\sim}{\alpha}^1 = \underset{\approx}{\mathbf{I}} + \theta \Delta p \left[(2/3)(C_{11}\underset{\approx}{\mathbf{N}}^{11} + C_{12}\underset{\approx}{\mathbf{N}}^{12}) + D_1 \underset{\approx}{\mathbf{I}}\right] \\
&\partial \underset{\sim}{\mathbf{R}}_{\alpha_1} / \partial \Delta \underset{\sim}{\alpha}^2 = \theta \Delta p \left[(2/3)(C_{12}\underset{\approx}{\mathbf{N}}^{11} + C_{22}\underset{\approx}{\mathbf{N}}^{12}) + (D_1 / C_{11}) C_{12} \underset{\approx}{\mathbf{I}}\right] \\
&\partial \underset{\sim}{\mathbf{R}}_{\alpha_1} / \Delta p = (3D_1 / 2C_{11}) \underset{\sim}{\mathbf{X}}^1 - \underset{\sim}{\mathbf{n}}^1
\end{aligned}$$

Function $\underset{\sim}{\mathbf{R}}_{\alpha 2}$

$$\partial \underset{\sim}{\mathbf{R}}_{\alpha_2}/\partial \Delta \underset{\sim}{\varepsilon}^e = -\theta \Delta p (\underset{\approx}{\mathbf{N}}^{21} + \underset{\approx}{\mathbf{N}}^{22}) \underset{\approx}{\mathbf{\Lambda}}$$
$$\partial \underset{\sim}{\mathbf{R}}_{\alpha_2}/\partial \Delta \underset{\sim}{\alpha}^1 = \theta \Delta p \left[(2/3)(C_{11}\underset{\approx}{\mathbf{N}}^{21} + C_{12}\underset{\approx}{\mathbf{N}}^{22}) + (D_2/C_{22})C_{12}\underset{\approx}{\mathbf{I}} \right]$$
$$\partial \underset{\sim}{\mathbf{R}}_{\alpha_2}/\partial \Delta \underset{\sim}{\alpha}^2 = \underset{\approx}{\mathbf{I}} + \theta \Delta p \left[(2/3)(C_{12}\underset{\approx}{\mathbf{N}}^{21} + C_{22}\underset{\approx}{\mathbf{N}}^{22}) + D_2 \underset{\approx}{\mathbf{I}} \right]$$
$$\partial \underset{\sim}{\mathbf{R}}_{\alpha_2}/\Delta p = (3D_2/2C_{22})\underset{\sim}{\mathbf{X}}^2 - \underset{\sim}{\mathbf{n}}^2$$

Function R_p

$$\partial R_p/\partial \Delta \underset{\sim}{\varepsilon}^e = -\theta \Delta t (n/k)\,(f/k)^{n-1}\, \underset{\approx}{\mathbf{\Lambda}}(A_1 \underset{\sim}{\mathbf{n}}^1 + A_2 \underset{\sim}{\mathbf{n}}^2)$$
$$\partial R_p/\partial \Delta \underset{\sim}{\alpha}^1 = (2/3)\theta \Delta t (n/k)\,(f/k)^{n-1}\,(C_{11}\underset{\sim}{\mathbf{n}}^1 + C_{12}\underset{\sim}{\mathbf{n}}^2)$$
$$\partial R_p/\partial \Delta \underset{\sim}{\alpha}^2 = (2/3)\theta \Delta t (n/k)\,(f/k)^{n-1}\,(C_{12}\underset{\sim}{\mathbf{n}}^1 + C_{22}\underset{\sim}{\mathbf{n}}^2)$$
$$\partial R_p/\Delta p = 1 + \theta \Delta t (n/k)\,(f/k)^{n-1}\, Qbe^{-bp}$$

It is worth noting that, after convergence of the Newton algorithm, the upper left part of the inverse of the Jacobian matrix is nothing but the term $\partial \underset{\sim}{\sigma}/\partial \underset{\sim}{\varepsilon}$ that is needed to evaluate the global consistent tangent matrix used by the finite element code at the structural level, as first pointed out in [SIM 97].

The 2M1C model with the described implicit integration scheme is implemented in the three FE codes Zebulon, Abaqus (θ=1) and Ansys (θ=1). The corresponding codes are respectively given in sections 6.6.4, 6.6.5 and 6.6.6.

6.1.2. *β models*

This section gives a short overview of the implementation of implicit integration schemes for a 2M1C model and a 2M2C model [REG 09b], the equations of which are summarized respectively in Tables 6.1 and 6.2. In this case, the evaluation of the "local" stresses on each mechanism takes into account a plastic accommodation between the two mechanisms, which are seen as two distinct phases in the material. This is represented by the scale transition variables $\underset{\sim}{\beta}^1$ and $\underset{\sim}{\beta}^2$. For the 2M1C model, the other variables are the elastic strain tensor $\underset{\sim}{\varepsilon}^e$, the kinematic variables for each phase $\underset{\sim}{\alpha}^1$ and $\underset{\sim}{\alpha}^2$, and the isotropic variable p. The latter is replaced by the two variables p^1 and p^2 for the 2M2C model. The residual vector is then obtained by assembling five tensorial equations ($\Delta\underset{\sim}{\varepsilon}^e$, $\Delta\underset{\sim}{\beta}^1$, $\Delta\underset{\sim}{\beta}^2$, $\Delta\underset{\sim}{\alpha}^1$, $\Delta\underset{\sim}{\alpha}^2$) and one Δp or two (Δp^1, Δp^2) scalar variables, as shown respectively in Tables 6.3 and 6.4.

$$\dot{\underset{\sim}{\varepsilon}} = \dot{\underset{\sim}{\varepsilon}}^e + (1-z)\dot{\underset{\sim}{\varepsilon}}^1 + z\dot{\underset{\sim}{\varepsilon}}^2$$

$$f = \left(J(\underset{\sim}{\sigma}^1 - \underset{\sim}{\mathbf{X}}^1)^2 + J(\underset{\sim}{\sigma}^2 - \underset{\sim}{\mathbf{X}}^2)^2\right)^{1/2} - R - R_0$$

$$R = Q(1 - \exp(-bp))$$

$$\underset{\sim}{\sigma}^1 = \underset{\sim}{\sigma} + A\left(\underset{\sim}{\beta} - \underset{\sim}{\beta}^1\right) \quad \underset{\sim}{\sigma}^2 = \underset{\sim}{\sigma} + A\left(\underset{\sim}{\beta} - \underset{\sim}{\beta}^2\right)$$

$$\underset{\sim}{\beta} = (1-z)\underset{\sim}{\beta}^1 + z\underset{\sim}{\beta}^2$$

$$\begin{pmatrix} \underset{\sim}{\mathbf{X}}^1 \\ \underset{\sim}{\mathbf{X}}^2 \end{pmatrix} = (2/3)\begin{pmatrix} C_{11} & C_{12} \\ C_{12} & C_{22} \end{pmatrix}\begin{pmatrix} \underset{\sim}{\alpha}^1 \\ \underset{\sim}{\alpha}^2 \end{pmatrix}$$

$$\dot{\underset{\sim}{\varepsilon}}^1 = \dot{p}\underset{\sim}{\mathbf{n}}^1 \quad \dot{\underset{\sim}{\varepsilon}}^2 = \dot{p}\underset{\sim}{\mathbf{n}}^2$$

$$\dot{p} = \langle f/K \rangle^n$$

$$\dot{\underset{\sim}{\beta}}^1 = \dot{p}\left(\underset{\sim}{\mathbf{n}}^1 - d_1\underset{\sim}{\beta}^1\right) \quad \dot{\underset{\sim}{\beta}}^2 = \dot{p}\left(\underset{\sim}{\mathbf{n}}^2 - d_2\underset{\sim}{\beta}^2\right)$$

$$\dot{\underset{\sim}{\alpha}}^1 = \dot{p}\left(\underset{\sim}{\mathbf{n}}^1 - \frac{3D_1}{2C_{11}}\underset{\sim}{\mathbf{X}}^1\right) \quad \dot{\underset{\sim}{\alpha}}^2 = \dot{p}\left(\underset{\sim}{\mathbf{n}}^2 - \frac{3D_2}{2C_{22}}\underset{\sim}{\mathbf{X}}^2\right)$$

Table 6.1. *Constitutive equations of the 2M1C model*

$$\dot{\underset{\sim}{\varepsilon}} = \dot{\underset{\sim}{\varepsilon}}^e + (1-z)\dot{\underset{\sim}{\varepsilon}}^1 + z\dot{\underset{\sim}{\varepsilon}}^2$$

$$f^1 = J(\underset{\sim}{\sigma}^1 - \underset{\sim}{\mathbf{X}}^1) - R^1 - R_{01} \quad f^2 = J(\underset{\sim}{\sigma}^2 - \underset{\sim}{\mathbf{X}}^2) - R^2 - R_{02}$$

$$R^1 = Q_1(1 - \exp(-b_1 p^1)) \quad R^2 = Q_2(1 - \exp(-b_2 p^2))$$

$$\underset{\sim}{\sigma}^1 = \underset{\sim}{\sigma} + A\left(\underset{\sim}{\beta} - \underset{\sim}{\beta}^1\right) \quad \underset{\sim}{\sigma}^2 = \underset{\sim}{\sigma} + A\left(\underset{\sim}{\beta} - \underset{\sim}{\beta}^2\right)$$

$$\underset{\sim}{\beta} = (1-z)\underset{\sim}{\beta}^1 + z\underset{\sim}{\beta}^2$$

$$\begin{pmatrix} \underset{\sim}{\mathbf{X}}^1 \\ \underset{\sim}{\mathbf{X}}^2 \end{pmatrix} = (2/3)\begin{pmatrix} C_{11} & C_{12} \\ C_{12} & C_{22} \end{pmatrix}\begin{pmatrix} \underset{\sim}{\alpha}^1 \\ \underset{\sim}{\alpha}^2 \end{pmatrix}$$

$$\dot{\underset{\sim}{\varepsilon}}^1 = \dot{p}^1\underset{\sim}{\mathbf{n}}^1 \quad \dot{\underset{\sim}{\varepsilon}}^2 = \dot{p}^2\underset{\sim}{\mathbf{n}}^2$$

$$\dot{p}^1 = \left\langle f^1/K_1 \right\rangle^{n_1} \quad \dot{p}^2 = \left\langle f^2/K_2 \right\rangle^{n_2}$$

$$\dot{\underset{\sim}{\beta}}^1 = \dot{p}^1\left(\underset{\sim}{\mathbf{n}}^1 - d_1\underset{\sim}{\beta}^1\right) \quad \dot{\underset{\sim}{\beta}}^2 = \dot{p}^2\left(\underset{\sim}{\mathbf{n}}^2 - d_2\underset{\sim}{\beta}^2\right)$$

$$\dot{\underset{\sim}{\alpha}}^1 = \dot{p}^1\left(\underset{\sim}{\mathbf{n}}^1 - \frac{3D_1}{2C_{11}}\underset{\sim}{\mathbf{X}}^1\right) \quad \dot{\underset{\sim}{\alpha}}^2 = \dot{p}^2\left(\underset{\sim}{\mathbf{n}}^2 - \frac{3D_2}{2C_{22}}\underset{\sim}{\mathbf{X}}^2\right)$$

Table 6.2. *Constitutive equations of the 2M2C model*

$$
\begin{aligned}
\underset{\sim}{\mathbf{R}}_e &= \Delta\underset{\sim}{\varepsilon}^e + \left((1-z)\underset{\sim}{\mathbf{n}}^1 + z\underset{\sim}{\mathbf{n}}^2\right)\Delta p - \Delta\underset{\sim}{\varepsilon}\\
R_p &= \Delta p - \left(\tfrac{f}{K}\right)^n \Delta t\\
\underset{\sim}{\mathbf{R}}_{\alpha 1} &= \Delta\underset{\sim}{\alpha}^1 - \left(\underset{\sim}{\mathbf{n}}^1 - \tfrac{3D_1}{2C_{11}}\underset{\sim}{\mathbf{X}}^1\right)\Delta p\\
\underset{\sim}{\mathbf{R}}_{\alpha 2} &= \Delta\underset{\sim}{\alpha}^2 - \left(\underset{\sim}{\mathbf{n}}^2 - \tfrac{3D_2}{2C_{22}}\underset{\sim}{\mathbf{X}}^2\right)\Delta p\\
\underset{\sim}{\mathbf{R}}_{\beta 1} &= \Delta\underset{\sim}{\beta}^1 - \left(\underset{\sim}{\mathbf{n}}^1 - d_1\underset{\sim}{\beta}^1\right)\Delta p\\
\underset{\sim}{\mathbf{R}}_{\beta 2} &= \Delta\underset{\sim}{\beta}^2 - \left(\underset{\sim}{\mathbf{n}}^2 - d_2\underset{\sim}{\beta}^2\right)\Delta p
\end{aligned}
$$

Table 6.3. *Equations of the residual for the 2M1C model*

$$
\begin{aligned}
\underset{\sim}{\mathbf{R}}_e &= \Delta\underset{\sim}{\varepsilon}^e + (1-z)\underset{\sim}{\mathbf{n}}^1\Delta p^1 + z\underset{\sim}{\mathbf{n}}^2\Delta p^2 - \Delta\underset{\sim}{\varepsilon}\\
R_{p^1} &= \Delta p^1 - \left(\tfrac{f_1}{K_1}\right)^{n_1} \Delta t\\
R_{p^2} &= \Delta p^2 - \left(\tfrac{f_2}{K_2}\right)^{n_2} \Delta t\\
\underset{\sim}{\mathbf{R}}_{\alpha 1} &= \Delta\underset{\sim}{\alpha}^1 - \left(\underset{\sim}{\mathbf{n}}^1 - \tfrac{3D_1}{2C_{11}}\underset{\sim}{\mathbf{X}}^1\right)\Delta p^1\\
\underset{\sim}{\mathbf{R}}_{\alpha 2} &= \Delta\underset{\sim}{\alpha}^2 - \left(\underset{\sim}{\mathbf{n}}^2 - \tfrac{3D_2}{2C_{22}}\underset{\sim}{\mathbf{X}}^2\right)\Delta p^2\\
\underset{\sim}{\mathbf{R}}_{\beta 1} &= \Delta\underset{\sim}{\beta}^1 - \left(\underset{\sim}{\mathbf{n}}^1 - d_1\underset{\sim}{\beta}^1\right)\Delta p^1\\
\underset{\sim}{\mathbf{R}}_{\beta 2} &= \Delta\underset{\sim}{\beta}^2 - \left(\underset{\sim}{\mathbf{n}}^2 - d_2\underset{\sim}{\beta}^2\right)\Delta p^2
\end{aligned}
$$

Table 6.4. *Equations of the residual for the 2M2C model*

6.1.2.1. *Jacobian matrix for the 2M1C model*

The derivative $\underset{\approx}{\mathbf{N}}^{ij} = \partial\underset{\sim}{\mathbf{n}}^i/\partial\underset{\sim}{\sigma}^j$ (i,j=1,2) will be presented in the computations of the derivation of the residual with respect to the unknowns

$$\underset{\approx}{\mathbf{N}}^{11} = \frac{\partial\underset{\sim}{\mathbf{n}}^1}{\partial\underset{\sim}{\sigma}^1} = \frac{\underset{\approx}{\mathbf{J}} - \underset{\sim}{\mathbf{n}}^1 \otimes \underset{\sim}{\mathbf{n}}^1}{\sqrt{\left(J(\underset{\sim}{\sigma}^1 - \underset{\sim}{\mathbf{X}}^1)\right)^2 + \left(J(\underset{\sim}{\sigma}^2 - \underset{\sim}{\mathbf{X}}^2)\right)^2}}$$

$$\underset{\approx}{\mathbf{N}}^{12} = \frac{\partial\underset{\sim}{\mathbf{n}}^1}{\partial\underset{\sim}{\sigma}^1} = \frac{\underset{\sim}{\mathbf{n}}^1 \otimes \underset{\sim}{\mathbf{n}}^2}{\sqrt{\left(J(\underset{\sim}{\sigma}^1 - \underset{\sim}{\mathbf{X}}^1)\right)^2 + \left(J(\underset{\sim}{\sigma}^2 - \underset{\sim}{\mathbf{X}}^2)\right)^2}}$$

$$\underset{\approx}{\mathbf{N}}^{21} = \frac{\partial\underset{\sim}{\mathbf{n}}^2}{\partial\underset{\sim}{\sigma}^1} = \frac{\underset{\sim}{\mathbf{n}}^2 \otimes \underset{\sim}{\mathbf{n}}^1}{\sqrt{\left(J(\underset{\sim}{\sigma}^1 - \underset{\sim}{\mathbf{X}}^1)\right)^2 + \left(J(\underset{\sim}{\sigma}^2 - \underset{\sim}{\mathbf{X}}^2)\right)^2}}$$

$$\underset{\approx}{\mathbf{N}}^{22} = \frac{\partial \underset{\sim}{\mathbf{n}}^2}{\partial \underset{\sim}{\sigma}^2} = \frac{\underset{\approx}{\mathbf{J}} - \underset{\sim}{\mathbf{n}}^2 \otimes \underset{\sim}{\mathbf{n}}^2}{\sqrt{\left(J(\underset{\sim}{\sigma}^1 - \underset{\sim}{\mathbf{X}}^1)\right)^2 + \left(J(\underset{\sim}{\sigma}^2 - \underset{\sim}{\mathbf{X}}^2)\right)^2}}$$

Term R_e

$$\partial \underset{\sim}{\mathbf{R}}_e / \partial \Delta \underset{\sim}{\varepsilon}^e = \underset{\approx}{\mathbf{I}} + \left((1-z)(\underset{\approx}{\mathbf{N}}^{11} + \underset{\approx}{\mathbf{N}}^{12}) + z(\underset{\approx}{\mathbf{N}}^{21} + \underset{\approx}{\mathbf{N}}^{22})\right) : \underset{\approx}{\mathbf{\Lambda}} \theta \Delta p$$

$$\partial \underset{\sim}{\mathbf{R}}_e / \Delta p = (1-z)\underset{\sim}{\mathbf{n}}^1 + z\underset{\sim}{\mathbf{n}}^2$$

$$\partial \underset{\sim}{\mathbf{R}}_e / \partial \Delta \underset{\sim}{\alpha}^1 = -(2/3)\theta \Delta p \left[(1-z)(C_{11}\underset{\approx}{\mathbf{N}}^{11} + C_{12}\underset{\approx}{\mathbf{N}}^{12}) + z(C_{11}\underset{\approx}{\mathbf{N}}^{21} + C_{12}\underset{\approx}{\mathbf{N}}^{22})\right]$$

$$\partial \underset{\sim}{\mathbf{R}}_e / \partial \Delta \underset{\sim}{\alpha}^2 = -(2/3)\theta \Delta p \left[(1-z)(C_{12}\underset{\approx}{\mathbf{N}}^{11} + C_{22}\underset{\approx}{\mathbf{N}}^{12}) + z(C_{12}\underset{\approx}{\mathbf{N}}^{21} + C_{22}\underset{\approx}{\mathbf{N}}^{22})\right]$$

$$\partial \underset{\sim}{\mathbf{R}}_e / \partial \Delta \underset{\sim}{\beta}^1 = -A\theta \Delta p \left[(1-z)(\underset{\approx}{\mathbf{N}}^{11} z - (1-z)\underset{\approx}{\mathbf{N}}^{12}) + z(\underset{\approx}{\mathbf{N}}^{21} z - (1-z)\underset{\approx}{\mathbf{N}}^{22})\right]$$

$$\partial \underset{\sim}{\mathbf{R}}_e / \partial \Delta \underset{\sim}{\beta}^2 = -A\theta \Delta p \left[-(1-z)(\underset{\approx}{\mathbf{N}}^{11} z - (1-z)\underset{\approx}{\mathbf{N}}^{12}) - z(\underset{\approx}{\mathbf{N}}^{21} z - (1-z)\underset{\approx}{\mathbf{N}}^{22})\right]$$

Term R_p

$$\partial R_p / \partial \Delta \underset{\sim}{\varepsilon}^e = -\theta \Delta t (n/K)(f/K)^{n-1}\, \underset{\approx}{\mathbf{\Lambda}} : (\underset{\sim}{\mathbf{n}}^1 + \underset{\sim}{\mathbf{n}}^2)$$

$$\partial R_p / \Delta p = 1 + \theta \Delta t (n/K)(f/K)^{n-1}\, Qbe^{-bp}$$

$$\partial R_p / \partial \Delta \underset{\sim}{\alpha}^1 = (2/3)\theta \Delta t (n/K)(f/K)^{n-1} (C_{11}\underset{\sim}{\mathbf{n}}^1 + C_{12}\underset{\sim}{\mathbf{n}}^2)$$

$$\partial R_p / \partial \Delta \underset{\sim}{\alpha}^2 = (2/3)\theta \Delta t (n/K)(f/K)^{n-1} (C_{12}\underset{\sim}{\mathbf{n}}^1 + C_{22}\underset{\sim}{\mathbf{n}}^2)$$

$$\partial R_p / \partial \Delta \underset{\sim}{\beta}^1 = A\theta \Delta t (n/K)(f/K)^{n-1} (z\underset{\sim}{\mathbf{n}}^1 - (1-z)\underset{\sim}{\mathbf{n}}^2)$$

$$\partial R_p / \partial \Delta \underset{\sim}{\beta}^2 = -Az\theta \Delta t (n/K)(f/K)^{n-1} (z\underset{\sim}{\mathbf{n}}^1 - (1-z)\underset{\sim}{\mathbf{n}}^2)$$

Term $\underset{\sim}{\mathbf{R}}_{\alpha_1}$

$$\partial\underset{\sim}{\mathbf{R}}_{\alpha_1}/\partial\Delta\underset{\sim}{\varepsilon}^e = -\theta\Delta p(\underset{\approx}{\mathbf{N}}^{11} + \underset{\approx}{\mathbf{N}}^{12}) : \underset{\approx}{\mathbf{\Lambda}}$$

$$\partial\underset{\sim}{\mathbf{R}}_{\alpha_1}/\Delta p = (3D_1/2C_{11})\underset{\sim}{\mathbf{X}}^1 - \underset{\sim}{\mathbf{n}}^1$$

$$\partial\underset{\sim}{\mathbf{R}}_{\alpha_1}/\partial\Delta\underset{\sim}{\alpha}^1 = \underset{\approx}{\mathbf{I}} + \theta\Delta p\left[(2/3)(C_{11}\underset{\approx}{\mathbf{N}}_{11} + C_{12}\underset{\approx}{\mathbf{N}}^{12}) + D_1\underset{\approx}{\mathbf{I}}\right]$$

$$\partial\underset{\sim}{\mathbf{R}}_{\alpha_1}/\partial\Delta\underset{\sim}{\alpha}^2 = \theta\Delta p\left[(2/3)(C_{12}\underset{\approx}{\mathbf{N}}_{11} + C_{22}\underset{\approx}{\mathbf{N}}^{12}) + (D_1/C_{11})C_{12}\underset{\approx}{\mathbf{I}}\right]$$

$$\partial\underset{\sim}{\mathbf{R}}_{\alpha_1}/\partial\Delta\underset{\sim}{\beta}^1 = Az\theta\Delta p(z\underset{\approx}{\mathbf{N}}^{11} - (1-z)\underset{\approx}{\mathbf{N}}^{12})$$

$$\partial\underset{\sim}{\mathbf{R}}_{\alpha_1}/\partial\Delta\underset{\sim}{\beta}^2 = -Az\theta\Delta p(z\underset{\approx}{\mathbf{N}}^{11} - (1-z)\underset{\approx}{\mathbf{N}}^{12})$$

Term $\underset{\sim}{\mathbf{R}}_{\alpha_2}$

$$\partial\underset{\sim}{\mathbf{R}}_{\alpha_2}/\partial\Delta\underset{\sim}{\varepsilon}^e = -\theta\Delta p(\underset{\approx}{\mathbf{N}}^{21} + \underset{\approx}{\mathbf{N}}^{22}) : \underset{\approx}{\mathbf{\Lambda}}$$

$$\partial\underset{\sim}{\mathbf{R}}_{\alpha_2}/\Delta p = (3D_2/2C_{22})\underset{\sim}{\mathbf{X}}^2 - \underset{\sim}{\mathbf{n}}^2$$

$$\partial\underset{\sim}{\mathbf{R}}_{\alpha_2}/\partial\Delta\underset{\sim}{\alpha}^1 = \theta\Delta p\left[(2/3)(C_{11}\underset{\approx}{\mathbf{N}}_{21} + C_{12}\underset{\approx}{\mathbf{N}}^{22}) + (D_2/C_{22})C_{12}\underset{\approx}{\mathbf{I}}\right]$$

$$\partial\underset{\sim}{\mathbf{R}}_{\alpha_2}/\partial\Delta\underset{\sim}{\alpha}^2 = \underset{\approx}{\mathbf{I}} + \theta\Delta p\left[(2/3)(C_{12}\underset{\approx}{\mathbf{N}}_{21} + C_{22}\underset{\approx}{\mathbf{N}}^{22}) + D_2\underset{\approx}{\mathbf{I}}\right]$$

$$\partial\underset{\sim}{\mathbf{R}}_{\alpha_2}/\partial\Delta\underset{\sim}{\beta}^1 = Az\theta\Delta p(z\underset{\approx}{\mathbf{N}}^{21} - (1-z)\underset{\approx}{\mathbf{N}}^{22})$$

$$\partial\underset{\sim}{\mathbf{R}}_{\alpha_2}/\partial\Delta\underset{\sim}{\beta}^2 = -Az\theta\Delta p(z\underset{\approx}{\mathbf{N}}^{21} - (1-z)\underset{\approx}{\mathbf{N}}^{22})$$

Term $\underset{\sim}{\mathbf{R}}_{\beta_1}$

$$\partial\underset{\sim}{\mathbf{R}}_{\beta_1}/\partial\Delta\underset{\sim}{\varepsilon}^e = -\theta\Delta p(\underset{\approx}{\mathbf{N}}^{11} + \underset{\approx}{\mathbf{N}}^{12}) : \underset{\approx}{\mathbf{\Lambda}}$$

$$\partial\underset{\sim}{\mathbf{R}}_{\beta_1}/\Delta p = d_1\underset{\sim}{\beta}^1 - \underset{\sim}{\mathbf{n}}^1$$

$$\partial\underset{\sim}{\mathbf{R}}_{\beta_1}/\partial\Delta\underset{\sim}{\alpha}^1 = \theta\Delta p(2/3)(C_{11}\underset{\approx}{\mathbf{N}}^{11} + C_{12}\underset{\approx}{\mathbf{N}}^{12})$$

$$\partial \underset{\sim}{\mathbf{R}}_{\beta_1} / \partial \Delta \underset{\sim}{\alpha}^2 = \theta \Delta p (2/3)(C_{12} \underset{\approx}{\mathbf{N}}^{11} + C_{22} \underset{\approx}{\mathbf{N}}^{12})$$

$$\partial \underset{\sim}{\mathbf{R}}_{\beta_1} / \partial \Delta \underset{\sim}{\beta}^1 = \underset{\approx}{\mathbf{I}} + \theta \Delta p \left[A(z \underset{\approx}{\mathbf{N}}^{11} - (1-z) \underset{\approx}{\mathbf{N}}^{12}) + d_1 \underset{\approx}{\mathbf{I}} \right]$$

$$\partial \underset{\sim}{\mathbf{R}}_{\beta_1} / \partial \Delta \underset{\sim}{\beta}^2 = -Az\theta \Delta p (z \underset{\approx}{\mathbf{N}}^{11} - (1-z) \underset{\approx}{\mathbf{N}}^{12})$$

Term $\underset{\sim}{\mathbf{R}}_{\beta 2}$

$$\partial \underset{\sim}{\mathbf{R}}_{\beta_2} / \partial \Delta \underset{\sim}{\varepsilon}^e = -\theta \Delta p (\underset{\approx}{\mathbf{N}}^{21} + \underset{\approx}{\mathbf{N}}^{22}) : \underset{\approx}{\mathbf{\Lambda}}$$

$$\partial \underset{\sim}{\mathbf{R}}_{\beta_2} / \Delta p = d_2 \underset{\sim}{\beta}^2 - \underset{\sim}{\mathbf{n}}^2$$

$$\partial \underset{\sim}{\mathbf{R}}_{\beta_2} / \partial \Delta \underset{\sim}{\alpha}^1 = \theta \Delta p (2/3)(C_{11} \underset{\approx}{\mathbf{N}}^{21} + C_{12} \underset{\approx}{\mathbf{N}}^{22})$$

$$\partial \underset{\sim}{\mathbf{R}}_{\beta_2} / \partial \Delta \underset{\sim}{\alpha}^2 = \theta \Delta p (2/3)(C_{12} \underset{\approx}{\mathbf{N}}^{21} + C_{22} \underset{\approx}{\mathbf{N}}^{22})$$

$$\partial \underset{\sim}{\mathbf{R}}_{\beta_2} / \partial \Delta \underset{\sim}{\beta}^1 = Az\theta \Delta p (z \underset{\approx}{\mathbf{N}}^{21} - (1-z) \underset{\approx}{\mathbf{N}}^{22})$$

$$\partial \underset{\sim}{\mathbf{R}}_{\beta_2} / \partial \Delta \underset{\sim}{\beta}^2 = \underset{\approx}{\mathbf{I}} + \theta \Delta p \left[-A(z \underset{\approx}{\mathbf{N}}^{21} - (1-z) \underset{\approx}{\mathbf{N}}^{22}) + d_2 \underset{\approx}{\mathbf{I}} \right]$$

6.1.2.2. *Jacobian matrix for the 2M2C model*

As in the 2M1C model, the derivative $\underset{\approx}{\mathbf{N}}^i = \partial \underset{\sim}{\mathbf{n}}^i / \partial \underset{\sim}{\sigma}$ of the normal $\underset{\sim}{\mathbf{n}}^i$ (i=1,2) will be presented in the computations:

$$\underset{\approx}{\mathbf{N}}^1 = \frac{\partial \underset{\sim}{\mathbf{n}}^1}{\partial \underset{\sim}{\sigma}^1} = \frac{1}{J^1} \left[\frac{3}{2} \underset{\approx}{\mathbf{J}}^1 - \underset{\sim}{\mathbf{n}}^1 \otimes \underset{\sim}{\mathbf{n}}^1 \right] \qquad \underset{\approx}{\mathbf{N}}^2 = \frac{\partial \underset{\sim}{\mathbf{n}}^2}{\partial \underset{\sim}{\sigma}^2} = \frac{1}{J^2} \left[\frac{3}{2} \underset{\approx}{\mathbf{J}}^2 - \underset{\sim}{\mathbf{n}}^2 \otimes \underset{\sim}{\mathbf{n}}^2 \right]$$

Term R_e

$$\partial \underset{\sim}{\mathbf{R}}_e / \partial \Delta \underset{\sim}{\varepsilon}^e = \underset{\approx}{\mathbf{I}} + (1-z) \underset{\approx}{\mathbf{N}}^1 : \underset{\approx}{\mathbf{\Lambda}} \theta \Delta p^1 + z \underset{\approx}{\mathbf{N}}^2 : \underset{\approx}{\mathbf{\Lambda}} \theta \Delta p^2$$

$$\partial \underset{\sim}{\mathbf{R}}_e / \Delta p^1 = (1-z) \underset{\sim}{\mathbf{n}}^1$$

$$\partial \underset{\sim}{\mathbf{R}}_e / \Delta p^2 = z \underset{\sim}{\mathbf{n}}^2$$

$$\partial \underset{\sim}{\mathbf{R}}_e / \partial \Delta \underset{\sim}{\alpha}^1 = -(2/3)\theta\left((1-z)C_{11}\underset{\approx}{\mathbf{N}}^1 \Delta p^1 + zC_{12}\underset{\approx}{\mathbf{N}}^2 \Delta p^2\right)$$

$$\partial \underset{\sim}{\mathbf{R}}_e / \partial \Delta \underset{\sim}{\alpha}^2 = -(2/3)\theta\left((1-z)C_{12}\underset{\approx}{\mathbf{N}}^1 \Delta p^1 + zC_{22}\underset{\approx}{\mathbf{N}}^2 \Delta p^2\right)$$

$$\partial \underset{\sim}{\mathbf{R}}_e / \partial \Delta \underset{\sim}{\beta}^1 = -Az(1-z)\underset{\approx}{\mathbf{N}}^1 \theta \Delta p^1 + Az(1-z)\underset{\approx}{\mathbf{N}}^2 \theta \Delta p^2$$

$$\partial \underset{\sim}{\mathbf{R}}_e / \partial \Delta \underset{\sim}{\beta}^2 = Az(1-z)\underset{\approx}{\mathbf{N}}^1 \theta \Delta p^1 - Az(1-z)\underset{\approx}{\mathbf{N}}^2 \theta \Delta p^2$$

Term R_{p1}

$$\partial R_{p1} / \partial \Delta \underset{\sim}{\varepsilon}^e = -\theta \Delta t (n_1/K_1)\left(f^1/K_1\right)^{n_1-1} \underset{\sim}{\mathbf{n}}^1 : \underset{\approx}{\mathbf{\Lambda}}$$

$$\partial R_{p1} / \Delta p^1 = 1 + \theta \Delta t (n_1/K_1)\left(f^1/K_1\right)^{n_1-1} Q_1 C_1 e^{-b_1 p^1}$$

$$\partial R_{p1} / \Delta p^2 = 0$$

$$\partial F_{p1} / \partial \Delta \underset{\sim}{\alpha}^1 = (2/3)\theta \Delta t (n_1/K_1)\left(f^1/K_1\right)^{n_1-1} C_{11} \underset{\sim}{\mathbf{n}}^1$$

$$\partial F_{p1} / \partial \Delta \underset{\sim}{\alpha}^2 = (2/3)\theta \Delta t (n_1/K_1)\left(f^1/K_1\right)^{n_1-1} C_{12} \underset{\sim}{\mathbf{n}}^1$$

$$\partial F_{p1} / \partial \Delta \underset{\sim}{\beta}^1 = Az\theta \Delta t (n_1/K_1)\left(f^1/K_1\right)^{n_1-1} \underset{\sim}{\mathbf{n}}^1$$

$$\partial F_{p1} / \partial \Delta \underset{\sim}{\beta}^2 = -Az\theta \Delta t (n_1/K_1)\left(f^1/K_1\right)^{n_1-1} \underset{\sim}{\mathbf{n}}^1$$

Term R_{p2}

$$\partial R_{p2} / \partial \Delta \underset{\sim}{\varepsilon}^e = -\theta \Delta t (n_2/K_2)\left(f^2/K_2\right)^{n_2-1} \underset{\sim}{\mathbf{n}}^2 : \underset{\approx}{\mathbf{\Lambda}}$$

$$\partial R_{p2} / \Delta p^1 = 0$$

$$\partial R_{p2} / \Delta p^2 = 1 + \theta \Delta t (n_2/K_2)\left(f^2/K_2\right)^{n_2-1} Q_2 C_2 e^{-b_2 p^2}$$

$$\partial R_{p2} / \partial \Delta \underset{\sim}{\alpha}^1 = (2/3)\theta \Delta t (n_2/K_2)\left(f^2/K_2\right)^{n_2-1} C_{12} \underset{\sim}{\mathbf{n}}^2$$

$$\partial R_{p2} / \partial \Delta \underset{\sim}{\alpha}^2 = (2/3)\theta \Delta t (n_2/K_2)\left(f^2/K_2\right)^{n_2-1} C_{22} \underset{\sim}{\mathbf{n}}^2$$

$$\partial R_{p2}/\partial \Delta \underset{\sim}{\beta}^1 = -A(1-z)\theta \Delta t(n_2/K_2)\left(f^2/K_2\right)^{n_2-1} \underset{\sim}{\mathbf{n}}^2$$

$$\partial R_{p2}/\partial \Delta \underset{\sim}{\beta}^2 = A(1-z)\theta \Delta t(n_2/K_2)\left(f^2/K_2\right)^{n_2-1} \underset{\sim}{\mathbf{n}}^2$$

Term $\underset{\sim}{\mathbf{R}}_{\alpha 1}$

$$\partial \underset{\sim}{\mathbf{R}}_{\alpha 1}/\partial \Delta \underset{\sim}{\varepsilon}^e = -\theta \Delta p^1 \underset{\approx}{\mathbf{N}}^1 : \underset{\approx}{\mathbf{\Lambda}}$$

$$\partial \underset{\sim}{\mathbf{R}}_{\alpha 1}/\Delta p^1 = (3D_1/2C_{11})\underset{\sim}{\mathbf{X}}^1 - \underset{\sim}{\mathbf{n}}^1$$

$$\partial \underset{\sim}{\mathbf{R}}_{\alpha 1}/\Delta p^2 = 0$$

$$\partial \underset{\sim}{\mathbf{R}}_{\alpha 1}/\partial \Delta \underset{\sim}{\alpha}^1 = \underset{\approx}{\mathbf{I}} + \theta \Delta p^1 \left((2/3)C_{11}\underset{\approx}{\mathbf{N}}_1 + D_1 \underset{\approx}{\mathbf{I}}\right)$$

$$\partial \underset{\sim}{\mathbf{R}}_{\alpha 1}/\partial \Delta \underset{\sim}{\alpha}^2 = \theta \Delta p^1 C_{12}\left((2/3)\underset{\approx}{\mathbf{N}}_1 + (D_1/C_{11})\underset{\approx}{\mathbf{I}}\right)$$

$$\partial \underset{\sim}{\mathbf{R}}_{\alpha 1}/\partial \Delta \underset{\sim}{\beta}^1 = Az\theta \Delta p^1 \underset{\approx}{\mathbf{N}}^1$$

$$\partial \underset{\sim}{\mathbf{R}}_{\alpha 1}/\partial \Delta \underset{\sim}{\beta}^2 = -Az\theta \Delta p^1 \underset{\approx}{\mathbf{N}}^1$$

Term $\underset{\sim}{\mathbf{R}}_{\alpha 2}$

$$\partial \underset{\sim}{\mathbf{R}}_{\alpha 2}/\partial \Delta \underset{\sim}{\varepsilon}^e = -\theta \Delta p^2 \underset{\approx}{\mathbf{N}}^2 : \underset{\approx}{\mathbf{\Lambda}}$$

$$\partial \underset{\sim}{\mathbf{R}}_{\alpha 2}/\Delta p^1 = 0$$

$$\partial \underset{\sim}{\mathbf{R}}_{\alpha 2}/\Delta p^2 = (3D_2/2C_{22})\underset{\sim}{\mathbf{X}}^2 - \underset{\sim}{\mathbf{n}}^2$$

$$\partial \underset{\sim}{\mathbf{R}}_{\alpha 2}/\partial \Delta \underset{\sim}{\alpha}^1 = \theta \Delta p^2 C_{12}\left((2/3)\underset{\approx}{\mathbf{N}}_2 + (D_2/C_{22})\underset{\approx}{\mathbf{I}}\right)$$

$$\partial \underset{\sim}{\mathbf{R}}_{\alpha 2}/\partial \Delta \underset{\sim}{\alpha}^2 = \underset{\approx}{\mathbf{I}} + \theta \Delta p^2 \left((2/3)C_{22}\underset{\approx}{\mathbf{N}}_2 + D_2 \underset{\approx}{\mathbf{I}}\right)$$

$$\partial \underset{\sim}{\mathbf{R}}_{\alpha 2}/\partial \Delta \underset{\sim}{\beta}^1 = -A(1-z)\theta \Delta p^2 \underset{\approx}{\mathbf{N}}^2$$

$$\partial \underset{\sim}{\mathbf{R}}_{\alpha 2}/\partial \Delta \underset{\sim}{\beta}^2 = A(1-z)\theta \Delta p^1 \underset{\approx}{\mathbf{N}}^1$$

Term $\underset{\sim}{\mathbf{R}}_{\beta 1}$

$$\partial \underset{\sim}{\mathbf{R}}_{\beta 1}/\partial \Delta \underset{\sim}{\varepsilon}^e = -\theta \Delta p^1 \underset{\approx}{\mathbf{N}}^1 : \underset{\approx}{\mathbf{\Lambda}}$$

$$\partial \underset{\sim}{\mathbf{R}}_{\beta 1}/\Delta p^1 = d_1 \underset{\sim}{\beta}^1 - \underset{\sim}{\mathbf{n}}^1$$

$$\partial \underset{\sim}{\mathbf{R}}_{\beta 1}/\Delta p^2 = 0$$

$$\partial \underset{\sim}{\mathbf{R}}_{\beta 1}/\partial \Delta \underset{\sim}{\alpha}^1 = \theta \Delta p^1 (2/3) C_{11} \underset{\approx}{\mathbf{N}}_1$$

$$\partial \underset{\sim}{\mathbf{R}}_{\beta 1}/\partial \Delta \underset{\sim}{\alpha}^2 = \theta \Delta p^1 (2/3) C_{12} \underset{\approx}{\mathbf{N}}_1$$

$$\partial \underset{\sim}{\mathbf{R}}_{\beta 1}/\partial \Delta \underset{\sim}{\beta}^1 = \underset{\approx}{\mathbf{I}} + \theta \Delta p^1 \left(Az \underset{\approx}{\mathbf{N}}^1 + d_1 \underset{\approx}{\mathbf{I}} \right)$$

$$\partial \underset{\sim}{\mathbf{R}}_{\beta 1}/\partial \Delta \underset{\sim}{\beta}^2 = -Az\theta \Delta p^1 \underset{\approx}{\mathbf{N}}^1$$

Term $\underset{\sim}{\mathbf{R}}_{\beta 2}$

$$\partial \underset{\sim}{\mathbf{R}}_{\beta 2}/\partial \Delta \underset{\sim}{\varepsilon}^e = -\theta \Delta p^2 \underset{\approx}{\mathbf{N}}^2 : \underset{\approx}{\mathbf{\Lambda}}$$

$$\partial \underset{\sim}{\mathbf{R}}_{\beta 2}/\Delta p^1 = 0$$

$$\partial \underset{\sim}{\mathbf{R}}_{\beta 2}/\Delta p^2 = d_2 \underset{\sim}{\beta}^2 - \underset{\sim}{\mathbf{n}}^2$$

$$\partial \underset{\sim}{\mathbf{R}}_{\beta 2}/\partial \Delta \underset{\sim}{\alpha}^1 = \theta \Delta p^2 (2/3) C_{12} \underset{\approx}{\mathbf{N}}_2$$

$$\partial \underset{\sim}{\mathbf{R}}_{\beta 2}/\partial \Delta \underset{\sim}{\alpha}^2 = \theta \Delta p^2 (2/3) C_{22} \underset{\approx}{\mathbf{N}}_2$$

$$\partial \underset{\sim}{\mathbf{R}}_{\beta 2}/\partial \Delta \underset{\sim}{\beta}^1 = -A(1-z)\theta \Delta p^2 \mathbf{N}^2$$

$$\partial \underset{\sim}{\mathbf{R}}_{\beta 2}/\partial \Delta \underset{\sim}{\beta}^2 = \underset{\approx}{\mathbf{I}} + \theta \Delta p^2 \left(A(1-z) \underset{\approx}{\mathbf{N}}^2 + d_2 \underset{\approx}{\mathbf{I}} \right)$$

6.2. Creep–plasticity interaction in a notched specimen

The present application introduces an axisymmetric notched specimen, which is simulated by means of a 2M2C model. As summarized in Table 3.2, the mechanism "1" is viscoplastic and the mechanism "2" is plastic.

The creep–plasticity interaction results from the coupling parameter C_{12}. Axisymmetric elements with reduced integration were used in the framework of an updated Lagrangian formulation to account for large deformations. The mesh is shown in Figure 6.1. A constant pressure of 200 MPa is applied on the "top" of the mesh. Two different rates (1 MPa.s^{-1} and 100 MPa.s^{-1}) are used for the loading stage, and the stress level is 635 MPa. Figure 6.2 shows the radial distribution of the plastic and the viscoplastic strains at the end of the creep period. It can be seen that for the smallest value of the coupling term (C_{12}=22500 MPa), viscoplasticity is more pronounced for the low loading rate, whereas plasticity is predominant for the high loading rate. For the larger coupling term (C_{12}=35000 MPa), plasticity is predominant, especially for the high loading rate.

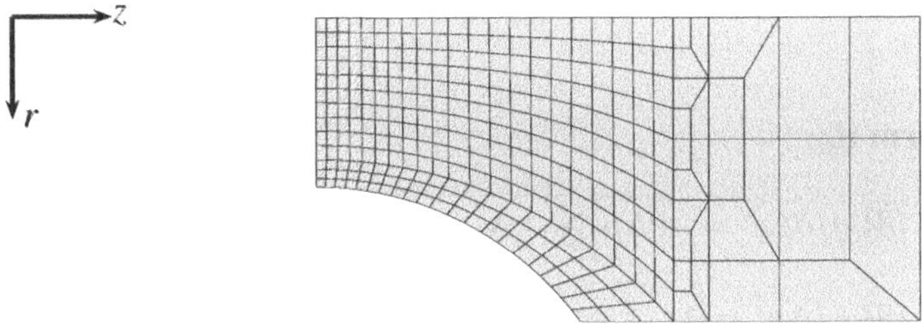

Figure 6.1. *Mesh of notched specimen used for the simulation of creep–plasticity interaction*

6.3. FE analysis of plane forging of polycarbonate specimens

In this section, FE simulations of plane-strain forging experiments at 25°C, under isothermal conditions, were performed on polycarbonate specimens, as in [SRI 10]. The initial shape of the specimen is a cylinder, the cross-section of which is a circle of diameter 12.7 mm. The cruciform final shape is imposed by split dies made of hardened tool steel. The forging experiment was performed up to a relative die displacement of 4.6 mm, with a displacement rate of 0.02 mm/s before the removal of the die at the same absolute displacement rate. Due to the symmetries of the problem, the FE modeling considers only one quarter of the system. The mesh includes 612 plane strain eight-node quadrilateral elements, whereas the die is modeled as a rigid body (Figure 6.3).

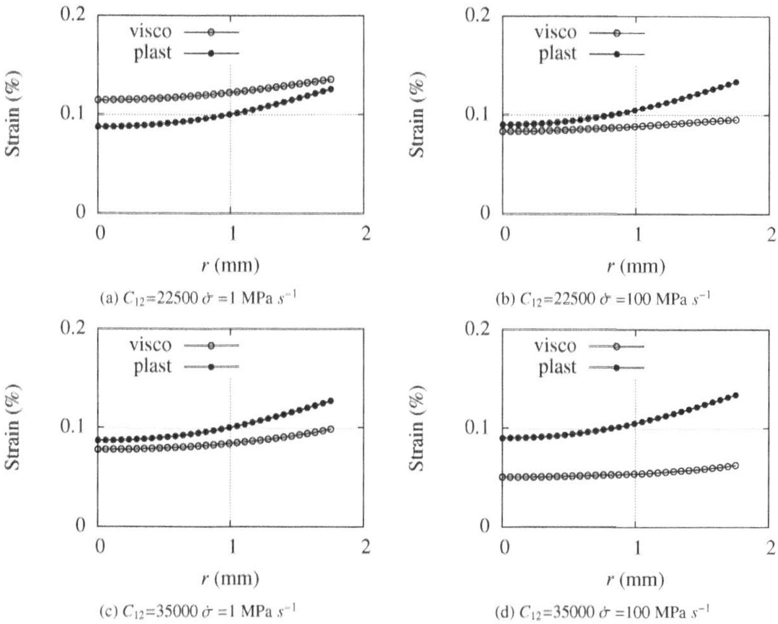

Figure 6.2. *Creep–plasticity interaction (notched specimen). Radial distribution of the plastic and the viscoplastic strains at the end of the creep stage.* $A_1 = A_2 = 1$, $E = 144{,}000$, $n = 10$, $K = 800$, $D_1 = D_2 = 100$, $R_{01} = 0$, $R_{02} = 140$ $C_{11} = 50{,}000$, $C_{22} = 30{,}000$ *(MPa, s)*

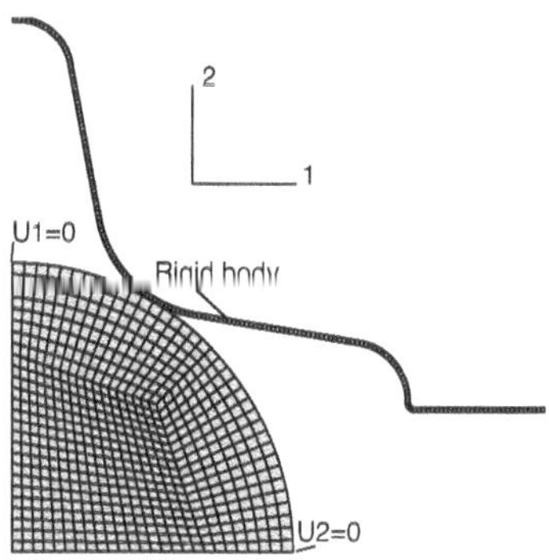

Figure 6.3. *FE simulation of plane-strain cruciform forging of PC: mesh and boundary conditions*

The contact between the die and the workpiece is assumed to be frictionless. In this specific case, the large deformation reached during the process produces a strong hardening, so that an updated expression is used for the isotropic variable R^1, where the additional term involves a negative parameter b'_1 in the exponential term. Namely:

$$\text{For} \quad I = 1 \quad R_{01} = Q_{11}(1 - \exp(-b_1 p^1) + Q_1(1 - \exp(-b'_1 p^1) \qquad [6.1]$$

$$\text{For} \quad I = 2, 3 \quad R_{0I} = Q_{II}(1 - \exp(-b_I p^I) \qquad [6.2]$$

Moreover, kinematic hardening is not taken into account even if loading-unloading stages are considered. The identified material parameters are displayed in Table 6.5. The comparison between the simulated and experimental stress-strain behavior is shown in Figure 6.4.

Mechanism 1	$R_{01} = 70$	$K_1 = 13$	$n_1 = 7$	$b_1 = 4.3$	$Q_{11} = -249$
Mechanism 2	$R_{02} = 15.5$	$K_2 = 45$	$n_2 = 7$	$b_2 = 9.95$	$Q_{22} = 2,079$
Mechanism 3	$R_{03} = 58$	$K_3 = 17$	$n_3 = 7$	$b_3 = 9.92$	$Q_{33} = 2,063$
Addititional parameters	$Q'_{11} = 33.$	$b'_1 = -5.61$			

Table 6.5. *Material parameters of the MM models used in the FE simulation of cruciform forging (MPa, s). The coupling parameters are set to zero*

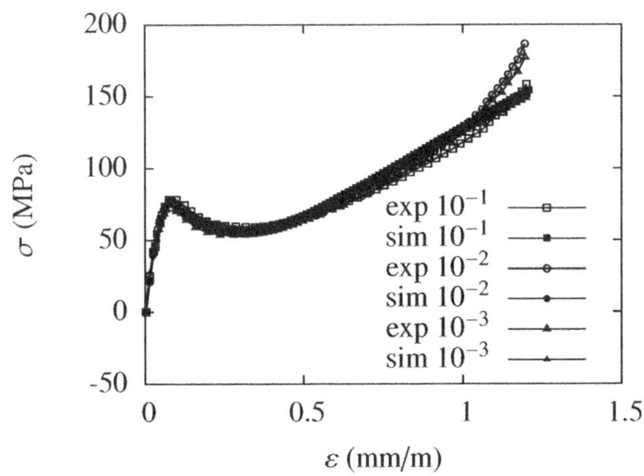

Figure 6.4. *FE simulation of plane-strain cruciform forging of PC. Experimental (taken from [SRI 10]) and simulated stress-strain curves*

Figure 6.6 compares the numerical simulation with the experimental shape taken from the work of [SRI 10] after die removal. A contour plot of the maximal component of the strain tensor at the end of the loading stage is shown in Figure 6.5.

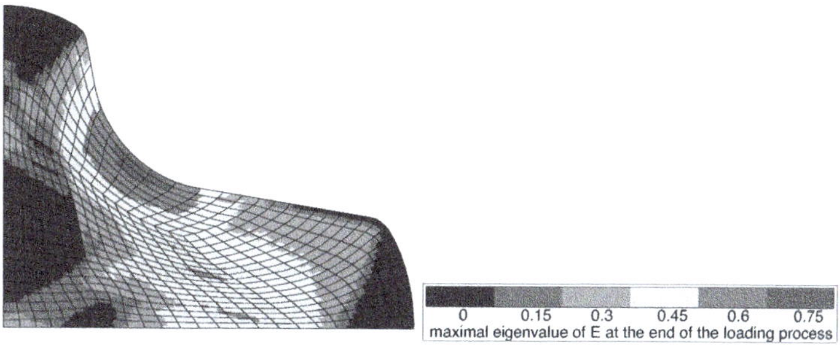

Figure 6.5. *FE simulation of plane-strain cruciform forging of PC. Contour plot of the maximal eigenstrain at the end of the loading stage. For a color version of the figure, see www.iste.co.uk/cailletaud/multi-mechanism.zip*

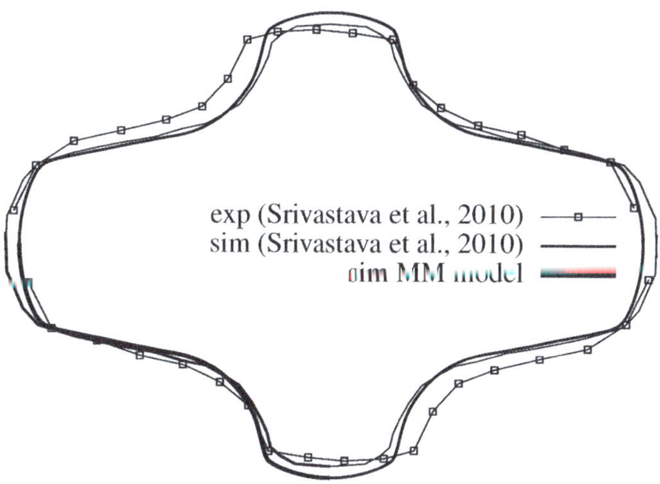

Figure 6.6. *FE simulation of plane-strain cruciform forging of PC. Comparison between experimentally deformed shape and FE prediction*

6.4. FE simulation of bulging of a 304SS sheet

Due to the strain-induced martensitic transformation during plastic deformation under cold working conditions, TRIP (transformation-induced plasticity) steels present interesting mechanical properties, combining high strength, ductility and toughness for large plastic deformations. These alloys are now classically used in industry, for instance for car bodies in automotive industry. The example of this section demonstrates how TRIP can be introduced in a MM model. The application is devoted to an austenitic stainless steel (304SS alloy) tested at low temperature.

The model has to take into account the phase transformation during cold working, and to introduce the individual properties of each phase. Then, an additional equation defines the kinetics of the transformation, and the resulting volume fraction of each phase operates as a weighting parameter to determine the influence of each mechanism. A bulging experiment performed at different temperatures has been used to validate the FE simulation.

Following previous attempts [VID 94, GAU 04], the local stress and strain tensors in each phase are defined in the framework of uniform field models by means of β-scale transition rules. One kinematic hardening and one isotropic hardening variable are introduced in each phase. The inelastic strains considered in the computations are intermediate (about 20 to 30%). This is the reason why transformation plasticity is not considered here. The amount of additional inelastic strain coming from either the Magee or Greenwood-Johnson effect is supposed to be small enough to be neglected in comparison to the amount of classical plasticity. It was assumed that the strain tensor is partitioned into the elastic part $\underset{\sim}{\varepsilon}^e$, the volumetric transformation $\underset{\sim}{\varepsilon}^{tr}$ and plastic strain $\underset{\sim}{\varepsilon}^p$:

$$\underset{\sim}{\varepsilon} = \underset{\sim}{\varepsilon}^e + \underset{\sim}{\varepsilon}^{tr} + \underset{\sim}{\varepsilon}^p \qquad [6.3]$$

– The elastic strain is defined by means of Hooke's law;

– The volumetric transformation is given by $\underset{\sim}{\varepsilon}^{tr} = z\delta\underset{\sim}{\mathbf{I}}$, where z is the volume fraction of the martensitic phase, δ is the intrinsic volume change between austenite and martensite and $\underset{\sim}{\mathbf{I}}$ is the second-rank identity tensor. An exponential law was used to fit the experimental kinetics results:

$$z = 1 - \exp\left(-\beta\left[1 - \exp(-\alpha\varepsilon)\right]\right)^m \qquad [6.4]$$

where ε is the equivalent strain, and , β and m are material parameters.

– Following [SAÏ 07], the inelastic strain results from the deformation process in both phases. Each inelastic strain is associated with a particular phase, austenite (ε^1) and martensite (ε^2):

$$\underset{\sim}{\varepsilon}^p = (1 - z)\underset{\sim}{\varepsilon}^1 + z\underset{\sim}{\varepsilon}^2 \qquad [6.5]$$

The model is identical to the version detailed in Chapter 4 and resumed in Table 3.4. The model is implemented using the θ-method solved by an implicit Newton scheme for the local integration. Simulations of the monotonic behavior of SS304 at low temperatures were performed to identify the parameters of the MM model. As shown in Figure 6.7, the simulations are in good agreement with the experiments for the tensile tests at various temperatures. The successful comparison includes the evolution of the volume fraction and the representation of the inflection points located in the strain range [5 to 10%] for which a significant growth of the martensite phase is observed. In order to assess the model reliability, a comparison is made between the FE simulations and an experimental database of bulging tests at various temperatures [SAÏ 06b]. The shape of the deformed sheet is studied, together with the applied force, for various boundary conditions. A numerical study is performed to demonstrate the validity of the computations, with a special attention to the friction coefficient. Typical contour plots performed by the FE simulation are also shown in Figure 6.8.

6.5. FE simulation of PA6 notched specimens

The purpose of this section is to demonstrate the way damage models can be combined with MM models. The case study relates to a semi-crystalline polymer, as in [JER 15]. Both Gurson-Tvergaard-Needleman (GTN) (see section 5.1) and continuum damage mechanics (CDM) (see section 5.4) approaches are used. Two assumptions are tested: (i) damage is described by a unique variable and (ii) each phase is associated with its own damage variable. Four models can then be distinguished. The comparison is based on creep tests performed on notched round bars made of PA6 semi-crystalline polymer to create a multiaxial stress state.

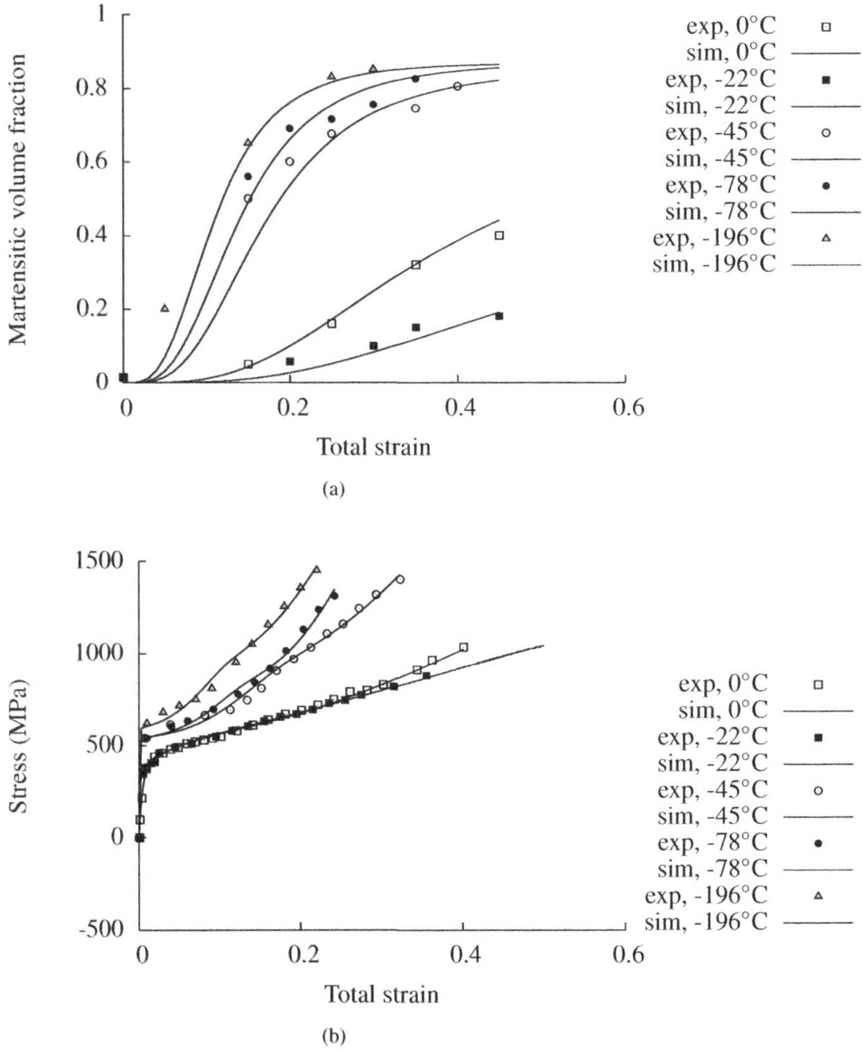

Figure 6.7. *(a) Evolution of the martensitic transformation induced by plastic deformation in 304SS, under different temperature conditions. (b) Stress-strain curves of the austenitic stainless steel under different temperature conditions [SAÏ 06b]*

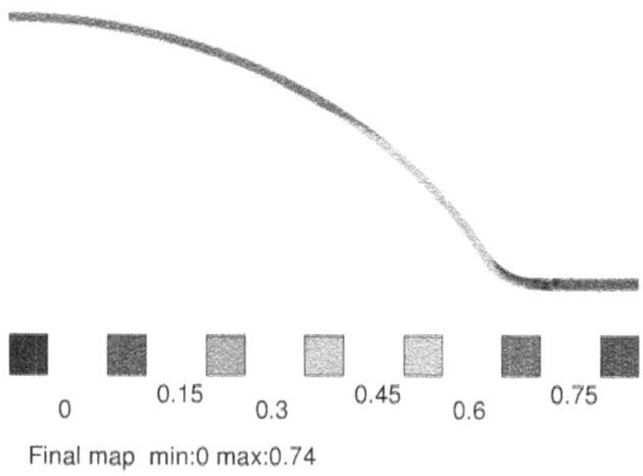

Figure 6.8. *Contour plot of the volume fraction of martensite T=–45° C at the end of the deformation process. For a color version of the figure, see www.iste.co.uk/cailletaud/multi-mechanism.zip*

The identification of the material parameter is performed according to the following procedure:

– tensile and creep tests performed on smooth specimens are first used to identify the material parameters of the "undamaged material" common to the different models. The strain rate is of 0.026 s^{-1} for the tensile test and the creep tests are performed at load levels of 71, 76, 79, 80 and 82 MPa;

– creep tests carried out on notched specimens are then used for the determination of the material parameters dedicated to damage modeling;

– numerical results are finally analyzed to check the relevance of the simulation and to select the best model. In the following, the first inelastic strain ($i = 1$) is associated with the amorphous phase, whereas the second one ($i = 2$) refers to the crystalline phase.

The two main steps of the identification process are:

1) The determination of the material parameters of the amorphous phase which deals with short-time effect. Since the crystalline phase does not produce creep strain during tensile tests, the only active mechanism is associated with the amorphous phase. The parameters n_1, K_1, Q_1, b_1 and R_{01} are then identified. The value of the yield stress is estimated graphically when the stress-strain curve deviates from the linear regime ($\simeq$ 40 MPa). The value of R_{01} is fixed to 20 MPa in order to activate the mechanism of the amorphous phase before the "apparent yield stress" of the PA6. Tensile tests show that the linear viscous effect is weak, so that n_1 is estimated at 2.5. The evaluation of the remaining parameters K_1, Q_1, b_1 is then performed through an optimization process.

2) The second step of the optimization procedure deals with the optimization of the parameters of the crystalline phase, which are supposed to represent the long-time effect encountered in the creep tests. To be in agreement with the time-dependent phenomena, the parameter n_2 is fixed to $2 \times n_1 \simeq 5$, meanwhile R_{02} is constrained at a small value, less than 10 MPa. In this step, the value of the parameters estimated in the first step can be adjusted to have a better representation of the experimental tensile curves.

The list of the calibrated coefficients for the MM models is given in Table 6.6 and the corresponding comparison between simulated responses and experimental tensile and creep data are shown in Figure 6.9.

	Isotropic hardening			Norton law	
amorphous	R_{01}	b_1	Q_1	n_1	K_1
	25	4.4	95	2.5	250
crystalline	R_{02}	b_2	Q_2	n_2	K_2
	3	3.8	110	5	800

Table 6.6. *Identified parameters for the "undamaged" PA6 (MPa, s)*

Since this first stage is only focused on the determination of undamaged model coefficients, the simulations were intentionally interrupted before necking occurs (after 6% for the tensile test and at the end of secondary creep for the creep tests). For the sake of clarity, the creep tests (Figure 6.9(b)-(f)) are shown in separate plots. Indeed, as the load level increases, the time of the end of secondary creep decreases.

The MM model is able to reproduce both short (tensile test) and long-term creep strain history of the PA6 thanks to the opportunity allowed by the MM model to separate the two different regimes in the constitutive equations. It is worth noting that only 10 parameters are used for the undamaged material. These parameters will be kept unchanged thereafter. Notched specimens are devoted to the characterization of the damage behavior and the identification of the remaining material parameters. The four models (GTN/CDM, unique/separate damage variable) will therefore be evaluated by comparing the corresponding predictions of the creep behavior of notched specimens with different notch radii (ρ=0.45 mm, 0.8 mm, 1.6 mm and 4 mm) to experimental data [CAY 13, JER 15].

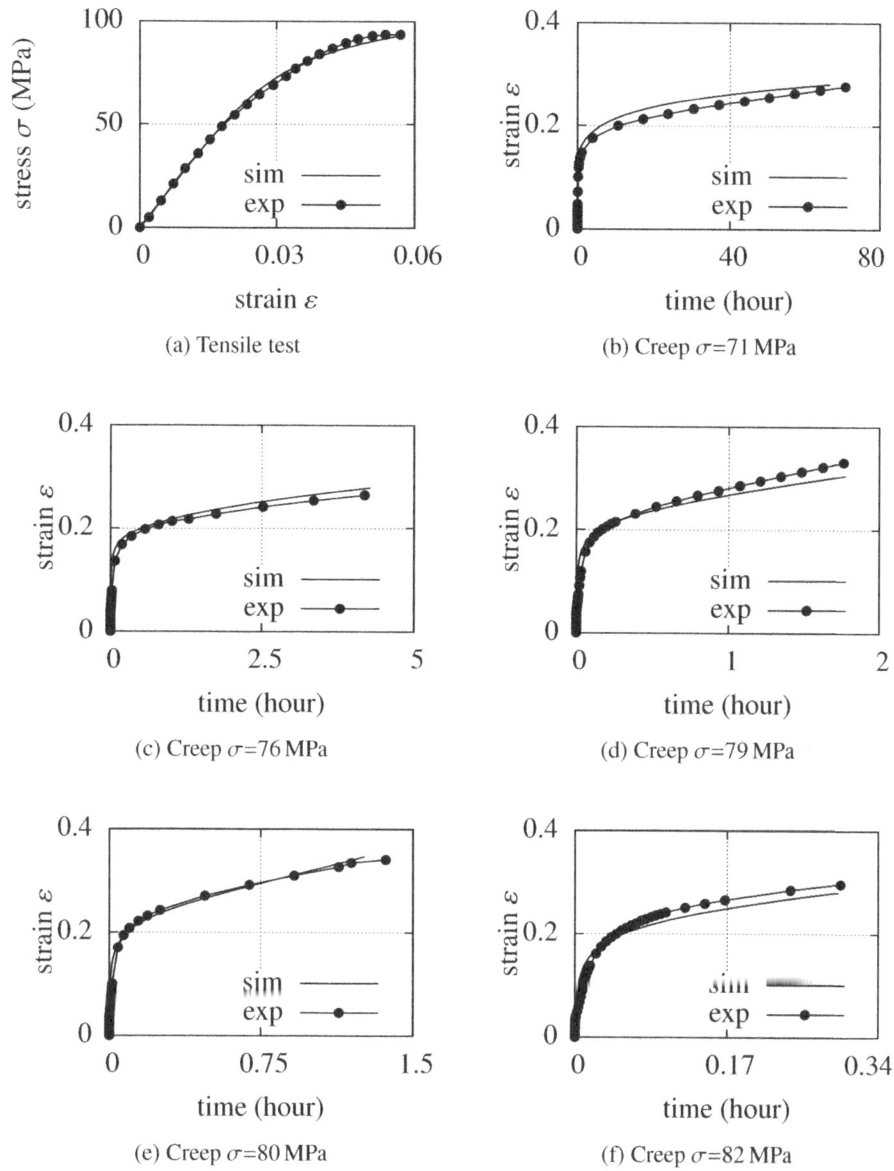

Figure 6.9. *Comparison of the experimental response and simulation of the tensile and creep tests*

The specimens with a sharper notch radius (ρ=0.45 mm, 0.8 mm) are used to check the accuracy of the models for high stress triaxiality ratios where the ductility loss is significant. The larger notch depths (ρ=1.6 mm and 4 mm) are used to check how the models can account for low stress triaxiality ratios.

For all the geometries, axisymmetric elements with reduced integration and an updated Lagrangian formulation are used. Only one-half of the geometry is meshed as shown in Figure 6.10.

A prescribed pressure is imposed on the top node set. Four creep levels (70 MPa, 74 MPa, 78 MPa and 80 MPa) were considered.

The material parameters are listed in Table 6.7. It was also noted that the general trends are captured by the models and the values of δ, the creep notch opening displacement (NOD) measured on the laser reflector, are also well reproduced. In terms of δ rate, the models underestimate the experimental values for the intermediate notch root radii (ρ=1.6 mm and 0.8 mm), but deliver a good prediction for ρ=4 mm and 0.45 mm except for the load level σ=78 MPa. The triggering of the tertiary creep corresponding to damage initiation responsible for an abrupt increase in strain is well simulated by the models with two damage variables. In order to investigate the creep rate, the normalized value ε is defined as

$$\varepsilon = \frac{\delta - \delta_{PL}}{\delta_0} \qquad [6.6]$$

where δ_{PL} is the NOD corresponding to the end of the preloading stage and δ_0 is the initial value. Figure 6.11 shows the creep rate for the different notch radii and the load level for the models using a variable damage per phase.

The summary of all these simulations demonstrates that:

– the onset of tertiary creep is underestimated by the CDM model with a unique damage variable, and overestimated by the GTN model with a unique damage variable;

– both CDM theory and GTN approach provide good estimations of the behavior under low triaxiality (ρ=1.6 mm and 4 mm) and high triaxiality (ρ=0.8 mm and 0.45 mm).

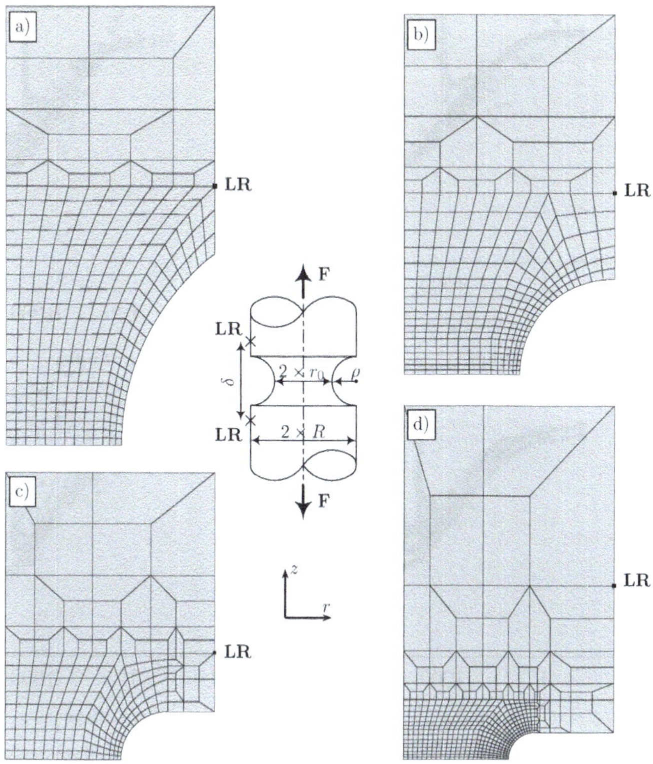

Figure 6.10. *Meshes of notched round bars with R=3.6 mm and r_0=1.8 mm for all specimens: (a) Notched round bars with ρ=4 mm, (b) ρ=1.6 mm, (c) ρ=0.8 mm, (d) ρ=0.45 mm, LR position of laser reflector*

GTN model, two damage variables
$q_1^a = 1.783$ $q_2^a = 1.34$ $\delta_a = 0$ $q_1^c = 2.68$ $q_2^c = 1.4$ $\delta_c = 0$
GTN model, unique damage variable
$q_1 = 1.77$ $q_2 = 1.25$ $\delta = 0$
CDM theory, two damage variables
$S_a = 6.9$ $s_a = 2.7$ $A_a = 0.025$ $S_c = 4.5$ $s_c = 3.0$ $A_c = 0.025$
CDM theory, unique damage variable
$S = 10.$ $s = 3.$ $A = 0.065$

Table 6.7. *Identified parameters for the damage behavior of the PA6 (MPa, s)*

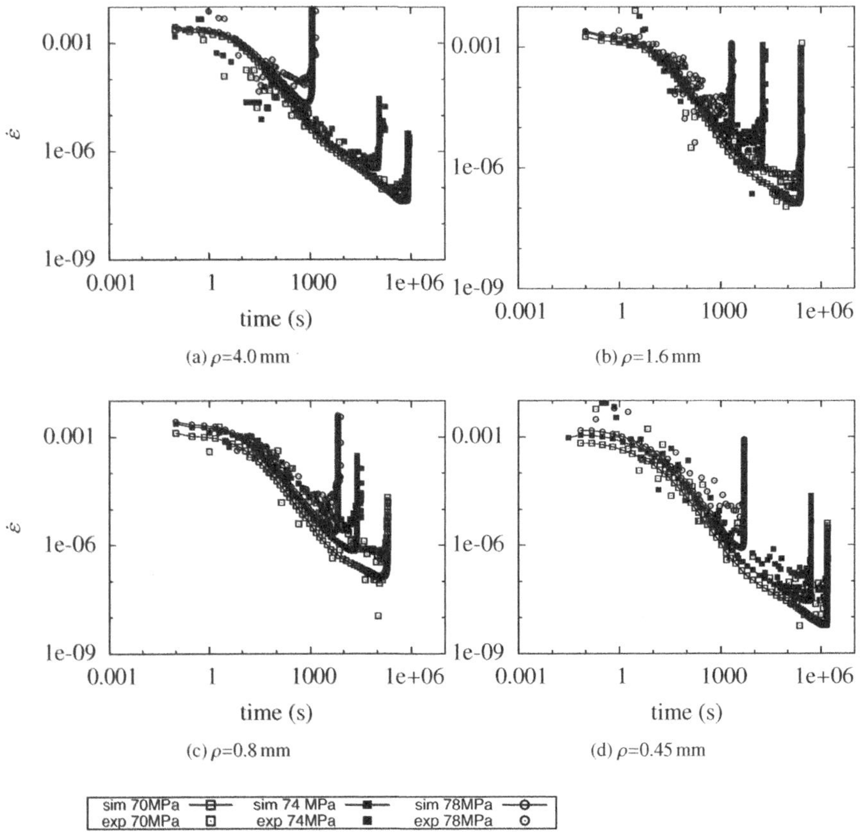

Figure 6.11. *Simulated (with models involving two damage variables) and experimental creep rates for the notched specimens*

The proposed models are valid for a large time interval ranging from 0.4 hour for the specimen with a notch radius ρ=4 mm subjected to a creep stress of 78 MPa, to 450 hours for the specimen with ρ=0.45 mm subjected to 70 MPa. In terms of creep rates, the GTN model and the CDM approach with two damage variables are very close to each other. Figure 6.11 shows the evolution of the creep rates for these two models together with the experimental results for the four notch radii (ρ=0.45 mm, 0.8 mm, 1.6 mm and 4 mm) and the different applied creep stresses (σ=70, 74, 80 MPa). A very good agreement between the experimental and simulated creep rates is obtained whatever the notch radius and whatever the applied creep stress in the primary, secondary and tertiary creep stages. Both GTN and CDM approaches are then reliable for the creep life prediction of materials like PA6.

Models with a unique damage variable are not considered anymore since they fail to reproduce the overall creep behavior of the notched specimens. The retained models introduce two damage variables. They are now scrutinized in terms of local variables (triaxiality ratio τ, damage variables at the phase level and overall damage). A qualitative comparison between experiments and the prediction of the GTN model is shown in Figure 6.12, which shows the contour plot of the porosity predicted at the beginning of the tertiary creep stage (108,000 s for ρ=4 mm, 504,000 s for ρ=0.45 mm) together with a cut of the specimen. As previously shown in [SAÏ 11b], white color characterizes the development of a substantial amount of damage.

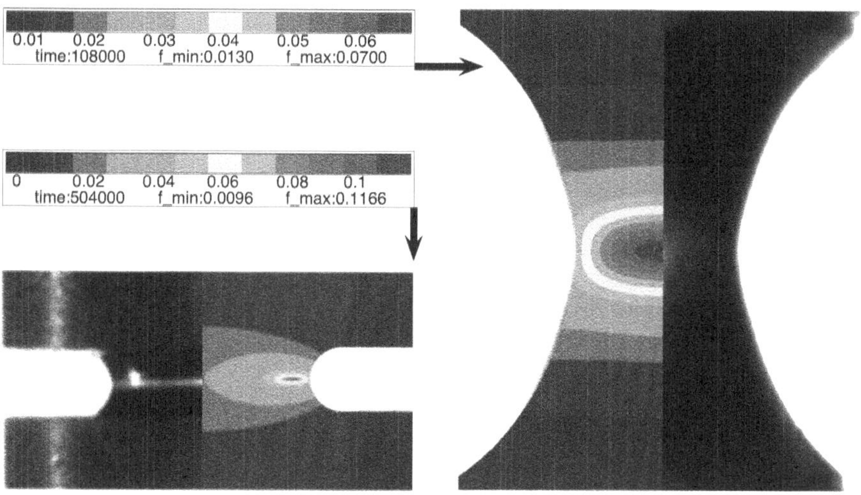

Figure 6.12. *Contour plot of the porosity predicted by the GTN model at the beginning of the tertiary creep stage (108,000 s for ρ=4 mm, 504,000 s for ρ=0.45 mm) [JER 15]. For a color version of the figure, see www.iste.co.uk/cailletaud/multi-mechanism.zip*

In the following, comparisons are performed for a creep stress of 74 MPa and two notch radii ρ=4.0 mm and ρ=0.45 mm corresponding to the lowest and the highest triaxiality ratio.

It was also shown that the trace of the strain tensors (amorphous, crystalline and macroscopic) predicted using the GTN approach is slightly higher than those simulated by the CDM theory. It can be noted in particular that value for the crystalline phase (Tr $(\underset{\sim}{\varepsilon}^c)$) is negligible at the beginning of the deformation stage, but increases rapidly and becomes predominant in comparison with Tr $(\underset{\sim}{\varepsilon}^a)$ and Tr $(\underset{\sim}{\varepsilon})$ for the GTN model in the case $\rho = 0.45$. Tr $(\underset{\sim}{\varepsilon}^c)$ also increases rapidly during the deformation process but is attenuated at the end of the deformation stage for the CDM theory.

The main concluding remarks can be summarized as follows:

– the volume change is directly driven by the hydrostatic stress;

– the simulations performed with the GTN model are in good agreement with the experimental results. The local damage in each phase is also correctly estimated thanks to the GTN approach deduced from the micromechanical analysis;

– the model based on the CDM theory is in good agreement with experiments for both large (low triaxiality) and small (high triaxiality) notch radii. Despite the fact that this model correctly reproduces the overall mechanical damage behavior, the analysis of the local variables such as damage at the phase level reveals some shortcomings of the model that need to be resolved.

6.6. Finite Element codes

6.6.1. *ZeBuLoN: explicit integration*

```
#include <Elasticity.h>
#include <Basic_nl_behavior.h>

@Class
dmunc_Explicit : BASIC_NL_BEHAVIOR, BASIC_SIMULATOR {
   @Name      dmunc_Explicit;
   @SubClass ELASTICITY elasticity;
   @tVarInt  eel, alpha1, alpha2;
   @sVarInt  evcum;
   @Coefs    A1, A2;
   @Coefs    D1, D2, C11, C22, C12;
   @Coefs    K, n, R0, Q, b;
   @tVarAux  epi,X1,X2;
};
```

```
@Derivative {
  TENSOR2 dev=0.;
  epi              = eto - eel;
  sig              = *elasticity*eel;
  X1               = (C11*alpha1 + C12*alpha2)/1.5;
  X2               = (C12*alpha1 + C22*alpha2)/1.5;
  TENSOR2 smx1d = deviator(sig - X1);
  TENSOR2 smx2d = deviator(sig - X2);
  double J1        = sqrt(1.5*(smx1d|smx1d));
  double J2        = sqrt(1.5*(smx2d|smx2d));
  double J         = pow(pow(J1,N) + pow(J2,N),(1./N));
  double R         = R0+Q*(1.-exp(-1.*b*evcum));
  double crit      = J-R;
  if(crit<=0.){devcum=0.;dalpha1=dalpha2;
    resolve_flux_grad(*elasticity, deel, deto);
}
    else{
        devcum           = pow((J-R)/K,n);
        TENSOR2 norm1 = (1.5*smx1d/J1)*pow(J1/J,N-1.);
        TENSOR2 norm2 = (1.5*smx2d/J2)*pow(J2/J,N-1.);
        dev        = devcum*(A1*norm1+A2*norm2); dalpha1
        = (norm1 - 1.5*(D1/C11)*X1)*devcum; dalpha2
        = (norm2 - 1.5*(D2/C22)*X2)*devcum;
        resolve_flux_grad(*elasticity, deel, deto, dev);
        }
}
```

6.6.2. *ABAQUS: explicit integration*

```
      SUBROUTINE UMAT(STRESS,STATEV,DDSDDE,SSE,SPD,SCD,
     + RPL,DDSDDT,DRPLDE,DRPLDT,
     + STRAN,DSTRAN,TIME,DTIME,TEMP,DTEMP,PREDEF,DPRED,CMNAME,
     + NDI,NSHR,NTENS,NSTATV,PROPS,NPROPS,COORDS,DROT,PNEWDT,
     + CELENT,DFGRD0,DFGRD1,NOEL,NPT,LAYER,KSPT,KSTEP,KINC)
C
      INCLUDE 'ABA_PARAM.INC'
C
      CHARACTER*80 CMNAME
C
     DIMENSION STRESS(NTENS),STATEV(NSTATV),
    + DDSDDE(NTENS,NTENS),DDSDDT(NTENS),DRPLDE(NTENS),
    + STRAN(NTENS),DSTRAN(NTENS),TIME(2),PREDEF(1),DPRED(1),
    + PROPS(NPROPS),COORDS(3),DROT(3,3),DFGRD0(3,3),DFGRD1(3,3)
```

```
C
   PARAMETER (M=3,ZERO=0.D0,ONE=1.D0,TWO=2.D0,THREE=3.D0)
C
      DIMENSION VN1(NTENS),VN2(NTENS),TN1(M,M),TN2(M,M),
     + VX1(NTENS),VX2(NTENS),TX1(M,M),TX2(M,M),
     + ALPHA1(NTENS),ALPHA2(NTENS),TA1(M,M),TA2(M,M),
     + DALPHA1(NTENS),DALPHA2(NTENS),
     + DPSTRAN(NTENS),DESTRAN(NTENS),DSTRESS(NTENS),
     + STR(M,M),DSTR(M,M),SMX1(M,M),SMX2(M,M),
     + DDS(NTENS,NTENS)
C
C MATERIAL PARAMETERS
C
      E=PROPS(1)
      XNU=PROPS(2)
      A1=PROPS(3)
      A2=PROPS(4)
      R0=PROPS(5)
      Q=PROPS(6)
      b=PROPS(7)
      xn=PROPS(8)
      xk=PROPS(9)
      C11=PROPS(10)
      C22=PROPS(11)
      C12=PROPS(12)
      D1=PROPS(13)
      D2=PROPS(14)
C
C RESTORE STATE VARIABLES
C
      p = STATEV(1)
      DO i =1, NTENS
         ALPHA1(i) = STATEV(1+i)
         ALPHA2(i) = STATEV(7+i)
      END DO
C
C  ELASTIC STIFNESS MATRIX
C
      EBULK3 = E/(ONE-TWO*XNU)
      EG2 = E/(ONE+XNU)
      EG = EG2/TWO
      ELAM = (EBULK3-EG2)/THREE
      DO I=1,NTENS
```

```
            DO J=1,NTENS
               DDS(I,J)=ZERO
             END DO
         END DO
         DO I = 1, M
            DO J = 1, M
               DDS(I,J) = ELAM
            END DO
            DDS(I,I) = EG2 + ELAM
         END DO
         DDS(4,4) = EG
         DDS(5,5) = EG
         DDS(6,6) = EG
C
C     WRITE STRESSES AND KINEMATIC VARIABLE IN MATRIX
                    FORM
C
         CALL COPY_VECTOR_TO_MATRIX(STRESS,STR)
         CALL COPY_VECT_TO_MATRIX_EPS(ALPHA1,TA1)
         CALL COPY_VECT_TO_MATRIX_EPS(ALPHA2,TA2)
C
C DEVIATORIC STRESS TENSOR
C
         CALL KDEVIA(STR,DSTR)
C
C COMPUTE THE TWO BACK STRESSES
C
         DO I=1,M
           DO J=1,M
            TX1(I,J)=(TWO/THREE)*(C11*TA1(I,J)+C12*TA2(I,J))
            TX2(I,J)=(TWO/THREE)*(C12*TA1(I,J)+C22*TA2(I,J))
           END DO
         END DO
C
C COMPUTE THE TWO EFFECTIVE STRESSES
C
         DO I=1,M
            DO J=1,M
               SMX1(I,J) = A1*DSTR(I,J)-TX1(I,J)
               SMX2(I,J) = A2*DSTR(I,J)-TX2(I,J)
            END DO
         END DO
```

```
C
C EQUIVALENT STRESSES
C
      CALL J_EQUIV(SMX1,PJ1)
      CALL J_EQUIV(SMX2,PJ2)
C
C YIELD CRITERION
C
      PJ=sqrt(PJ1*PJ1+PJ2*PJ2)
      ZY = PJ - Q*(ONE-exp(-B*p))-R0
C
      IF (ZY.LE.ZERO) THEN
C ELASTIC STEP
         DLAMBDA = ZERO
         DO K = 1,NTENS
           DPSTRAN(K)=ZERO
           DALPHA1(K)=ZERO
           DALPHA2(K)=ZERO
           DESTRAN(K)=DSTRAN(K)
         END DO
      ELSE
C PLASTIC STEP
         DO I=1,M
           DO J=1,M
              TN1(I,J) = (THREE/TWO)*SMX1(I,J)/PJ
              TN2(I,J) = (THREE/TWO)*SMX2(I,J)/PJ
           END DO
          END DO
          CALL COPY_MATRIX_TO_VECT_EPS(TN1,VN1)
          CALL COPY_MATRIX_TO_VECT_EPS(TN2,VN2)
          CALL COPY_MATRIX_TO_VECT_EPS(TX1,VX1)
          CALL COPY_MATRIX_TO_VECT_EPS(TX2,VX2)
          DLAMBDA = exp(xn*log(ZY/xk))*DTIME
          DO K = 1,NTENS
            DPSTRAN(K)=DLAMBDA*(A1*VN1(K)+A2*VN2(K))
            DALPHA1(K)=DLAMBDA*(VN1(K)-1.5*D1*VX1(K)/C11)
            DALPHA2(K)=DLAMBDA*(VN2(K)-1.5*D2*VX2(K)/C22)
            DESTRAN(K)=DSTRAN(K)-DPSTRAN(K)
          END DO
      END IF
C
C STRESS INCREMENT
```

```
C
      CALL MULT6_6(DDS,DESTRAN,DSTRESS)
C
C UPDATE STRESS AND INTERNAL VARIABLES
C
      DO K = 1,NTENS
         STRESS(K) = STRESS(K) + DSTRESS(K)
      END DO
       p = p + DLAMBDA
       DO K = 1,NTENS
          ALPHA1(K)  = ALPHA1(K) + DALPHA1(K)
          ALPHA2(K)  = ALPHA2(K) + DALPHA2(K)
       END DO
C
C UPDATE STATE VARIABLES
C
       STATEV(1) = p
       DO i =1, NTENS
          STATEV(1+i) = ALPHA1(i)
          STATEV(7+i) = ALPHA2(i)
       END DO
C
C TANGENT OPERATOR
C
      DO I=1,NTENS
         DO J=1,NTENS
           DDSDDE(I,J) = DDS(I,J)
          END DO
      END DO
C
      RETURN
      END
****************************************************
**         MULTIPLY 6X6 MATRIX WITH 6X1 VECTOR     *
****************************************************
      SUBROUTINE MULT6_6(DM1,DM2,DM)
      INCLUDE 'ABA_PARAM.INC'
      DIMENSION DM1(6,6),DM2(6),DM(6)
      DO I=1,6
         DM(I)=0.0
         DO K=1,6
            DM(I)=DM(I)+DM1(I,K)*DM2(K)
         END DO
```

```
      END DO
      RETURN
      END
****************************************
**            EFFECTIVE STRESS          *
**   (CONTRACTED MATRIX CALCULATION)    *
****************************************
      SUBROUTINE J_EQUIV(EFF1,VAL1)
      INCLUDE 'ABA_PARAM.INC'
      PARAMETER (M=3)
      DIMENSION EFF1(M,M)
      X=0.0
      DO I=1,M
        DO J=1,M
          X=X+EFF1(I,J)*EFF1(I,J)
        END DO
      END DO
      VAL1=SQRT(3.0*X/2.0)
      RETURN
      END
********************************************
**          DOT PRODUCT OF TWO VECTORS      *
********************************************
      SUBROUTINE DOTPROD(DM1,DM2,DM)
      INCLUDE 'ABA_PARAM.INC'
      DIMENSION DM1(6),DM2(6)
      DM=0.0
      DO K=1,6
        DM=DM+DM1(K)*DM2(K)
      END DO
      RETURN
      END
****************************************
**   DEVIATORIC STRESS DETERMINATION   **
****************************************
      SUBROUTINE KDEVIA(STRSS,DEVITO)
      INCLUDE 'ABA_PARAM.INC'
      PARAMETER (M=3)
      DIMENSION STRSS(M,M),DEVITO(M,M)
      TRACE=0.0
      DO I=1,M
        TRACE=TRACE+STRSS(I,I)
      END DO
```

```
      DO I=1,M
        DO J=1,M
          IF(I .EQ. J) THEN
           DEVITO(I,J)=STRSS(I,J)-TRACE/3.
          ELSE
           DEVITO(I,J)=STRSS(I,J)
          END IF
        END DO
      END DO
      RETURN
      END
*********************************************
** PUT VECTOR IN A MATRIX FORM             **
** STRESS LIKE VARIABLE                    **
*********************************************
      SUBROUTINE COPY_VECTOR_TO_MATRIX(VECT,TENS)
      INCLUDE 'ABA_PARAM.INC'
      DIMENSION VECT(6),TENS(3,3)
      TENS(1,1) = VECT(1)
      TENS(2,2) = VECT(2)
      TENS(3,3) = VECT(3)
      TENS(1,2) = VECT(4)
      TENS(2,1) = VECT(4)
      TENS(1,3) = VECT(5)
      TENS(3,1) = VECT(5)
      TENS(2,3) = VECT(6)
      TENS(3,2) = VECT(6)
      END
*********************************************
** PUT VECTOR IN A MATRIX FORM             **
** STRAIN LIKE VARIABLE IN ABAQUS NOTATION **
*********************************************
      SUBROUTINE COPY_VECT_TO_MATRIX_EPS(VECT,TENS)
      INCLUDE 'ABA_PARAM.INC'
      DIMENSION VECT(6),TENS(3,3)
      TWO=2.0
      TENS(1,1) = VECT(1)
      TENS(2,2) = VECT(2)
      TENS(3,3) = VECT(3)
      TENS(1,2) = VECT(4)/TWO
      TENS(2,1) = VECT(4)/TWO
      TENS(1,3) = VECT(5)/TWO
      TENS(3,1) = VECT(5)/TWO
```

```
      TENS(2,3) = VECT(6)/TWO
      TENS(3,2) = VECT(6)/TWO
      END
******************************************
** PUT MATRIX IN A VECTOR FORM          **
** STRAIN LIKE VARIABLE IN ABAQUS NOTATION **
******************************************
      SUBROUTINE COPY_MATRIX_TO_VECT_EPS(TENS,VECT)
      INCLUDE 'ABA_PARAM.INC'
      DIMENSION TENS(3,3),VECT(6)
      TWO=2.0
      VECT(1)=TENS(1,1)
      VECT(2)=TENS(2,2)
      VECT(3)=TENS(3,3)
      VECT(4)=TWO*TENS(1,2)
      VECT(5)=TWO*TENS(1,3)
      VECT(6)=TWO*TENS(2,3)
      END
```

6.6.3. *ANSYS: explicit integration*

```
      subroutine usermat(
     & matId, elemId,kDomIntPt, kLayer, kSectPt,
     & ldstep,isubst,keycut,
     & nDirect,nShear,ncomp,nStatev,nProp,
     & Time,dTime,Temp,dTemp,
     & stress,ustatev,dsdePl,sedEl,sedPl,epseq,
     & Strain,dStrain, epsPl, prop, coords,
     & var0, defGrad_t, defGrad,
     & tsstif, epsZZ,
     & var1, var2, var3, var4, var5,
     & var6, var7, var8)
C
#include "impcom.inc"
C
      INTEGER
     & matId, elemId,
     & kDomIntPt, kLayer, kSectPt,
     & ldstep,isubst,keycut,
     & nDirect,nShear,ncomp,nStatev,nProp
      DOUBLE PRECISION
     & Time, dTime, Temp, dTemp,
     & sedEl, sedPl, epseq, epsZZ
```

```
       DOUBLE PRECISION
      & stress (ncomp ), ustatev (nStatev),
      & dsdePl (ncomp,ncomp),
      & Strain (ncomp ), dStrain (ncomp ),
      & epsPl (ncomp ), prop (nProp ),
      & coords (3),
      & defGrad (3,3), defGrad_t(3,3),
      & tsstif (2)
c --- User defined part
      INTEGER I,J,K,M
      DOUBLE PRECISION ZERO,ONE,TWO,THREE
      PARAMETER (M=3,ZERO=0.D0,ONE=1.D0,TWO=2.D0,THREE=3.D0)
C
       DOUBLE PRECISION VN1(ncomp),VN2(ncomp),TN1(M,M),TN2(M,M),
      + VX1(ncomp),VX2(ncomp),TX1(M,M),TX2(M,M),
      + ALPHA1(ncomp),ALPHA2(ncomp),TA1(M,M),TA2(M,M),
      + DALPHA1(ncomp),DALPHA2(ncomp),
      + DPSTRAN(ncomp),DESTRAN(ncomp),DSTRESS(ncomp),
      + STR(M,M),DSTR(M,M),SMX1(M,M),SMX2(M,M),
      + DDS(ncomp,ncomp)
c --- Temporary variables for solution purpose
       DOUBLE PRECISION var0, var1, var2, var3, var4,var5,
      & var6, var7, var8
c ------------------------------
       DOUBLE PRECISION E,XNU,A1,A2,R0,Q,b,xn,xk,
      &                  C11,C22,C12,D1,D2
       DOUBLE PRECISION  EG,EG2,ELAM,EBULK3
       DOUBLE PRECISION p,PJ1,PJ2,PJ,ZY,DLAMBDA
       EXTERNAL usermat3d, usermatps, usermatbm, usermat1d
C
C MATERIAL PARAMETERS
C
       E=prop(1)
       XNU=prop(2)
       A1=prop(3)
       A2=prop(4)
       R0=prop(5)
       Q=prop(6)
       b=prop(7)
       xn=prop(8)
       xk=prop(9)
       C11=prop(10)
       C22=prop(11)
```

```
      C12=prop(12)
      D1=prop(13)
      D2=prop(14)
C
C     RECOVER STATE VARIABLES
C
      p = ustatev(1)
      DO i =1, ncomp
         ALPHA1(i) = ustatev(1+i)
         ALPHA2(i) = ustatev(7+i)
      END DO
C
C   ELASTIC STIFNESS MATRIX
C
      EBULK3 = E/(ONE-TWO*XNU)
      EG2 = E/(ONE+XNU)
      EG = EG2/TWO
      ELAM = (EBULK3-EG2)/THREE
      DO I=1,ncomp
          DO J=1,ncomp
             DDS(I,J)=ZERO
           END DO
      END DO
      DO I = 1, M
          DO J = 1, M
             DDS(I,J) = ELAM
          END DO
          DDS(I,I) = EG2 + ELAM
      END DO
      DDS(4,4) = EG
      DDS(5,5) = EG
      DDS(6,6) = EG
C
C  WRITE STRESSES AND KINEMATIC VARIABLE IN MATRIX FORM
C
      CALL COPY_VECTOR_TO_MATRIX(STRESS,STR)
      CALL COPY_VECT_TO_MATRIX_EPS(ALPHA1,TA1)
      CALL COPY_VECT_TO_MATRIX_EPS(ALPHA2,TA2)
C
C DEVIATORIC STRESS TENSOR
C
      CALL KDEVIA(STR,DSTR)
```

```
C
C COMPUTE THE TWO BACK STRESSES
C
      DO I=1,M
        DO J=1,M
         TX1(I,J)=(TWO/THREE)*(C11*TA1(I,J)+C12*TA2(I,J))
         TX2(I,J)=(TWO/THREE)*(C12*TA1(I,J)+C22*TA2(I,J))
        END DO
      END DO
C
C COMPUTE THE TWO EFFECTIVE STRESSES
C
      DO I=1,M
         DO J=1,M
           SMX1(I,J) = A1*DSTR(I,J)-TX1(I,J)
           SMX2(I,J) = A2*DSTR(I,J)-TX2(I,J)
         END DO
      END DO
C
C EQUIVALENT STRESSES
C
      CALL J_EQUIV(SMX1,PJ1)
      CALL J_EQUIV(SMX2,PJ2)
C
      PJ=sqrt(PJ1*PJ1+PJ2*PJ2)
      ZY = PJ - Q*(ONE-exp(-B*p))-R0
C
C
      IF (ZY.LE.ZERO) THEN
C ELASTIC STEP
         DLAMBDA = ZERO
         DO K = 1,ncomp
           DPSTRAN(K)=ZERO
           DALPHA1(K)=ZERO
           DALPHA2(K)=ZERO
           DESTRAN(K)=dStrain(K)
         END DO
      ELSE
C PLASTIC STEP
              WRITE(6,*) 'ZY PJ ',ZY,PJ
         DO I=1,M
           DO J=1,M
              TN1(I,J) = (THREE/TWO)*SMX1(I,J)/PJ
```

```
            TN2(I,J) = (THREE/TWO)*SMX2(I,J)/PJ
          END DO
         END DO
         CALL COPY_MATRIX_TO_VECT_EPS(TN1,VN1)
         CALL COPY_MATRIX_TO_VECT_EPS(TN2,VN2)
         CALL COPY_MATRIX_TO_VECT_EPS(TX1,VX1)
         CALL COPY_MATRIX_TO_VECT_EPS(TX2,VX2)
         DLAMBDA = exp(xn*log(ZY/xk))*dTime
         DO K = 1,ncomp
           DPSTRAN(K)=DLAMBDA*(A1*VN1(K)+A2*VN2(K))
           DALPHA1(K)=DLAMBDA*(VN1(K)-1.5*D1*VX1(K)/C11)
           DALPHA2(K)=DLAMBDA*(VN2(K)-1.5*D2*VX2(K)/C22)
           DESTRAN(K)=dStrain(K)-DPSTRAN(K)
         END DO
      END IF
C
C
      CALL MULT6_6(DDS,DESTRAN,DSTRESS)
C
C UPDATE STRESS AND INTERNAL VARIABLES
C
      DO K = 1,ncomp
         STRESS(K) = STRESS(K) + DSTRESS(K)
      END DO
       p = p + DLAMBDA
       DO K = 1,ncomp
          ALPHA1(K)  = ALPHA1(K) + DALPHA1(K)
          ALPHA2(K)  = ALPHA2(K) + DALPHA2(K)
       END DO
C
C UPDATE STATE VARIABLES
C
       ustatev(1) = p
       DO i =1, ncomp
          ustatev(1+i) = ALPHA1(i)
          ustatev(7+i) = ALPHA2(i)
       END DO
C
C     TANGENT OPERATOR
C
      DO I=1,ncomp
       DO J=1,ncomp
            dsdePl(I,J) = DDS(I,J)
```

```
        END DO
      END DO
C
      return
C
      end
*************************************************
**         MULTIPLY 6X6 MATRIX WITH 6X1 VECTOR      *
*************************************************
      SUBROUTINE MULT6_6(DM1,DM2,DM)
      INTEGER I,K
      DOUBLE PRECISION DM1(6,6),DM2(6),DM(6)
      DO I=1,6
         DM(I)=0.0d0
         DO K=1,6
            DM(I)=DM(I)+DM1(I,K)*DM2(K)
         END DO
      END DO
      RETURN
      END
*************************************
**           EFFECTIVE STRESS           *
**   (CONTRACTED MATRIX CALCULATION)    *
*************************************
      SUBROUTINE J_EQUIV(EFF1,VAL1)
      INTEGER I,J
      DOUBLE PRECISION EFF1(3,3)
      DOUBLE PRECISION VAL1,TWO,THREE
      TWO=2.0d0
      THREE=3.0d0
      VAL1=0.0d0
      DO I=1,3
        DO J=1,3
           VAL1=VAL1+EFF1(I,J)*EFF1(I,J)
        END DO
      END DO
      VAL1=SQRT(THREE*VAL1/TWO)
      RETURN
      END
*****************************************
**         DOT PRODUCT OF TWO VECTORS        *
*****************************************
      SUBROUTINE DOTPROD(DM1,DM2,DM)
```

```
      INTEGER K
      DOUBLE PRECISION DM1(6),DM2(6)
      DOUBLE PRECISION DM
      DM=0.0d0
      DO K=1,6
         DM=DM+DM1(K)*DM2(K)
      END DO
      RETURN
      END
****************************************
**   DEVIATORIC STRESS CALCULATION    **
****************************************
      SUBROUTINE KDEVIA(STRSS,DEVITO)
      INTEGER I,J
      DOUBLE PRECISION STRSS(3,3),DEVITO(3,3)
      DOUBLE PRECISION TRACE,THREE
      THREE=3.0d0
      TRACE=0.0d0
      DO I=1,3
         TRACE=TRACE+STRSS(I,I)
      END DO
      DO I=1,3
        DO J=1,3
          IF(I .EQ. J) THEN
           DEVITO(I,J)=STRSS(I,J)-TRACE/THREE
          ELSE
           DEVITO(I,J)=STRSS(I,J)
          END IF
        END DO
      END DO
      RETURN
      END
*********************************************
** PUT VECTOR IN A MATRIX FORM             **
** STRESS LIKE VARIABLE                    **
*********************************************
      SUBROUTINE COPY_VECTOR_TO_MATRIX(VECT,TENS)
      DOUBLE PRECISION TENS(3,3),VECT(6)
      TENS(1,1) = VECT(1)
      TENS(2,2) = VECT(2)
      TENS(3,3) = VECT(3)
      TENS(1,2) = VECT(4)
      TENS(2,1) = VECT(4)
```

```
      TENS(1,3) = VECT(5)
      TENS(3,1) = VECT(5)
      TENS(2,3) = VECT(6)
      TENS(3,2) = VECT(6)
      END
C
*********************************************
** PUT VECTOR IN A MATRIX FORM             **
** STRAIN LIKE VARIABLE IN ABAQUS NOTATION **
*********************************************
      SUBROUTINE COPY_VECT_TO_MATRIX_EPS(VECT,TENS)
      DOUBLE PRECISION TENS(3,3),VECT(6)
      DOUBLE PRECISION TWO
      TWO=2.0d0
C
      TENS(1,1) = VECT(1)
      TENS(2,2) = VECT(2)
      TENS(3,3) = VECT(3)
      TENS(1,2) = VECT(4)/TWO
      TENS(2,1) = VECT(4)/TWO
      TENS(1,3) = VECT(5)/TWO
      TENS(3,1) = VECT(5)/TWO
      TENS(2,3) = VECT(6)/TWO
      TENS(3,2) = VECT(6)/TWO
      END
*********************************************
** PUT MATRIX IN A VECTOR FORM             **
** STRAIN LIKE VARIABLE IN ABAQUS NOTATION **
*********************************************
      SUBROUTINE COPY_MATRIX_TO_VECT_EPS(TENS,VECT)
      DOUBLE PRECISION TENS(3,3),VECT(6)
      DOUBLE PRECISION TWO
      TWO=2.0d0
C
      VECT(1)=TENS(1,1)
      VECT(2)=TENS(2,2)
      VECT(3)=TENS(3,3)
      VECT(4)=TWO*TENS(1,2)
      VECT(5)=TWO*TENS(1,3)
      VECT(6)=TWO*TENS(2,3)
      END
C
      subroutine usermat3d
```

```
      end
C
      subroutine usermat1d
      end
C
      subroutine usermatbm
      end
C
      subroutine usermatps
      end
```

6.6.4. *ZeBuLoN: implicit integration*

```
#include <Elasticity.h>
#include <Basic_nl_behavior.h>

@Class
dmunc_Implicit : BASIC_NL_BEHAVIOR, BASIC_SIMULATOR {
   @Name     dmunc_Implicit;
   @SubClass ELASTICITY elasticity;
   @tVarInt  eel, alpha1, alpha2;
   @sVarInt  evcum;
   @Coefs    A1, A2;
   @Coefs    D1, D2, C11, C22, C12;
   @Coefs    K, n, R0, Q, b;
   @tVarAux  epi,X1,X2;
   @Implicit
};

@CalcGradF {
     double  dv_df;
     TENSOR2 df_fs;
     SMATRIX dn1_ds1;
     SMATRIX dn1_ds2;
     SMATRIX dn2_ds1;
     SMATRIX dn2_ds2;

     ELASTICITY& E=*elasticity;
     sig             = E*eel;
//
     double tdv      = theta*devcum;
//
     X1              = (C11*alpha1 + C12*alpha2)/1.5;
```

```
      X2                = (C12*alpha1 + C22*alpha2)/1.5;
      TENSOR2 smx1d = deviator(sig - X1);
      TENSOR2 smx2d = deviator(sig - X2);
      double J1         = sqrt(1.5*(smx1d|smx1d));
      double J2         = sqrt(1.5*(smx2d|smx2d));
      double J          = pow(pow(J1,2.) + pow(J2,2.),
                            (1./2.));
      double R          = R0+Q*(1.-exp(-1.*b*evcum));
      double crit       = J-R;
//
       f_vec_eel           -= deto;
//
       if(crit>0.){
         TENSOR2 norm1       = 1.5*smx1d/J;
         TENSOR2 norm2       = 1.5*smx2d/J;
         dn1_ds1             = unit32;
         dn1_ds1            -= norm1^norm1;
         dn1_ds1            *= tdv/J;
         dn1_ds2             = norm1^norm2;
         dn1_ds2            *= tdv/J;
         dn2_ds2             = unit32;
         dn2_ds2            -= norm2^norm2;
         dn2_ds2            *= tdv/J;
         dn2_ds1             = norm2^norm1;
         dn2_ds1            *= tdv/J;
         dv_df               = tdt*n*pow(crit/K,n-1.)/K;
         df_fs               = dv_df*(A1*norm1+A2*norm2);
//
         f_vec_eel          += (A1*norm1+A2*norm2)*devcum;
         f_vec_evcum        -= dt*pow(crit/K,n);
         f_vec_alpha1       -= devcum*(norm1-1.5*D1*X1/C11);
         f_vec_alpha2       -= devcum*(norm2-1.5*D2*X2/C22);
//
         deel_deel          += (A1*(dn1_ds1+dn1_ds2)+A2*
                                     (dn2_ds1+dn2_ds2))*E;
         deel_devcum         = A1*norm1+A2*norm2;
         deel_dalpha1        = A1*(C11*dn1_ds1+C12*dn1_ds2);
         deel_dalpha1       += A2*(C11*dn2_ds1+C12*dn2_ds2);
         deel_dalpha1       *= (-2./3.);
         deel_dalpha2        = A1*(C12*dn1_ds1+C22*dn1_ds2);
         deel_dalpha2       += A2*(C12*dn2_ds1+C22*dn2_ds2);
         deel_dalpha2       *= (-2./3.);
//
```

```
          devcum_deel       -= df_fs*E;
          devcum_devcum     += dv_df*Q*b*exp(-1.*b*evcum);
          devcum_dalpha1     = (2./3.)*dv_df*(C11*norm1+C12*
                                   norm2);
          devcum_dalpha2     = (2./3.)*dv_df*(C12*norm1+C22*
                                   norm2);
//
         dalpha1_deel      -= (dn1_ds1+dn1_ds2)*E;
         dalpha1_devcum     =(3./2.)*D1*X1/C11-norm1;
         dalpha1_dalpha1   *= (1.+D1*tdv);
         dalpha1_dalpha1   +=(2./3.)*(C11*dn1_ds1+C12*dn1_ds2);
         dalpha1_dalpha2   *= D1*C12/C11*tdv;
         dalpha1_dalpha2   +=(2./3.)*(C12*dn1_ds1+C22*dn1_ds2);
//
         dalpha2_deel      -= (dn2_ds1+dn2_ds2)*E;
         dalpha2_devcum     =(3./2.)*D2*X2/C22-norm2;
         dalpha2_dalpha1   *= D2*C12/C22*tdv;
         dalpha2_dalpha1   +=(2./3.)*(C11*dn2_ds1+C12*dn2_ds2);
         dalpha2_dalpha2   *= (1.+D2*tdv);
         dalpha2_dalpha2   +=(2./3.)*(C12*dn2_ds1+C22*dn2_ds2);
//
                    }
}
```

6.6.5. *ABAQUS: implicit integration*

```
      SUBROUTINE UMAT(STRESS,STATEV,DDSDDE,SSE,SPD,SCD,
     1 RPL,DDSDDT,DRPLDE,DRPLDT,
     2 STRAN,DSTRAN,TIME,DTIME,TEMP,DTEMP,PREDEF,DPRED,CMNAME,
     3 NDI,NSHR,NTENS,NSTATV,PROPS,NPROPS,COORDS,DROT,PNEWDT,
     4 CELENT,DFGRD0,DFGRD1,NOEL,NPT,LAYER,KSPT,KSTEP,KINC)
C
      INCLUDE 'ABA_PARAM.INC'
C
      CHARACTER*80 CMNAME
C
      DIMENSION STRESS(NTENS),STATEV(NSTATV),
     1 DDSDDE(NTENS,NTENS),DDSDDT(NTENS),DRPLDE(NTENS),
     2 STRAN(NTENS),DSTRAN(NTENS),TIME(2),PREDEF(1),DPRED(1),
     3 PROPS(NPROPS),COORDS(3),DROT(3,3),DFGRD0(3,3),DFGRD1(3,
C
      PARAMETER
     + (M=3,ZERO=0.D0,ONE=1.D0,TWO=2.D0,THREE=3.D0,UD=1.5D0)
```

```
C
      DIMENSION VN1(NTENS),VN2(NTENS),TN1(M,M),TN2(M,M),
     + VX1(NTENS),VX2(NTENS),TX1(M,M),TX2(M,M),
     + ALPHA1(NTENS),ALPHA2(NTENS),TA1(M,M),TA2(M,M),
     + DSTRESS(NTENS),DESTRAN(NTENS),ESTRAN(NTENS),
     + DALPHA1(NTENS),DALPHA2(NTENS),
     + STR(M,M),DSTR(M,M),SMX1(M,M),SMX2(M,M),
     + DDS(NTENS,NTENS)
C-----------------------------------------------------------------
      DIMENSION UNIT32(6,6),XIDEN66(6,6),INDX(NSTATV),
     X XJAC(NSTATV,NSTATV),DELTA_X(NSTATV),F_VEC(NSTATV)
      DIMENSION XNV1_XNV1(NTENS,NTENS),XNV1_XNV2(NTENS,NTENS),
     X XNV2_XNV1(NTENS,NTENS),XNV2_XNV2(NTENS,NTENS),
     X XNV(NTENS),DDS_XNV(NTENS),PCN11(NTENS),PCN22(NTENS),
     X DN1DS1(NTENS,NTENS),DN1DS2(NTENS,NTENS),
     X DN2DS1(NTENS,NTENS),DN2DS2(NTENS,NTENS),
     X TEMPO_A_FE(NTENS,NTENS),TEMPO_A_FE_DDS(NTENS,NTENS),
     X TEMPO_AC_FE_1(NTENS,NTENS),TEMPO_AC_FE_2(NTENS,NTENS),
     X CN11(NTENS,NTENS),CN12(NTENS,NTENS),CN21(NTENS,NTENS),
     X CN22(NTENS,NTENS),
     X XN11_N12(NTENS,NTENS),XN21_N22(NTENS,NTENS),
     X XN11_N12_E(NTENS,NTENS),XN21_N22_E(NTENS,NTENS),
     X XJAC_FE(NTENS,NTENS)
     LOGICAL CONVERGENCE
      data
      XIDEN66 /1.0E0, 0.0E0, 0.0E0, 0.0E0, 0.0E0, 0.0E0,
     x         0.0E0, 1.0E0, 0.0E0, 0.0E0, 0.0E0, 0.0E0,
     x         0.0E0, 0.0E0, 1.0E0, 0.0E0, 0.0E0, 0.0E0,
     x         0.0E0, 0.0E0, 0.0E0, 1.0E0, 0.0E0, 0.0E0,
     x         0.0E0, 0.0E0, 0.0E0, 0.0E0, 1.0E0, 0.0E0,
     x         0.0E0, 0.0E0, 0.0E0, 0.0E0, 0.0E0, 1.0E0/
      data UNIT32/1.0E0, 0.5E0, 0.5E0,0.0 E0,0.0E0,0.0E0,
     x 0.5E0, 1.0E0, 0.5E0, 0.0E0,0.0E0, 0.0E0,
     x 0.5E0, 0.5E0, 1.0E0, 0.0E0,0.0E0, 0.0E0,
     x 0.0E0, 0.0E0, 0.0E0, 1.5E0, 0.0E0, 0.0E0,
     x 0.0E0, 0.0E0, 0.0E0, 0.0E0, 1.5E0, 0.0E0,
     x 0.0E0, 0.0E0, 0.0E0, 0.0E0, 0.0E0, 1.5E0 /
C
      TOLER=1.E-8
      MAXITER=100
C
      E=PROPS(1)
      XNU=PROPS(2)
```

```
      A1=PROPS(3)
      A2=PROPS(4)
      R0=PROPS(5)
      Q=PROPS(6)
      b=PROPS(7)
      xn=PROPS(8)
      xk=PROPS(9)
      C11=PROPS(10)
      C22=PROPS(11)
      C12=PROPS(12)
      D1=PROPS(13)
      D2=PROPS(14)
C
      DO I =1, NTENS
         ESTRAN(I) = STATEV(I)
         ALPHA1(I) = STATEV(NTENS+I)
         ALPHA2(I) = STATEV(2*NTENS+I)
      END DO
      p = STATEV(NSTATV)
C
C  ELASTIC STIFNESS
C
      EBULK3 = E/(ONE-TWO*XNU)
      EG2 = E/(ONE+XNU)
      EG = EG2/TWO
      ELAM = (EBULK3-EG2)/THREE
      DO I=1,NTENS
         DO J=1,NTENS
            DDS(I,J)=ZERO
         END DO
      END DO
      DO I = 1, M
         DO J = 1, M
            DDS(I,J) = ELAM
         END DO
         DDS(I,I) = EG2 + ELAM
      END DO
      DDS(4,4) = EG
      DDS(5,5) = EG
      DDS(6,6) = EG
C
C    OBTAIN TRIAL (ELASTIC) STRESS
C
```

```
      CALL KMLT1(DDS,DSTRAN,DSTRESS)
      DO I=1,NTENS
        STRESS(I)=STRESS(I)+DSTRESS(I)
      END DO
C
C WRITE STRESSES AND KINEMATIC VARIABLE IN MATRIX FORM
C
      CALL COPY_VECTOR_TO_MATRIX(STRESS,STR)
      CALL COPY_VECT_TO_MATRIX_EPS(ALPHA1,TA1)
      CALL COPY_VECT_TO_MATRIX_EPS(ALPHA2,TA2)
C
C DEVIATORIC STRESS TENSOR
C
      CALL KDEVIA(STR,DSTR)
C
C COMPUTE THE TWO BACK STRESSES
C
      DO I=1,M
        DO J=1,M
         TX1(I,J)=(TWO/THREE)*(C11*TA1(I,J)+C12*TA2(I,J))
         TX2(I,J)=(TWO/THREE)*(C12*TA1(I,J)+C22*TA2(I,J))
        END DO
      END DO
C
      DO I=1,M
         DO J=1,M
            SMX1(I,J) = A1*DSTR(I,J)-TX1(I,J)
            SMX2(I,J) = A2*DSTR(I,J)-TX2(I,J)
         END DO
      END DO

      CALL J_EQUIV(SMX1,PJ1)
      CALL J_EQUIV(SMX2,PJ2)
C
      PJ=sqrt(PJ1*PJ1+PJ2*PJ2)
C
      ZY = PJ - Q*(ONE-exp(-b*p))-R0
C
***********************************
**    BEGIN IMPLICIT INTEGRATION   **
***********************************
      IF (ZY.GT.ZERO) THEN
c
```

```
        DO I=1,NSTATV
          DELTA_X(I)=ZERO
        END DO
        DO I=1,M
          DO J=1,M
              TN1(I,J) = UD*SMX1(I,J)/PJ
              TN2(I,J) = UD*SMX2(I,J)/PJ
          END DO
         END DO
         CALL COPY_MATRIX_TO_VECT_EPS(TN1,VN1)
         CALL COPY_MATRIX_TO_VECT_EPS(TN2,VN2)
         CALL COPY_MATRIX_TO_VECT_EPS(TX1,VX1)
         CALL COPY_MATRIX_TO_VECT_EPS(TX2,VX2)
c
        DLAMBDA = exp(xn*log(ZY/xk))*DTIME
c
        CALL DYADICPROD(VN1,VN1,XNV1_XNV1)
        CALL DYADICPROD(VN1,VN2,XNV1_XNV2)
        CALL DYADICPROD(VN2,VN1,XNV2_XNV1)
        CALL DYADICPROD(VN2,VN2,XNV2_XNV2)
c
        DO I=1,NTENS
          DO J=1,NTENS
           DN1DS1(I,J) = (UNIT32(I,J)-XNV1_XNV1(I,J))/PJ
           DN1DS2(I,J) = XNV1_XNV2(I,J)/PJ
           DN2DS1(I,J) = XNV2_XNV1(I,J)/PJ
           DN2DS2(I,J) = (UNIT32(I,J)-XNV2_XNV2(I,J))/PJ
           TEMPO_A_FE(I,J)=A1*(DN1DS1(I,J)+DN1DS2(I,J))
     X                    +A2*(DN2DS1(I,J)+DN2DS2(I,J))
           XN11_N12(I,J)=DN1DS1(I,J)+DN1DS2(I,J)
           XN21_N22(I,J)=DN2DS1(I,J)+DN2DS2(I,J)
           CN11(I,J)=C11*DN1DS1(I,J)+C12*DN1DS2(I,J)
           CN12(I,J)=C11*DN2DS1(I,J)+C12*DN2DS2(I,J)
           CN21(I,J)=C12*DN1DS1(I,J)+C22*DN1DS2(I,J)
           CN22(I,J)=C12*DN2DS1(I,J)+C22*DN2DS2(I,J)
           TEMPO_AC_FE_1(I,J)=A1*CN11(I,J)+A2*CN12(I,J)
           TEMPO_AC_FE_2(I,J)=A1*CN21(I,J)+A2*CN22(I,J)
          END DO
        END DO
c
        CALL MULT_T4_T4(DDS,TEMPO_A_FE,TEMPO_A_FE_DDS)
        CALL MULT_T4_T4(DDS,XN11_N12,XN11_N12_E)
        CALL MULT_T4_T4(DDS,XN21_N22,XN21_N22_E)
```

```
c
      TERM=DTIME*xn*exp((xn-ONE)*log(ZY/xk))
      DO I=1,NTENS
        XNV(I)=A1*VN1(I)+A2*VN2(I)
        PCN11(I)=C11*VN1(I)+C12*VN2(I)
        PCN22(I)=C12*VN1(I)+C22*VN2(I)
      END DO
      CALL KMLT1(DDS,XNV,DDS_XNV)
C
C BEGIN OF NEWTON ITERATIONS
C
      CONVERGENCE=.FALSE.
      NITMAX=0
      DO 500 ITER=1,MAXITER
      IF (.NOT.CONVERGENCE) THEN
        NITMAX=NITMAX+1
        DELTAP=DELTA_X(NSTATV)
C
c============== FILL F_vec BEGIN ==========================
        DO I=1,NTENS
          F_VEC(I)=DELTA_X(I)+XNV(I)*DELTAP-DSTRAN(I)
        END DO
C       WRITE(*,*)NITMAX,' FV3 ',F_VEC(3), ' DSTR3 ',
         DSTRAN(3)
        DO I=1,NTENS
          J=I+NTENS
          F_VEC(J)=DELTA_X(J)-DELTAP*(VN1(I)-UD*D1*VX1
          (I)/C11)
        END DO
        DO I=1,NTENS
          J=I+2*NTENS
          F_VEC(J)=DELTA_X(J)-DELTAP*(VN2(I)-UD*D2*VX2
          (I)/C22)
        END DO
        F_VEC(NSTATV)=DELTAP-DLAMBDA
c============== FILL F_vec END ============================
C
C============== FILL XJAC BEGIN ===========================
C    DEEL_DEEL
        DO I=1,NTENS
          DO J=1,NTENS
            XJAC(I,J) = XIDEN66(I,J)+DELTAP*TEMPO_A_FE_
            DDS(I,J)
```

```
            END DO
          END DO
C     DEEL_DALPHA1
          DO I=1,NTENS
            DO J=1,NTENS
              XJAC(I,NTENS+J) = -TEMPO_AC_FE_1(I,J)*DEL
              TAP/UD
            END DO
          END DO
C     DEEL_DALPHA2
          DO I=1,NTENS
            DO J=1,NTENS
              XJAC(I,2*NTENS+J) = -TEMPO_AC_FE_2(I,J)*
              DELTAP/UD
            END DO
          END DO
C     DEEL_DLAMBDA
          DO I=1,NTENS
             XJAC(I,NSTATV) = XNV(I)
          END DO
C     DALPHA1_DEEL
          DO I=1,NTENS
            DO J=1,NTENS
              XJAC(I+NTENS,J) = -DELTAP*XN11_N12_E(I,J)
            END DO
          END DO
C     DALPHA1_DALPHA1
          DO I=1,NTENS
            DO J=1,NTENS
              XJAC(I+NTENS,NTENS+J)=
     X  (ONE+D1*DELTAP)*XIDEN66(I,J)+CN11(I,J)*DELTAP/UD
            END DO
          END DO
C     DALPHA1_DALPHA2
          DO I=1,NTENS
            DO J=1,NTENS
              XJAC(I+NTENS,2*NTENS+J) =
     X  (CN21(I,J)/UD+C12*D1*XIDEN66(I,J)/C11)*DELTAP
            END DO
          END DO
C     DALPHA1_DELTAP
         DO I=1,NTENS
           XJAC(I+NTENS,NSTATV) = -VN1(I)+UD*D1*VX1(I)/C11
```

```
            END DO
C     DALPHA2_DEEL
             DO I=1,NTENS
               DO J=1,NTENS
                 XJAC(I+2*NTENS,J) = -DELTAP*XN21_N22_E(I,J)
               END DO
             END DO
C     DALPHA2_DALPHA1
             DO I=1,NTENS
               DO J=1,NTENS
                 XJAC(I+2*NTENS,NTENS+J) =
     X   (CN12(I,J)/UD+C12*D2*XIDEN66(I,J)/C22)*DELTAP
               END DO
             END DO
C     DALPHA2_DALPHA2
             DO I=1,NTENS
               DO J=1,NTENS
                 XJAC(I+2*NTENS,2*NTENS+J)=
     X  (ONE+D2*DELTAP)*XIDEN66(I,J)+CN22(I,J)*DELTAP/UD
               END DO
             END DO
C     DALPHA2_DELTAP
            DO I=1,NTENS
             XJAC(I+2*NTENS,NSTATV)= -VN2(I)+UD*D2*VX2(I)/C22
            END DO
c      DELTAP_DEEL
             DO J=1,NTENS
               XJAC(NSTATV,J)=-TERM*DDS_XNV(J)
             END DO
C
c      DELTAP_DALPHA1
             DO J 1,NTENS
               XJAC(NSTATV,J+NTENS)=TERM*PCN11(J)/UD
             END DO
c      DELTAP_DALPHA2
             DO J=1,NTENS
               XJAC(NSTATV,J+2*NTENS)=TERM*PCN22(J)/UD
             END DO
c      DELTAP_DELTAP
             XJAC(NSTATV,NSTATV)=ONE+TERM*Q*b*exp(-b*p)
C
C============ FILL XJAC END  ========================
C
```

```
C           STORE THE TOP LEFT BLOCK OF THE JACOBIAN
C           FOR THE TANGENT OPERATOR DETERMINATION
C
            DO I=1,NTENS
              DO J=1,NTENS
                XJAC_FE(I,J)=XJAC(I,J)
              END DO
            END DO
C
            CALL LUDCMP(XJAC,NSTATV,NSTATV,INDX)
            CALL LUBKSB(XJAC,NSTATV,NSTATV,INDX,F_VEC)
            DO I=1,NSTATV
              DELTA_X(I)=DELTA_X(I)-F_VEC(I)
            END DO
            FNORM=ZERO
            DO I=1,NSTATV
              FNORM=FNORM+F_VEC(I)*F_VEC(I)
            END DO
            FNORM=SQRT(FNORM)
            IF (FNORM.LT.TOLER) CONVERGENCE=.TRUE.
          ELSE
            GOTO 600
          END IF
 500      CONTINUE
 600    CONTINUE
        IF (.NOT.CONVERGENCE) THEN
          WRITE(*,*)'Warning Divergence'
        END IF
        END IF
***********************************
**   END   IMPLICIT INTEGRATION   **
***********************************
C
        IF (ZY.LE.ZERO) THEN
          DLAMBDA = ZERO
          DO K = 1,NTENS
            DESTRAN(K)=DSTRAN(K)
            DALPHA1(K)=ZERO
            DALPHA2(K)=ZERO
          END DO
        ELSE
            DLAMBDA = DELTA_X(NSTATV)
            DO K = 1,NTENS
```

```
            DESTRAN(K)=DELTA_X(K)
            DALPHA1(K)=DELTA_X(K+NTENS)
            DALPHA2(K)=DELTA_X(K+2*NTENS)
          END DO
          DO K = 1,NTENS
            STRESS(K) = STRESS(K) - EG2*DLAMBDA*XNV(K)
          END DO
      END IF
C
      DO K = 1,NTENS
        ESTRAN(K) = ESTRAN(K) + DESTRAN(K)
        ALPHA1(K)  = ALPHA1(K) + DALPHA1(K)
        ALPHA2(K)  = ALPHA2(K) + DALPHA2(K)
      END DO
      p = p + DLAMBDA
C
C UPDATE STATE VARIABLES
C
       DO I =1, NTENS
          STATEV(I) = ESTRAN(I)
          STATEV(NTENS+I) = ALPHA1(I)
          STATEV(2*NTENS+I) = ALPHA2(I)
       END DO
       STATEV(NSTATV) = p
C
C     DETERMINE TANGENT OPERATOR
C
       IF ((ZY.LT.ZERO).OR.(DLAMBDA.LE.ZERO)) THEN
         DO I=1,NTENS
          DO J=1,NTENS
            DDSDDE(I,J) = DDS(I,J)
          END DO
         END DO
      ELSE
         CALL MATINV (XJAC_FE,NTENS)
         CALL MULT_T4_T4(DDS,XJAC_FE,DDSDDE)
      END IF
C
      RETURN
      END
*************************************************************
**          MULTIPLY 6X6 MATRIX WITH 6X1 VECTOR            *
*************************************************************
```

```
      SUBROUTINE KMLT1(DM1,DM2,DM)
      INCLUDE 'ABA_PARAM.INC'
      DIMENSION DM1(6,6),DM2(6),DM(6)
      DO I=1,6
         DM(I)=0.0
         DO K=1,6
            DM(I)=DM(I)+DM1(I,K)*DM2(K)
         END DO
      END DO
      RETURN
      END
***************************************
**           EFFECTIVE STRESS          *
**   (CONTRACTED MATRIX CALCULATION)   *
***************************************
      SUBROUTINE J_EQUIV(EFF1,VAL1)
      INCLUDE 'ABA_PARAM.INC'
      PARAMETER (M=3)
      DIMENSION EFF1(M,M)
      X=0.0
      DO I=1,M
        DO J=1,M
          X=X+EFF1(I,J)*EFF1(I,J)
        END DO
      END DO
      VAL1=SQRT(3.0*X/2.0)
      RETURN
      END
********************************************
**         DOT PRODUCT OF TWO VECTORS      *
********************************************
      SUBROUTINE DOTPROD(DM1,DM2,DM)
      INCLUDE 'ABA_PARAM.INC'
      DIMENSION DM1(6),DM2(6)
      DM=0.0
      DO K=1,6
         DM=DM+DM1(K)*DM2(K)
      END DO
      RETURN
      END
****************************************
**   DEVIATORIC STRESS CALCULATION     **
****************************************
```

```
      SUBROUTINE KDEVIA(STRSS,DEVITO)
      INCLUDE 'ABA_PARAM.INC'
      PARAMETER (M=3)
      DIMENSION STRSS(M,M),DEVITO(M,M)
      TRACE=0.0
      DO I=1,M
         TRACE=TRACE+STRSS(I,I)
      END DO
      DO I=1,M
        DO J=1,M
          IF(I .EQ. J) THEN
           DEVITO(I,J)=STRSS(I,J)-TRACE/3.
          ELSE
           DEVITO(I,J)=STRSS(I,J)
          END IF
        END DO
      END DO
      RETURN
      END

C**************************************************
      SUBROUTINE MULT_T4_T4(A,B,C)
      INCLUDE 'ABA_PARAM.INC'
      PARAMETER (M=6)
      DIMENSION A(M,M),B(M,M),C(M,M)
      DO I=1,M
            DO J=1, M
                C(I,J)=0.0
                    DO K=1, M
                        C(I,J)=C(I,J)+A(I,K)*B(K,J)
                    END DO
            END DO
      END DO
      END
C**************************************************
      SUBROUTINE DYADICPROD(DM1,DM2,DM3)
      INCLUDE 'ABA_PARAM.INC'
      DIMENSION DM1(6),DM2(6),DM3(6,6)
      DO I=1,6
       DO J=1,6
        DM3(I,J) = DM1(I)*DM2(J)
       END DO
      END DO
```

```
      END
******************************************
** PUT VECTOR IN A MATRIX FORM          **
** STRESS LIKE VARIABLE                 **
******************************************
      SUBROUTINE COPY_VECTOR_TO_MATRIX(VECT,TENS)
      INCLUDE 'ABA_PARAM.INC'
      DIMENSION VECT(6),TENS(3,3)
      TENS(1,1) = VECT(1)
      TENS(2,2) = VECT(2)
      TENS(3,3) = VECT(3)
      TENS(1,2) = VECT(4)
      TENS(2,1) = VECT(4)
      TENS(1,3) = VECT(5)
      TENS(3,1) = VECT(5)
      TENS(2,3) = VECT(6)
      TENS(3,2) = VECT(6)
      END
******************************************
** PUT VECTOR IN A MATRIX FORM          **
** STRAIN LIKE VARIABLE IN ABAQUS NOTATION **
******************************************
      SUBROUTINE COPY_VECT_TO_MATRIX_EPS(VECT,TENS)
      INCLUDE 'ABA_PARAM.INC'
      DIMENSION VECT(6),TENS(3,3)
      TWO=2.0
      TENS(1,1) = VECT(1)
      TENS(2,2) = VECT(2)
      TENS(3,3) = VECT(3)
      TENS(1,2) = VECT(4)/TWO
      TENS(2,1) = VECT(4)/TWO
      TENS(1,3) = VECT(5)/TWO
      TENS(3,1) = VECT(5)/TWO
      TENS(2,3) = VECT(6)/TWO
      TENS(3,2) = VECT(6)/TWO
      END
******************************************
** PUT MATRIX IN A VECTOR FORM          **
** STRAIN LIKE VARIABLE IN ABAQUS NOTATION **
******************************************
      SUBROUTINE COPY_MATRIX_TO_VECT_EPS(TENS,VECT)
      INCLUDE 'ABA_PARAM.INC'
      DIMENSION TENS(3,3),VECT(6)
```

```
      TWO=2.0
      VECT(1)=TENS(1,1)
      VECT(2)=TENS(2,2)
      VECT(3)=TENS(3,3)
      VECT(4)=TWO*TENS(1,2)
      VECT(5)=TWO*TENS(1,3)
      VECT(6)=TWO*TENS(2,3)
      END
*****************************************************
** FROM NUMERICAL RECIPES                          **
** DECOMPOSE THE MATRIX A(N,N) INTO                **
** AN LOWER AND UPPER TRIANGULAR MATRIX            **
** RETURN THIS MATRICES IN A                       **
** INPUT PARAMETERS:                               **
** NP         : PHYSICAL DIMENSION OF MATRIX       **
** N          : ACTUAL DIMENSION OF MATRIX         **
** A(NP,NP)   : MATRIX TO DECOMPOSE                **
** OUTPUT PARAMETERS:                              **
** INDX       : VECTOR RECORDING                   **
**              THE NUMBERS OF ROW CHANGES         **
*****************************************************
      SUBROUTINE LUDCMP(A,N,NP,INDX)
      INCLUDE 'ABA_PARAM.INC'
      PARAMETER (NMAX=100,TINY=1.0E-20)
      DIMENSION A(NP,NP),INDX(N),VV(NMAX)
C
      D=1.
      DO 12 I=1,N
       AAMAX=0.
       DO 11 J=1,N
        IF (ABS(A(I,J)).GT.AAMAX) AAMAX=ABS(A(I,J))
11      CONTINUE
        IF (AAMAX.EQ.0.) STOP 'Singular matrix.'
        VV(I)=1./AAMAX
12    CONTINUE
      DO 19 J=1,N
       IF (J.GT.1) THEN
       DO 14 I=1,J-1
        SUM=A(I,J)
        IF (I.GT.1)THEN
        DO 13 K=1,I-1
         SUM=SUM-A(I,K)*A(K,J)
13      CONTINUE
```

```
          A(I,J)=SUM
          ENDIF
14      CONTINUE
        ENDIF
        AAMAX=0.
        DO 16 I=J,N
          SUM=A(I,J)
          IF (J.GT.1)THEN
          DO 15 K=1,J-1
            SUM=SUM-A(I,K)*A(K,J)
15        CONTINUE
          A(I,J)=SUM
          ENDIF
          DUM=VV(I)*ABS(SUM)
          IF (DUM.GE.AAMAX) THEN
            IMAX=I
            AAMAX=DUM
          ENDIF
16      CONTINUE
        IF (J.NE.IMAX)THEN
          DO 17 K=1,N
            DUM=A(IMAX,K)
            A(IMAX,K)=A(J,K)
            A(J,K)=DUM
17        CONTINUE
          D=-D
          VV(IMAX)=VV(J)
        ENDIF
        INDX(J)=IMAX
        IF(J.NE.N)THEN
        IF(A(J,J).EQ.0.)A(J,J)=TINY
          DUM=1./A(J,J)
          DO 18 I=J+1,N
            A(I,J)=A(I,J)*DUM
18        CONTINUE
        ENDIF
19    CONTINUE
      IF(A(N,N).EQ.0.)A(N,N)=TINY
      RETURN
      END
*******************************************
** FROM NUMERICAL RECIPES                **
** SOLVE THE SYSTEM A(N,N)*X(N) = B(N),  **
```

```
** USING THE DECOMPOSITION OF                          **
** A(N,N) WITH SUBROUTINE LUDCMP.                      **
** THE SOLUTION X(N) IS RETURNED IN B(N)               **
*******************************************************
      SUBROUTINE LUBKSB(A,N,NP,INDX,B)
C
      INCLUDE 'ABA_PARAM.INC'
      DIMENSION A(NP,NP),B(N),INDX(N)
C
      II=0
      DO 12 I=1,N
       LL=INDX(I)
       SUM=B(LL)
       B(LL)=B(I)
       IF (II.NE.0)THEN
        DO 11 J=II,I-1
         SUM=SUM-A(I,J)*B(J)
11      CONTINUE
       ELSE IF (SUM.NE.0.) THEN
        II=I
       ENDIF
       B(I)=SUM
12    CONTINUE
      DO 14 I=N,1,-1
       SUM=B(I)
       DO 13 J=I+1,N
        SUM=SUM-A(I,J)*B(J)
13     CONTINUE
       B(I)=SUM/A(I,I)
14    CONTINUE
      RETURN
      END
*******************************************************
** MATINV FROM NUMERICAL RECIPES                       **
*******************************************************
      SUBROUTINE MATINV(A,N)
C
      INCLUDE 'ABA_PARAM.INC'
      DIMENSION A(N,N),IK(N),JK(N)
C
      IFAIL=0
      DO 100 K=1,N
       AMAX=0.
```

```
21      DO 30 I=K,N
        DO 30 J=K,N
        IF(ABS(AMAX)-ABS(A(I,J)))24,24,30
24      AMAX=A(I,J)
        IK(K)=I
        JK(K)=J
30      CONTINUE
        IF(AMAX)41,32,41
32      IFAIL=1
        GO TO 140
41      I=IK(K)
        IF(I-K)21,51,43
43      DO 50 J=1,N
        SAVE=A(K,J)
        A(K,J)=A(I,J)
50      A(I,J)=-SAVE
51      J=JK(K)
        IF(J-K)21,61,53
53      DO 60 I=1,N
        SAVE=A(I,K)
        A(I,K)=A(I,J)
60      A(I,J)=-SAVE
61      DO 70 I=1,N
        IF(I-K)63,70,63
63      A(I,K)=-A(I,K)/AMAX
70      CONTINUE
        DO 80 I=1,N
        DO 80 J=1,N
        IF(I-K)74,80,74
74      IF(J-K)75,80,75
75      A(I,J)=A(I,J)+A(I,K)*A(K,J)
80      CONTINUE
        DO 90 J=1,N
        IF(J-K)83,90,83
83      A(K,J)=A(K,J)/AMAX
90      CONTINUE
        A(K,K)=1./AMAX
100     CONTINUE
        DO 130 L=1,N
        K=N-L+1
        J=IK(K)
        IF(J-K)111,111,105
105     DO 110 I=1,N
```

```
             SAVE=A(I,K)
             A(I,K)=-A(I,J)
 110             A(I,J)=SAVE
 111             I=JK(K)
             IF(I-K)130,130,113
 113       DO 120 J=1,N
           SAVE=A(K,J)
           A(K,J)=-A(I,J)
 120       A(I,J)=SAVE
 130      CONTINUE
 140     CONTINUE
       RETURN
       END
C*************************************************
```

6.6.6. *ANSYS: implicit integration*

```
       subroutine usermat(
      & matId, elemId,kDomIntPt, kLayer, kSectPt,
      & ldstep,isubst,keycut,
      & nDirect,nShear,ncomp,nStatev,nProp,
      & Time,dTime,Temp,dTemp,
      & stress,ustatev,dsdePl,sedEl,sedPl,epseq,
      & Strain,dStrain, epsPl, prop, coords,
      & var0, defGrad_t, defGrad,
      & tsstif, epsZZ,
      & var1, var2, var3, var4, var5,
      & var6, var7, var8)
c
#include "impcom.inc"
       INTEGER
      & matId, elemId,
      & kDomIntPt, kLayer, kSectPt,
      & ldstep,isubst,keycut,
      & nDirect,nShear,ncomp,nStatev,nProp
       DOUBLE PRECISION
      & Time, dTime, Temp, dTemp,
      & sedEl, sedPl, epseq, epsZZ
       DOUBLE PRECISION
      & stress (ncomp ), ustatev (nStatev),
      & dsdePl (ncomp,ncomp),
      & Strain (ncomp ), dStrain (ncomp ),
      & epsPl (ncomp ), prop (nProp ),
```

```
     & coords (3),
     & defGrad (3,3), defGrad_t(3,3),
     & tsstif (2)
c
c***************** User defined part
C
  INTEGER I,J,K,M,ITER,MAXITER,NITMAX
  DOUBLE PRECISION ZERO,ONE,TWO,THREE,UD,TOLER,FNORM
  PARAMETER (M=3,ZERO=0.D0,ONE=1.D0,TWO=2.D0,THREE=3.
  D0,UD=1.5D0)
C
  DOUBLE PRECISION VN1(ncomp),VN2(ncomp),TN1(M,M),TN2(M,M),
  + VX1(ncomp),VX2(ncomp),TX1(M,M),TX2(M,M),ESTRAN(ncomp),
  + ALPHA1(ncomp),ALPHA2(ncomp),TA1(M,M),TA2(M,M),
  + DALPHA1(ncomp),DALPHA2(ncomp),
  + DPSTRAN(ncomp),DESTRAN(ncomp),DSTRESS(ncomp),
  + STR(M,M),DSTR(M,M),SMX1(M,M),SMX2(M,M),
  + DDS(ncomp,ncomp)
c --- temporary variables for solution purpose
  DOUBLE PRECISION var0, var1, var2, var3, var4, var5,
  & var6, var7, var8
c
  DOUBLE PRECISION E,XNU,A1,A2,R0,Q,b,xn,xk,
  &                    C11,C22,C12,D1,D2
  DOUBLE PRECISION  EG,EG2,ELAM,EBULK3
  DOUBLE PRECISION p,PJ1,PJ2,PJ,ZY,DLAMBDA,TERM,DELTAP
  EXTERNAL usermat3d, usermatps, usermatbm, usermat1d
C
  INTEGER INDX(nStatev)
  DOUBLE PRECISION UNIT32(6,6),XIDEN66(6,6),
  X XJAC(nStatev,nStatev),DELTA_X(nStatev),F_VEC(nStatev)
  DOUBLE PRECISION XNV1_XNV1(ncomp,ncomp),XNV1_XNV2(ncomp,ncom
  X  XNV2_XNV1(ncomp,ncomp),XNV2_XNV2(ncomp,ncomp),
  X XNV(ncomp),DDS_XNV(ncomp),
  X PCN11(ncomp),PCN22(ncomp),
  X DN1DS1(ncomp,ncomp),DN1DS2(ncomp,ncomp),
  X DN2DS1(ncomp,ncomp),DN2DS2(ncomp,ncomp),
  X TEMPO_A_FE(ncomp,ncomp),TEMPO_A_FE_DDS(ncomp,ncomp),
  X TEMPO_AC_FE_1(ncomp,ncomp),TEMPO_AC_FE_2(ncomp,ncomp),
  X CN11(ncomp,ncomp),CN12(ncomp,ncomp),CN21(ncomp,ncomp),
  X CN22(ncomp,ncomp),
  X XN11_N12(ncomp,ncomp),XN21_N22(ncomp,ncomp),
  X XN11_N12_E(ncomp,ncomp),XN21_N22_E(ncomp,ncomp),
```

```
      X XJAC_FE(ncomp,ncomp)
       LOGICAL CONVERGENCE
       data XIDEN66/1.0d0,0.0d0,0.0d0,0.0d0, 0.0d0, 0.0d0,
      x  0.0d0, 1.0d0, 0.0d0, 0.0d0, 0.0d0, 0.0d0,
      x  0.0d0, 0.0d0, 1.0d0, 0.0d0, 0.0d0, 0.0d0,
      x  0.0d0, 0.0d0, 0.0d0, 1.0d0, 0.0d0, 0.0d0,
      x  0.0d0, 0.0d0, 0.0d0, 0.0d0, 1.0d0, 0.0d0,
      x  0.0d0, 0.0d0, 0.0d0, 0.0d0, 0.0d0, 1.0d0/
       data UNIT32/1.0d0, 0.5d0, 0.5d0,0.0d0,0.0d0,0.0d0,
      x  -0.5d0, 1.0d0, -0.5d0,  0.0d0,  0.0d0,  0.0d0,
      x  -0.5d0, -0.5d0, 1.0d0,  0.0d0,  0.0d0,  0.0d0,
      x  0.0d0, 0.0d0, 0.0d0,  1.5d0,  0.0d0,  0.0d0,
      x  0.0d0, 0.0d0, 0.0d0,  0.0d0,  1.5d0,  0.0d0,
      x  0.0d0, 0.0d0, 0.0d0,  0.0d0,  0.0d0,  1.5d0/
       TOLER=1.D-12
       MAXITER=100
C
       E=prop(1)
       XNU=prop(2)
       A1=prop(3)
       A2=prop(4)
       R0=prop(5)
       Q=prop(6)
       b=prop(7)
       xn=prop(8)
       xk=prop(9)
       C11=prop(10)
       C22=prop(11)
       C12=prop(12)
       D1=prop(13)
       D2=prop(14)
C
C     RECOVER STATE VARIABLES
C
       DO i =1, ncomp
         ESTRAN(i) = ustatev(i)
         ALPHA1(i) = ustatev(ncomp+i)
         ALPHA2(i) = ustatev(2*ncomp+i)
       END DO
       p = ustatev(nStatev)
C
C   ELASTIC STIFNESS
C
```

```
      EBULK3 = E/(ONE-TWO*XNU)
      EG2 = E/(ONE+XNU)
      EG = EG2/TWO
      ELAM = (EBULK3-EG2)/THREE
      DO I=1,ncomp
         DO J=1,ncomp
            DDS(I,J)=ZERO
         END DO
      END DO
      DO I = 1, M
         DO J = 1, M
            DDS(I,J) = ELAM
         END DO
         DDS(I,I) = EG2 + ELAM
      END DO
      DDS(4,4) = EG
      DDS(5,5) = EG
      DDS(6,6) = EG
C
C     OBTAIN TRIAL (ELASTIC) STRESS
C
      CALL KMLT1(DDS,dStrain,DSTRESS)
      DO I=1,ncomp
         stress(I)=stress(I)+DSTRESS(I)
      END DO
C

C    WRITE STRESSES AND KINEMATIC VARIABLE IN MATRIX FORM
C
      CALL COPY_VECTOR_TO_MATRIX(STRESS,STR)
      CALL COPY_VECT_TO_MATRIX_EPS(ALPHA1,TA1)
      CALL COPY_VECT_TO_MATRIX_EPS(ALPHA2,TA2)
C
C DEVIATORIC STRESS TENSOR
C
      CALL KDEVIA(STR,DSTR)
C
C COMPUTE THE TWO BACK STRESSES
C
      DO I=1,M
        DO J=1,M
          TX1(I,J)=(TWO/THREE)*(C11*TA1(I,J)+C12*TA2(I,J))
          TX2(I,J)=(TWO/THREE)*(C12*TA1(I,J)+C22*TA2(I,J))
```

```
        END DO
      END DO
C
      DO I=1,M
         DO J=1,M
            SMX1(I,J) = A1*DSTR(I,J)-TX1(I,J)
            SMX2(I,J) = A2*DSTR(I,J)-TX2(I,J)
         END DO
      END DO

      CALL J_EQUIV(SMX1,PJ1)
      CALL J_EQUIV(SMX2,PJ2)
C
      PJ=sqrt(PJ1*PJ1+PJ2*PJ2)
      ZY = PJ - Q*(ONE-exp(-B*p))-R0
C
***********************************
**   BEGIN IMPLICIT INTEGRATION  **
***********************************
      IF (ZY.GT.ZERO) THEN
c
         DO I=1,nStatev
            DELTA_X(I)=ZERO
         END DO
         DO I=1,M
            DO J=1,M
               TN1(I,J) = UD*SMX1(I,J)/PJ
               TN2(I,J) = UD*SMX2(I,J)/PJ
            END DO
          END DO
          CALL COPY_MATRIX_TO_VECT_EPS(TN1,VN1)
          CALL COPY_MATRIX_TO_VECT_EPS(TN2,VN2)
          CALL COPY_MATRIX_TO_VECT_EPS(TX1,VX1)
          CALL COPY_MATRIX_TO_VECT_EPS(TX2,VX2)
c
         DLAMBDA = exp(xn*log(ZY/xk))*DTIME
c
         CALL DYADICPROD(VN1,VN1,XNV1_XNV1)
         CALL DYADICPROD(VN1,VN2,XNV1_XNV2)
         CALL DYADICPROD(VN2,VN1,XNV2_XNV1)
         CALL DYADICPROD(VN2,VN2,XNV2_XNV2)
c
         DO I=1,ncomp
```

```
            DO J=1,ncomp
              DN1DS1(I,J) = (UNIT32(I,J)-XNV1_XNV1(I,J))/PJ
              DN1DS2(I,J) = XNV1_XNV2(I,J)/PJ
              DN2DS1(I,J) = XNV2_XNV1(I,J)/PJ
              DN2DS2(I,J) = (UNIT32(I,J)-XNV2_XNV2(I,J))/PJ
              TEMPO_A_FE(I,J)=A1*(DN1DS1(I,J)+DN1DS2(I,J))
     X                       +A2*(DN2DS1(I,J)+DN2DS2(I,J))
              XN11_N12(I,J)=DN1DS1(I,J)+DN1DS2(I,J)
              XN21_N22(I,J)=DN2DS1(I,J)+DN2DS2(I,J)
              CN11(I,J)=C11*DN1DS1(I,J)+C12*DN1DS2(I,J)
              CN12(I,J)=C11*DN2DS1(I,J)+C12*DN2DS2(I,J)
              CN21(I,J)=C12*DN1DS1(I,J)+C22*DN1DS2(I,J)
              CN22(I,J)=C12*DN2DS1(I,J)+C22*DN2DS2(I,J)
              TEMPO_AC_FE_1(I,J)=A1*CN11(I,J)+A2*CN12(I,J)
              TEMPO_AC_FE_2(I,J)=A1*CN21(I,J)+A2*CN22(I,J)
            END DO
          END DO
c
          CALL MULT_T4_T4(DDS,TEMPO_A_FE,TEMPO_A_FE_DDS)
          CALL MULT_T4_T4(DDS,XN11_N12,XN11_N12_E)
          CALL MULT_T4_T4(DDS,XN21_N22,XN21_N22_E)
c
          TERM=DTIME*xn*exp((xn-ONE)*log(ZY/xk))
          DO I=1,ncomp
            XNV(I)=A1*VN1(I)+A2*VN2(I)
            PCN11(I)=C11*VN1(I)+C12*VN2(I)
            PCN22(I)=C12*VN1(I)+C22*VN2(I)
          END DO
          CALL KMLT1(DDS,XNV,DDS_XNV)
C
C BEGIN OF NEWTON ITERATIONS
C
          CONVERGENCE=.FALSE.
          NITMAX=0
          DO 500 ITER=1,MAXITER
          IF (.NOT.CONVERGENCE) THEN
            NITMAX=NITMAX+1
            DELTAP=DELTA_X(nStatev)
C
c============ FILL F_vec BEGIN =======================
            DO I=1,ncomp
              F_VEC(I)=DELTA_X(I)+XNV(I)*DELTAP-dStrain(I)
            END DO
```

```
            DO I=1,ncomp
               J=I+ncomp
               F_VEC(J)=DELTA_X(J)-DELTAP*(VN1(I)-UD*D1*VX1
               (I)/C11)
            END DO
            DO I=1,ncomp
               J=I+2*ncomp
               F_VEC(J)=DELTA_X(J)-DELTAP*(VN2(I)-UD*D2*VX2
               (I)/C22)
            END DO
            F_VEC(nStatev)=DELTAP-DLAMBDA
C============= FILL F_vec END ============================
C
C============= FILL XJAC BEGIN ==========================
C    DEEL_DEEL
            DO I=1,ncomp
              DO J=1,ncomp
                XJAC(I,J) = XIDEN66(I,J)+DELTAP*TEMPO_A_FE_
                DDS(I,J)
              END DO
            END DO
C    DEEL_DALPHA1
            DO I=1,ncomp
              DO J=1,ncomp
                XJAC(I,ncomp+J) = -TEMPO_AC_FE_1(I,J)*
                DELTAP/UD
              END DO
            END DO
C    DEEL_DALPHA2
            DO I=1,ncomp
              DO J=1,ncomp
                XJAC(I,2*ncomp+J) = -TEMPO_AC_FE_2(I,J)*
                DELTAP
                /UD
              END DO
            END DO
C    DEEL_DLAMBDA
            DO I=1,ncomp
               XJAC(I,nStatev) = XNV(I)
            END DO
C---
C    DALPHA1_DEEL
            DO I=1,ncomp
```

```
          DO J=1,ncomp
            XJAC(I+ncomp,J) = -DELTAP*XN11_N12_E(I,J)
          END DO
        END DO
C   DALPHA1_DALPHA1
        DO I=1,ncomp
          DO J=1,ncomp
            XJAC(I+ncomp,ncomp+J)=
   X   (ONE+D1*DELTAP)*XIDEN66(I,J)+CN11(I,J)*
   DELTAP/UD
          END DO
        END DO
C   DALPHA1_DALPHA2
        DO I=1,ncomp
          DO J=1,ncomp
            XJAC(I+ncomp,2*ncomp+J) =
   X   (CN21(I,J)/UD+C12*D1*XIDEN66(I,J)/C11)*DELTAP
          END DO
        END DO
C   DALPHA1_DELTAP
        DO I=1,ncomp
          XJAC(I+ncomp,nStatev) = -VN1(I)+UD*D1*VX1(I)
          /C11
        END DO
C   DALPHA2_DEEL
        DO I=1,ncomp
          DO J=1,ncomp
           XJAC(I+2*ncomp,J) = -DELTAP*XN21_N22_E(I,J)
          END DO
        END DO
C   DALPHA2_DALPHA1
        DO I=1,ncomp
          DO J=1,ncomp
            XJAC(I+2*ncomp,ncomp+J) =
   X   (CN12(I,J)/UD+C12*D2*XIDEN66(I,J)/C22)*DELTAP
          END DO
        END DO
C   DALPHA2_DALPHA2
        DO I=1,ncomp
          DO J=1,ncomp
            XJAC(I+2*ncomp,2*ncomp+J)=
   X (ONE+D2*DELTAP)*XIDEN66(I,J)+CN22(I,J)*DELTAP/UD
          END DO
```

```
            END DO
C     DALPHA2_DELTAP
            DO I=1,ncomp
              XJAC(I+2*ncomp,nStatev)=-VN2(I)+UD*D2*VX2(I)
              /C22
            END DO
c      DELTAP_DEEL
            DO J=1,ncomp
              XJAC(nStatev,J)=-TERM*DDS_XNV(J)
            END DO
c      DELTAP_DALPHA1
            DO J=1,ncomp
              XJAC(nStatev,J+ncomp)=TERM*PCN11(J)/UD
            END DO
c      DELTAP_DALPHA2
            DO J=1,ncomp
              XJAC(nStatev,J+2*ncomp)=TERM*PCN22(J)/UD
            END DO
c      DELTAP_DELTAP
            XJAC(nStatev,nStatev)=ONE+TERM*Q*b*exp(-b*p)
C============= FILL XJAC END  =========================
C
C           STORE THE TOP LEFT BLOCK OF THE JACOBIAN
C           FOR THE TANGENT OPERATOR DETERMINATION
C
            DO I=1,ncomp
              DO J=1,ncomp
               XJAC_FE(I,J)=XJAC(I,J)
              END DO
            END DO
C
            CALL LUDCMP(XJAC,nStatev,nStatev,INDX)
            CALL LUBKSB(XJAC,nStatev,nStatev,INDX,F_VEC)
            DO I=1,nStatev
              DELTA_X(I)=DELTA_X(I)-F_VEC(I)
            END DO
            FNORM = ZERO
            DO I=1,nStatev
              FNORM=FNORM+F_VEC(I)*F_VEC(I)
            END DO
            FNORM=DSQRT(FNORM)
            IF (FNORM.LT.TOLER) CONVERGENCE=.TRUE.
          ELSE
```

```
            GOTO 600
          END IF
 500      CONTINUE
 600    CONTINUE
        IF (.NOT.CONVERGENCE) THEN
          WRITE(*,*) 'Warning Divergence '
        END IF
        END IF
**********************************
**   END   IMPLICIT INTEGRATION   **
**********************************
C
        IF (ZY.LE.ZERO) THEN
          DLAMBDA = ZERO
          DO K = 1,ncomp
            DESTRAN(K)=dStrain(K)
            DALPHA1(K)=ZERO
            DALPHA2(K)=ZERO
          END DO
        ELSE
          DLAMBDA = DELTA_X(nStatev)
           DO K = 1,ncomp
             DESTRAN(K)=DELTA_X(K)
             DALPHA1(K)=DELTA_X(K+ncomp)
             DALPHA2(K)=DELTA_X(K+2*ncomp)
           END DO
           DO K = 1,ncomp
             STRESS(K) = STRESS(K) - EG2*DLAMBDA*XNV(K)
           END DO
        END IF
C
        DO K = 1,ncomp
          ESTRAN(K) = ESTRAN(K) + DESTRAN(K)
          ALPHA1(K)  = ALPHA1(K) + DALPHA1(K)
          ALPHA2(K)  = ALPHA2(K) + DALPHA2(K)
        END DO
        p = p + DLAMBDA
C
C UPDATE STATE VARIABLES
C
         DO i =1, ncomp
            ustatev(i) = ESTRAN(i)
            ustatev(ncomp+i) = ALPHA1(i)
```

```
            ustatev(2*ncomp+i) = ALPHA2(i)
          END DO
          ustatev(nStatev) = p
C
C     DETERMINE TANGENT OPERATOR
C
          IF ((ZY.LT.ZERO).OR.(DLAMBDA.LE.ZERO)) THEN
            DO I=1,ncomp
             DO J=1,ncomp
               dsdePl(I,J) = DDS(I,J)
             END DO
            END DO
        ELSE
            CALL MATINV (XJAC_FE,ncomp)
            CALL MULT_T4_T4(DDS,XJAC_FE,dsdePl)
        END IF
C
        return
        end
**************************************************
**          MULTIPLY 6X6 MATRIX WITH 6X1 VECTOR          *
**************************************************
        SUBROUTINE KMLT1(DM1,DM2,DM)
        INTEGER I,K
        DOUBLE PRECISION DM1(6,6),DM2(6),DM(6)
        DO I=1,6
           DM(I)=0.0d0
           DO K=1,6
              DM(I)=DM(I)+DM1(I,K)*DM2(K)
           END DO
        END DO
        RETURN
        END
*************************************
**          EFFECTIVE STRESS           *
**   (CONTRACTED MATRIX CALCULATION)   *
*************************************
        SUBROUTINE J_EQUIV(EFF1,VAL1)
C
        INTEGER I,J
        DOUBLE PRECISION EFF1(3,3)
        DOUBLE PRECISION VAL1,TWO,THREE
C
```

```
      TWO=2.0d0
      THREE=3.0d0
      VAL1=0.0d0
      DO I=1,3
        DO J=1,3
          VAL1=VAL1+EFF1(I,J)*EFF1(I,J)
        END DO
      END DO
      VAL1=SQRT(THREE*VAL1/TWO)
      RETURN
      END
*******************************************
**          DOT PRODUCT OF TWO VECTORS        *
*******************************************
      SUBROUTINE DOTPROD(DM1,DM2,DM)
C
      INTEGER K
      DOUBLE PRECISION DM1(6),DM2(6)
      DOUBLE PRECISION DM
C
      DM=0.0d0
      DO K=1,6
         DM=DM+DM1(K)*DM2(K)
      END DO
      RETURN
      END
****************************************
**    DEVIATORIC STRESS CALCULATION     **
****************************************
      SUBROUTINE KDEVIA(STRSS,DEVITO)
      INTEGER I,J
      DOUBLE PRECISION STRSS(3,3),DEVITO(3,3)
      DOUBLE PRECISION TRACE,THREE
C
      THREE=3.0d0
      TRACE=0.0d0
      DO I=1,3
         TRACE=TRACE+STRSS(I,I)
      END DO
C
      DO I=1,3
        DO J=1,3
          IF(I .EQ. J) THEN
```

```
            DEVITO(I,J)=STRSS(I,J)-TRACE/THREE
          ELSE
            DEVITO(I,J)=STRSS(I,J)
          END IF
        END DO
      END DO
      RETURN
      END
********************************************
** PUT VECTOR IN A MATRIX FORM            **
** STRESS LIKE VARIABLE                   **
********************************************
      SUBROUTINE COPY_VECTOR_TO_MATRIX(VECT,TENS)
      DOUBLE PRECISION TENS(3,3),VECT(6)
C
      TENS(1,1) = VECT(1)
      TENS(2,2) = VECT(2)
      TENS(3,3) = VECT(3)
      TENS(1,2) = VECT(4)
      TENS(2,1) = VECT(4)
      TENS(1,3) = VECT(5)
      TENS(3,1) = VECT(5)
      TENS(2,3) = VECT(6)
      TENS(3,2) = VECT(6)
      END
C
********************************************
** PUT VECTOR IN A MATRIX FORM            **
** STRAIN LIKE VARIABLE IN ANSYS NOTATION **
********************************************
      SUBROUTINE COPY_VECT_TO_MATRIX_EPS(VECT,TENS)
      DOUBLE PRECISION TENS(3,3),VECT(6)
      DOUBLE PRECISION TWO
      TWO=2.0d0
C
      TENS(1,1) = VECT(1)
      TENS(2,2) = VECT(2)
      TENS(3,3) = VECT(3)
      TENS(1,2) = VECT(4)/TWO
      TENS(2,1) = VECT(4)/TWO
      TENS(1,3) = VECT(5)/TWO
      TENS(3,1) = VECT(5)/TWO
      TENS(2,3) = VECT(6)/TWO
```

```
      TENS(3,2) = VECT(6)/TWO
      END
*******************************************
** PUT MATRIX IN A VECTOR FORM          **
** STRAIN LIKE VARIABLE IN ANSYS NOTATION **
*******************************************
      SUBROUTINE COPY_MATRIX_TO_VECT_EPS(TENS,VECT)
      DOUBLE PRECISION TENS(3,3),VECT(6)
      DOUBLE PRECISION TWO
      TWO=2.0d0
C
      VECT(1)=TENS(1,1)
      VECT(2)=TENS(2,2)
      VECT(3)=TENS(3,3)
      VECT(4)=TWO*TENS(1,2)
      VECT(5)=TWO*TENS(1,3)
      VECT(6)=TWO*TENS(2,3)
      END
C
      SUBROUTINE MULT_T4_T4(A,B,C)
      INTEGER I,J,K,M
      PARAMETER (M=6)
      DOUBLE PRECISION A(M,M),B(M,M),C(M,M)
      DO I=1,M
            DO J=1, M
               C(I,J)=0.0D0
                  DO K=1, M
                     C(I,J)=C(I,J)+A(I,K)*B(K,J)
                  END DO
            END DO
      END DO
      END
C
      SUBROUTINE DYADICPROD(DM1,DM2,DM3)
      INTEGER I,J
      DOUBLE PRECISION DM1(6),DM2(6),DM3(6,6)
      DO I=1,6
       DO J=1,6
        DM3(I,J) = DM1(I)*DM2(J)
       END DO
      END DO
      END
*******************************************
```

```
** FROM NUMERICAL RECIPES                                   **
** DECOMPOSE THE MATRIX A(N,N) INTO                         **
** AN LOWER AND UPPER TRIANGULAR MATRIX                     **
** RETURN THIS MATRICES IN A                                **
** INPUT PARAMETERS:                                        **
** NP          : PHYSICAL DIMENSION OF MATRIX               **
** N           : ACTUAL DIMENSION OF MATRIX                 **
** A(NP,NP)    : MATRIX TO DECOMPOSE                        **
** OUTPUT PARAMETERS:                                       **
** INDX        : VECTOR RECORDING                           **
**               THE NUMBERS OF ROW CHANGES                 **
**************************************************************
      SUBROUTINE LUDCMP(A,N,NP,INDX)
C
      INTEGER NMAX,I,J,K,N,IMAX
      DOUBLE PRECISION TINY, D, AAMAX,SUM,DUM
      PARAMETER (NMAX=100,TINY=1.0D-20)
      DOUBLE PRECISION A(NP,NP),VV(NMAX)
      INTEGER INDX(N)
C
      D=1.0d0
      DO 12 I=1,N
       AAMAX=0.0d0
       DO 11 J=1,N
        IF (ABS(A(I,J)).GT.AAMAX) AAMAX=ABS(A(I,J))
11      CONTINUE
        IF (AAMAX.EQ.0.) THEN
            WRITE(*,*) 'Singular matrix.'
        END IF
        VV(I)=1./AAMAX
12    CONTINUE
      DO 19 J=1,N
       IF (J.GT.1) THEN
       DO 14 I=1,J-1
        SUM=A(I,J)
        IF (I.GT.1)THEN
        DO 13 K=1,I-1
         SUM=SUM-A(I,K)*A(K,J)
13      CONTINUE
        A(I,J)=SUM
        ENDIF
14     CONTINUE
       ENDIF
```

```
         AAMAX=0.
         DO 16 I=J,N
           SUM=A(I,J)
           IF (J.GT.1)THEN
           DO 15 K=1,J-1
             SUM=SUM-A(I,K)*A(K,J)
15         CONTINUE
           A(I,J)=SUM
           ENDIF DUM=VV(I)*ABS(SUM)
           IF (DUM.GE.AAMAX) THEN
             IMAX=I
             AAMAX=DUM
           ENDIF
16       CONTINUE
         IF (J.NE.IMAX)THEN
           DO 17 K=1,N
             DUM=A(IMAX,K)
             A(IMAX,K)=A(J,K)
             A(J,K)=DUM
17         CONTINUE
           D=-D
           VV(IMAX)=VV(J)
         ENDIF
         INDX(J)=IMAX
         IF(J.NE.N)THEN
         IF(A(J,J).EQ.0.)A(J,J)=TINY
           DUM=1./A(J,J)
           DO 18 I=J+1,N
             A(I,J)=A(I,J)*DUM
18         CONTINUE
         ENDIF
19     CONTINUE
       IF(A(N,N).EQ.0.)A(N,N)=TINY
       RETURN
       END
*********************************************
** FROM NUMERICAL RECIPES                  **
** SOLVE THE SYSTEM A(N,N)*X(N) = B(N),    **
** USING THE DECOMPOSITION OF              **
** A(N,N) WITH SUBROUTINE LUDCMP.          **
** THE SOLUTION X(N) IS RETURNED IN B(N)   **
*********************************************
```

```
      SUBROUTINE LUBKSB(A,N,NP,INDX,B)
C
      INTEGER N,NP,I,J,II
      DOUBLE PRECISION SUM
      DOUBLE PRECISION A(NP,NP),B(N)
      INTEGER INDX(N)
C
      II=0
      DO 12 I=1,N
       LL=INDX(I)
       SUM=B(LL)
       B(LL)=B(I)
       IF (II.NE.0)THEN
        DO 11 J=II,I-1
         SUM=SUM-A(I,J)*B(J)
11      CONTINUE
       ELSE IF (SUM.NE.0.) THEN
        II=I
       ENDIF
       B(I)=SUM
12    CONTINUE
      DO 14 I=N,1,-1
       SUM=B(I)
       DO 13 J=I+1,N
        SUM=SUM-A(I,J)*B(J)
13     CONTINUE
       B(I)=SUM/A(I,I)
14    CONTINUE
      RETURN
      END
C**************************************************
C
*********************************************
** MATINV FROM NUMERICAL RECIPES           **
*********************************************
      SUBROUTINE MATINV(A,N)
C
      INTEGER I,J,K,M,N,IFAIL
      DOUBLE PRECISION A(N,N),IK(N),JK(N)
      DOUBLE PRECISION AAMAX
C
      IFAIL=0
      DO 100 K=1,N
```

```
      AMAX=0.
21        DO 30 I=K,N
      DO 30 J=K,N
      IF(ABS(AMAX)-ABS(A(I,J)))24,24,30
24          AMAX=A(I,J)
      IK(K)=I
      JK(K)=J
30        CONTINUE
      IF(AMAX)41,32,41
32        IFAIL=1
          GO TO 140
41        I=IK(K)
          IF(I-K)21,51,43
43        DO 50 J=1,N
      SAVE=A(K,J)
      A(K,J)=A(I,J)
50        A(I,J)=-SAVE
51      J=JK(K)
        IF(J-K)21,61,53
53          DO 60 I=1,N
            SAVE=A(I,K)
            A(I,K)=A(I,J)

60          A(I,J)=-SAVE
61          DO 70 I=1,N
             IF(I-K)63,70,63
63           A(I,K)=-A(I,K)/AMAX
70          CONTINUE
            DO 80 I=1,N
             DO 80 J=1,N
        IF(I-K)74,80,74
74              IF(J-K)75,80,75
75              A(I,J)=A(I,J)+A(I,K)*A(K,J)
80           CONTINUE
            DO 90 J=1,N
       IF(J-K)83,90,83
83             A(K,J)=A(K,J)/AMAX
90          CONTINUE
            A(K,K)=1./AMAX
100       CONTINUE
            DO 130 L=1,N
             K=N-L+1
             J=IK(K)
```

```
                  IF(J-K)111,111,105
105         DO 110 I=1,N
            SAVE=A(I,K)
            A(I,K)=-A(I,J)
110             A(I,J)=SAVE
111             I=JK(K)
            IF(I-K)130,130,113
113       DO 120 J=1,N
          SAVE=A(K,J)
          A(K,J)=-A(I,J)
120       A(I,J)=SAVE
130      CONTINUE
140     CONTINUE
      RETURN
      END
C***********************************************************
      subroutine usermat3d
      end
      subroutine usermat1d
      end
      subroutine usermatbm
      end
      subroutine usermatps
      end
```

Bibliography

[AND 87] ANDRIEU E., Influence de l'environnement sur la propagation des fissures dans un superalliage base nickel, PhD Thesis, École Nationale Supérieure des Mines de Paris, 1987.

[ARM 66] ARMSTRONG P., FREDERICK C., A mathematical representation of the multiaxial Bauschinger effect, G.E.C.B., Report RD/B/N 731, 1966.

[AUB 03] AUBIN V., BULTHÉ A.-L., DEGALLAIX S. *et al.*, "Ratcheting behavior of a duplex stainless steel: characterization and modeling", *Proceedings of the 9th International Conference on the Mechanical Behavior of Materials*, Geneva, Switzerland, pp. 1–6, 25–29 May, 2003.

[BAN 04] BANABIC D., KUWABARA T., BALAN T. *et al.*, "An anisotropic yield criterion for sheet metals", *Journal of Materials Processing Technology*, vol. 157–158, pp. 462–465, 2004.

[BAR 05] BARLAT F., ARETZ H., YOON J. *et al.*, "Linear transformation-based anisotropic yield functions", *International Journal of Plasticity*, vol. 21, pp. 1009–1039, 2005.

[BAU 09] BAUDET C., GRANDIDIER J., CANGÉMI L., "A two-phase model for the diffuso-mechanical behaviour of semicrystalline polymers in gaseous environment", *International Journal of Solids and Structures*, vol. 46, pp. 1389–1401, 2009.

[BEN 87] BENALLAL A., MARQUIS D., "Constitutive equations for nonproportional cyclic elasto-viscoplasticity", *Journal of Engineering Materials and Technology*, vol. 109, pp. 326–336, 1987.

[BER 79] BERVEILLER M., ZAOUI A., "An extension of the self-consistent scheme to plastically-flowing polycrystals", *Journal of the Mechanics and Physics of Solids*, vol. 26, pp. 325–344, 1979.

[BES 03] BESSON J., GUILLEMER-NEEL C., "An extension of the Green and Gurson models to kinematic hardening", *Mechanics of Materials*, vol. 35, pp. 1–18, 2003.

[BES 09a] BESSON J., "Damage of ductile materials deforming under multiple plastic or viscoplastic mechanisms", *International Journal of Plasticity*, vol. 25, pp. 2204–2221, 2009.

[BES 09b] BESSON J., CAILLETAUD G., CHABOCHE J.-L. *et al.*, *Nonlinear Mechanics of Materials*, Springer, 2009.

[BIS 51] BISHOP J., HILL R., "A theory of the plastic distortion of a polycrystalline aggregate under combined stresses", *Philosophical Magazine*, vol. 42, pp. 414–427, 1951.

[BOD 87] BODNER S., "Unified constitutive equations for plastic deformation and creep of engineering alloys", in MILLER A.K. (ed.), *Unified Constitutive Equations for Creep and Plasticity*, Elsevier Applied Science, 1987.

[BOU 09] BOUDIFA M., SAANOUNI K., CHABOCHE J., "A micromechanical model for inelastic ductile damage prediction in polycrystalline metals for metal forming", *International Journal of Mechanical Sciences*, vol. 51, pp. 273–302, 2009.

[BRO 04] BRON F., BESSON J., "A yield function for anisotropic materials application to aluminum alloys", *International Journal of Plasticity*, vol. 20, pp. 937–963, 2004.

[BRU 08] BRUSSELLE-DUPEND N., CANGÉMI L., "A two-phase model for the mechanical behaviour of semicrystalline polymers. Part I: large strains multiaxial validation on HDPE", *Mechanics of Materials*, vol. 40, pp. 743–760, 2008.

[BUD 62] BUDIANSKY T., WU T., "Theoretical prediction of plastic strains of polycrystals", in ROSENBERG R.M. (ed.), *4th U.S. National Congress on Applied Mechanics*, vol. 2, American Society of Mechanical Engineers, New York, pp. 1175–1185, 1962.

[BUI 69] BUI H., Etude de l'évolution de la frontière du domaine élastique avec l'écrouissage et relation de comportement élastoplastique des métaux cubiques, PhD Thesis, University Pierre and Marie Curie, Paris, 1969.

[BUR 87] BURLET H., CAILLETAUD G., "Modelling of cyclic plasticity in finite element codes", in DESAI C.S., KREMPL E. (ed.) *Proceedings of 2nd International Conference on Constitutive Laws for Engineering Materials; Theory and Application*, Elsevier, Tucson, AZ, pp. 1157–1164, 1987.

[CAI 84] CAILLETAUD G., KACZMAREK H., POLICELLA H., "Some elements on multiaxial behavior of 316 stainless steel at room temperature", *Mechanics of Materials*, vol. 3, pp. 333–347, 1984.

[CAI 87] CAILLETAUD G., Une approche micromécanique phénoménologique du comportement inélastique des métaux, PhD Thesis, University Pierre and Marie Curie, Paris, 1987.

[CAI 94] CAILLETAUD G., PILVIN P., "Utilisation de modèles polycristallins pour le calcul par éléments finis", *Revue Européenne des Éléments Finis*, vol. 3, pp. 515–541, 1994.

[CAI 95] CAILLETAUD G., SAÏ K., "Study of plastic/viscoplastic models with various inelastic mechanisms", *International Journal of Plasticity*, vol. 11, pp. 991–1005, 1995.

[CAI 08] CAILLETAUD G., SAÏ K., "A polycrystalline model for the description of ratchetting: effect of intergranular and intragranular hardening", *Materials Science and Engineering A*, vol. 480, pp. 24–39, 2008.

[CAY 13] CAYZAC H., SAÏ K., LAIARINANDRASANA L., "Damage based constitutive relationships in semi-crystalline polymer by using multi-mechanisms model", *International Journal of Plasticity*, vol. 51, pp. 47–64, 2013.

[CAZ 04] CAZACU O., BARLAT F., "A criterion for description of anisotropy and yield differential effects in pressure-insensitive metals", *International Journal of Plasticity*, vol. 20, pp. 2027–2045, 2004.

[CAZ 06] CAZACU O., PLUNKETT B., BARLAT F., "Orthotropic yield criterion for hexagonal closed packed metals", *International Journal of Plasticity*, vol. 22, pp. 1171–1194, 2006.

[CHA 75] CHABOCHE J., "Viscoplastic constitutive equations for the description of cyclic and anisotropic behaviour of metals", *17th Polish Conference on Mechanics of Solids*, vol. 25, pp. 33–42, 1975.

[CHA 77] CHABOCHE J., "Viscoplastic constitutive equations for the the description of cyclic and anisotropic behaviour of metals", *Bulletin de l'Acad. des Sciences série Sciences et Technologie*, vol. 25, pp. 33–42, 1977.

[CHA 79] CHABOCHE J., DANG-VAN K., CORDIER G., "Modelization of strain memory effect on the cyclic hardening of 316 stainless steel", *SMIRT-5, Division L.*, Berlin, 1979.

[CHA 86] CHABOCHE J., "Time-independent constitutive theories for cyclic plasticity", *International Journal of Plasticity*, vol. 2, pp. 149–188, 1986.

[CHA 87] CHABOCHE J., "Continuum damage mechanics: present state and future trends", *Nuclear Engineering and Design*, vol. 105, pp. 19–33, 1987.

[CHA 89] CHABOCHE J., "Constitutive equations for cyclic plasticity and cyclic viscoplasticity", *International Journal of Plasticity*, vol. 5, pp. 247–302, 1989.

[CHA 91] CHABOCHE J., "On some modifications of kinematic hardening to improve the description of ratchetting effects", *International Journal of Plasticity*, vol. 7, pp. 661–678, 1991.

[CHA 06] CHABOCHE J., BOUDIFA M., SAANOUNI K., "A CDM approach of ductile damage with plastic compressibility", *International Journal of Fracture*, vol. 137, pp. 51–75, 2006.

[CHA 08] CHABOCHE J., "A review of some plasticity and viscoplasticity constitutive theories", *International Journal of Plasticity*, vol. 24, pp. 1642–1693, 2008.

[CHE 04] CHEN X., JIAO R., "Modified kinematic hardening rule for multiaxial ratcheting prediction", *International Journal of Plasticity*, vol. 20, pp. 871–898, 2004.

[CLE 12] CLERI F., "A two phase model of large strain plasticity in covalent nanostructures", *International Journal of Plasticity*, vol. 37, pp. 31–52, 2012.

[CON 89] CONTESTI E., CAILLETAUD G., "Description of creep plasticity interaction with non-unified constitutive equations: application to an austenitic stainless steel", *Nuclear Engineering and Design*, vol. 116, pp. 265–280, 1989.

[DAF 76] DAFALIAS Y., POPOV E., "Rate-independent cyclic plasticity in a plastic internal variables formalism", *Mechanics Research Communications*, vol. 3, pp. 33–38, 1976.

[DEB 08] DE BUHAN P., HASSEN G., "Multiphase approach as a generalized homogenization procedure for modelling the macroscopic behaviour of soils reinforced by linear inclusions", *European Journal of Mechanics – A/Solids*, vol. 27, pp. 662–679, 2008.

[DEL 84] DELOBELLE P., OYTANA C., "Experimental study of the flow rules of a 316 stainless steel at high and low stress", *Nuclear Engineering and Design*, vol. 83, pp. 333–339, 1984.

[DEL 95] DELOBELLE P., ROBINET P., BOCHER L., "Experimental study and phenomenological modelization of ratchet under uniaxial and biaxial loading on an austenitic stainless steel", *International Journal of Plasticity*, vol. 11, pp. 295–330, 1995.

[ELS 12] EL SAYED T., GÜRSES E., SIDDIQ A., "A phenomenological two-phase constitutive model for porous shape memory alloys", *Computational Materials Science*, vol. 60, pp. 44–52, 2012.

[FAL 98] FALESKOG J., GAO X., SHIH C., "Cell model for nonlinear fracture analysis – I. Micromechanics calibration", *International Journal of Fracture*, vol. 89, pp. 355–373, 1998.

[GAL 07] GALLÉE S., MANACH P., THUILLIER S., "Mechanical behavior of a metastable austenitic stainless steel under simple and complex loading paths", *Materials Science and Engineering A*, vol. 466, pp. 47–55, 2007.

[GAL 10] GALLÉE S., PILVIN P., "Deep drawing simulation of a metastable austenitic stainless steel using a two-phase model", *Journal of Materials Processing Technology*, vol. 210, pp. 835–843, 2010.

[GAU 04] GAUTIER E., CAILLETAUD C., "N-phase modeling applied to phase transformations in steels: a coupled kinetics-mechanics approach", *ICHMM-2004 International Conference on Heterogeneous Material Mechanics*, Oral communication, Chongqing, Chine, 2004.

[GER 83] GERMAIN P., NGUYEN Q., SUQUET P., "Continuum thermodynamics", *Journal of Applied Mechanics*, vol. 50, pp. 1010–1020, 1983.

[GEY 99] GEYER P., Comportement élasto-viscoplastique de tubes en Zircaloy-4: Approche expérimentale et modélisation micromécanique, PhD Thesis, Ecole Nationale Supérieure des Mines de Paris, 1999.

[GHO 91] GHONEM H., ZHENG D., "Depth of intergranular oxygen diffusion during environment-dependent fatigue crack growth in alloy 718", *Materials Science and Engineering A*, vol. 150, pp. 151–160, 1991.

[GOT 77] GOTOH M., "A theory of plastic anisotropy based on yield function of fourth order (plane stress state) – I", *International Journal of Mechanical Sciences*, vol. 19, pp. 505–512, 1977.

[GRO 09] GROSJEAN C., Etude de l'anisotropie de comportement en fluage thermique de tubes gaine et de tubes guide en alliages de Zirconium. Développements expérimentaux et résultats, PhD Thesis, INP Toulouse, 2009.

[GUR 77] GURSON A., "Continuum theory of ductile rupture by void nucleation and growth, part I: yield criteria and flow rules for porous ductile media", *Journal of Engineering Materials and Technology*, vol. 99, pp. 2–15, 1977.

[HAL 75] HALPHEN B., NGUYEN Q., "Sur les matériaux standards généralisés", *Journal de Mécanique*, vol. 14, pp. 39–63, 1975.

[HAS 92a] HASSAN T., KYRIAKIDES S., "Ratcheting in cyclic plasticity, part I: uniaxial behavior", *International Journal of Plasticity*, vol. 8, pp. 91–116, 1992.

[HAS 92b] HASSAN T., CORONA E., KYRIAKIDES S., "Ratcheting in cyclic plasticity, part II: multiaxial behavior", *International Journal of Plasticity*, vol. 8, pp. 117–146, 1992.

[HAS 94a] Hassan T., Kyriakides S., "Ratcheting of cyclically hardening and softening materials: I. Uniaxial behavior", *International Journal of Plasticity*, vol. 10, pp. 149–184, 1994.

[HAS 94b] Hassan T., Kyriakides S., "Ratcheting of cyclically hardening and softening materials: II. Multiaxial behavior", *International Journal of Plasticity*, vol. 10, pp. 185–212, 1994.

[HAS 08] Hassan T., Taleb L., Krishna S., "Influence of non-proportional loading on ratcheting responses and simulations by two recent cyclic plasticity models", *International Journal of Plasticity*, vol. 24, pp. 1863–1889, 2008.

[HAU 76] Hausselt J., Blum W., "Dynamic recovery during and after steady state deformation of Al-11wt%Zn", *Acta Metallurgica*, vol. 24, pp. 1027–1039, 1976.

[HEN 87] Henshall G., Miller A., Tanaka J., "Modeling cyclic deformation with the MATMOD-BSSOL unified constitutive equations", *2nd Conference on Low-Cycle Fatigue and Elasto-Plastic Behaviour of Materials*, Munich, pp. 184–191, 1987.

[HIL 65] Hill R., "Continuum micro-mechanisms of elastoplastic polycrystals", *Journal of the Mechanics and Physics of Solids*, vol. 13, pp. 89–101, 1965.

[HIL 76] Hill R., "Plastic analysis of pressurized cylinders under axial load", *International Journal of Mechanical Sciences*, vol. 18, pp. 145–148, 1976.

[HOS 72] Hosford W., "A generalized isotropic yield criterion", *Journal of Applied Mechanics*, vol. 39, pp. 607–609, 1972.

[HOS 10] Hosseininia E., Farzaneh O., "A simplified two-phase macroscopic model for reinforced soils", *Geotextiles and Geomembranes*, vol. 28, pp. 85–92, 2010.

[HU 05] Hu W., "An orthtropic yield criterion in a 3-D general stress state", *International Journal of Plasticity*, vol. 21, pp. 1771–1796, 2005.

[JER 14] Jeridi M., Saï K., Chouchene H. *et al.*, "Multi-mechanism modeling of amorphous polymers", *Mechanics Research Communication*, vol. 56, pp. 136–142, 2014.

[JER 15] Jeridi M., Laiarinandrasana L., Saï K., "Comparative study of continuum damage mechanics and Mechanics of porous media based on multi-mechanism model on polyamide 6 semi-crystalline polymer", *International Journal of Solids and Structures*, vol. 53, pp. 12–27, 2015.

[JIA 96] Jiang Y., Kurath P., "Characteristics of the Armstrong–Frederick type plasticity models", *International Journal of Plasticity*, vol. 12, pp. 387–415, 1996.

[JIA 08] Jiang Y., Zhang J., "Benchmark experiments and characteristic cyclic plasticity deformation", *International Journal of Plasticity*, vol. 24, pp. 1481–1515, 2008.

[KAR 74] Karlsson B., Sundström B., "Inhomogeneity in plastic deformation of two-phase steels", *Materials Science and Engineering*, vol. 16, pp. 161–168, 1974.

[KAR 93] Karafillis A., Boyce M., "A general anisotropic yield criterion using bounds and a transformation weighting tensor", *Journal of the Mechanics and Physics of Solids*, vol. 41, pp. 1859–1886, 1993.

[KOC 76] KOCKS U., "Laws for work-hardening and low-temperature creep", *Journal of Engineering Materials and Technology*, vol. 98, pp. 76–85, 1976.

[KOI 60] KOITER W., "General theorems for elastic-plastic solids", SNEDDON I.N., HILL R. (eds), *Progress in Solid Mechanics*, North Holland, Amsterdam, 1960.

[KRE 86] KREMPL E., MCMAHON J., YAO D., "Viscoplasticity based on overstress with a differential growth law for the equilibrium stress", *Mechanics of Materials*, vol. 5, pp. 35–48, 1986.

[KRÖ 61] KRÖNER E., "Zur plastischen Verformung des Vielkristalls", *Acta Metallurgica*, vol. 9, pp. 155–161, 1961.

[KUB 85] KUBIN L., ESTRIN Y., "The Portevin–Le Chatelier effect in deformation with constant stress rate", *Acta Metallurgica*, vol. 33, pp. 397–407, 1985.

[LAD 99] LADEVEZE P., *Nonlinear Computational Structural Mechanics – New Approaches and Non-Incremental Methods of Calculation*, Springer-Verlag, 1999.

[LAM 78] LAMBA H., SIDEBOTTOM O., "Cyclic plasticity for nonproportional paths, parts I and II", *Journal of Engineering Materials and Technology*, vol. 100, pp. 96–111, 1978.

[LEM 90] LEMAITRE J., CHABOCHE J.-L., *Mechanics of Solid Materials*, Cambridge University Press, Cambridge, UK, 1990.

[LON 09] LONGUET A., ROBERT Y., AEBY-GAUTIER E. *et al.*, "A multiphase mechanical model for Ti-6Al-4V: application to the modeling of laser assisted processing", *Computational Materials Science*, vol. 46, pp. 761–766, 2009.

[LUB 91] LUBLINER J., "A simple model of generalized plasticity", *International Journal of Solids and Structures*, vol. 28, pp. 769–778, 1991.

[LUB 93] LUBLINER J., TAYLOR R., AURICCHIO F., "A new model of generalized plasticity and its numerical implementation", *International Journal of Solids and Structures*, vol. 30, pp. 3171–3184, 1993.

[MAL 72] MALININ N., KHADJINSKY G., "Theory of creep with anisotropic hardening", *International Journal of Mechanical Sciences*, vol. 14, pp. 235–246, 1972.

[MAN 65] MANDEL J., "Généralisation de la théorie de plasticité de W.T. Koiter", *International Journal of Solids and Structures*, vol. 1, pp. 273–295, 1965.

[MAR 89] MARQUIS D., Phénoménologie et thermodynamique: couplages entre thermoélasticité, plasticité, vieillissement et endommagement, PhD Thesis, University Paris 6, 1989.

[MAZ 06] MAZIÈRE M., BESSON J., FOREST S. *et al.*, "Simulation de l'effet Portevin–Le Chatelier dans un superalliage pour disque de turbomachine", *8e colloque national en calcul des structures*, pp. 193–198, 21–25 May, Giens, 2006.

[MAZ 10] MAZIÈRE M., BESSON J., FOREST S. *et al.*, "Numerical aspects in the finite element simulation of the Portevin–Le Chatelier effect", *Computer Methods in Applied Mechanics and Engineering*, vol. 199, pp. 734–754, 2010.

[MEB 03] Mebarki N., Relationship between microstructure and mechanical properties of tempered martensitic steel, PhD Thesis, Ecole Nationale Supérieure des Mines de Paris, 2003.

[MEC 81] Mecking H., Kocks U., "Kinetics of flow and strain-hardening", *Acta Metallurgica*, vol. 29, pp. 1865–1875, 1981.

[MIL 76] Miller A., "An inelastic constitutive model for monotonic, cyclic, and creep deformation: part I. Equations development and analytical procedures", *Journal of Engineering Materials and Technology*, vol. 98, pp. 97–105, 1976.

[MIS 13] Mises R.V., "Mechanik der festen Körper in plastisch-deformablen Zustand", *Mechanics of Solid Bodies in the Plastically-deformable, Gottinger Nachrichten*, vol. 4, pp. 582–592, 1913.

[MOH 00] Mohr O., "Welche Umstande bedingen die Elastizitatsgrenze und den Bruch eines Materials?", *Zeitschrift des Vereins Deutscher Ingenieure Band*, vol. 44, pp. 1524–1530, 1900.

[MOL 99] Molinari A., "Extensions of the self-consistent tangent model", *Modelling and Simulation in Materials Science and Engineering*, vol. 7, pp. 683–697, 1999.

[MRO 67] Mroz Z., "On the description of anisotropic work-hardening", *Journal of the Mechanics and Physics of Solids*, vol. 15, pp. 163–175, 1967.

[MUR 82] Murakami S., Ohno N., "A constitutive equation of creep based on the concept of a creep-hardening surface", *International Journal of Solids and Structures*, vol. 18, pp. 597–609, 1982.

[NOR 29] Norton F., *Creep of Steel at High Temperatures*, McGraw-Hill Book Co., New York, 1929.

[NOU 89] Nouailhas D., "Unified modelling of cyclic viscoplasticity: application to austenitic stainless steels", *International Journal of Plasticity*, vol. 5, pp. 501–520, 1989.

[NOU 92] Nouailhas D., Cailletaud G., "Comparaison de divers critères anisotropes pour monocristaux cubiques à faces centrées CFC", *Comptes Rendus de l'Académie des Sciences – Paris*, vol. 315, pp. 1573–1579, 1992.

[OHA 83] Ohashi Y., Kawai M., Simizu H., "Effects of prior creep on subsequent plasticity of type 316 stainless steel at elevated temperature", *Journal of Engineering Materials and Technology*, vol. 105, pp. 257–263, 1983.

[OHA 86] Ohashi Y., Kawai M., Momose T., "Effect of prior plasticity on subsequent creep of type 316 stainless steel at elevated temperature", *Journal of Engineering Materials and Technology*, vol. 108, pp. 68–74, 1986.

[OHN 82] Ohno N., "A constitutive model of cyclic plasticity with a non-hardening strain region", *Journal of Applied Mechanics*, vol. 49, pp. 721–727, 1982.

[OHN 93] Ohno N., Wang J., "Kinematic hardening rules with critical state of dynamic recovery, part I: formulation and basis features for ratchetting behaviour", *International Journal of Plasticity*, vol. 9, pp. 375–390, 1993.

[PEN 72] Penning P., "Mathematics of the Portevin–Le Chatelier effect", *Acta Metallurgica*, vol. 20, no. 10, pp. 1169–1175, 1972.

[POR 00] Portier L., Calloch S., Marquis D. *et al.*, "Ratchetting under tension-torsion loading: experiments and modelling", *International Journal of Plasticity*, vol. 16, pp. 303–335, 2000.

[PRA 56] Prager W., "A new method of analyzing stresses and strains in work hardening plastic solids", *Journal of Applied Mechanics*, vol. 23, p. 493, 1956.

[PRI 08] Priser M., Analyses multi-échelles du comportement en fluage d'alliages de zirconium, PhD Thesis, Université de Bretagne-Sud, 2008.

[REG 09a] Regrain C., Comportement, endommagement et fissuration par fluage du Polyamide 6: étude expérimentale et modélisation, PhD Thesis, École Nationale Supérieure des Mines de Paris, 2009.

[REG 09b] Regrain C., Laiarinandrasana L., Toillon S. *et al.*, "Multi-mechanism models for semi-crystalline polymer: constitutive relations and finite element implementation", *International Journal of Plasticity*, vol. 25, pp. 1253–1279, 2009.

[ROB 95] Robinet P., Etude expérimentale et modélisation du comportement viscoplastique anisotrope du Zircaloy-4 dans deux états métallurgiques, PhD Thesis, Université de Franche-Comté, 1995.

[ROU 85a] Rousselier G., Etude comparative de modèles de comportement pour la simulation d'essais en traction-compression sur tubes en acier inoxydable, Document EDF, Annexe du rapport no. 8 du GIS Rupture à chaud, 1985.

[ROU 85b] Rousset M., Surfaces seuils en plasticité - détermination automatique et modélisation, PhD Thesis, University Pierre and Marie Curie, Paris 6, 1985.

[ROU 00] Roumagnac P., Guillemer-Néel C., Saanouni K. *et al.*, "Mechanical behaviour and deformation mechanicsms of Ni-Ti shape memory alloys in tension", *The European Physical Journal Applied Physics*, vol. 10, pp. 109–122, 2000.

[SAA 88] Saanouni K., Sur l'analyse de la fissuration des milieux élasto-viscoplastiques par la théorie de l'endommagement continu, PhD Thesis, Université de Technologie de Compiégne, 1988.

[SAÏ 93] Saï K., Modèles à grand nombre de variables internes et méthodes numériques associées, PhD Thesis, École Nationale Supérieure des Mines de Paris, 1993.

[SAÏ 04] Saï K., Aubourg V., Cailletaud G. *et al.*, "Physical basis for model with various inelastic mechanisms for nickel base superalloy", *Materials Science and Technology*, vol. 20, pp. 747–755, 2004.

[SAÏ 06a] Saï K., Cailletaud G., Forest S., "Micro-mechanical modeling of the inelastic behavior of directionally solidified materials", *Mechanics of Materials*, vol. 38, pp. 203–217, 2006.

[SAÏ 06b] Saï K., Cailletaud G., Tourki Z. *et al.*, "N-phase modeling of phase transformation in 304 stainless steel, bulging: finite element simulation and experimental validation", *12th International Symposium on Plasticity and its Applications*, Halifax, Nova Scotia, Canada, 2006.

[SAÏ 07] SAÏ K., CAILLETAUD G., "Multi-mechanism models for the description of ratchetting: effect of the scale transition rule and of the coupling between hardening variables", *International Journal of Plasticity*, vol. 23, pp. 1589–1617, 2007.

[SAÏ 11a] SAÏ K., "Multi-mechanism models: present state and future trends", *International Journal of Plasticity*, vol. 27, pp. 250–281, 2011.

[SAÏ 11b] SAÏ K., LAIARINANDRASANA L., BEN NACEUR I. *et al.*, "Multi-mechanism damage-plasticity model for semi-crystalline polymer: creep damage of notched specimen of PA6", *Materials Science and Engineering A*, vol. 528, pp. 1087–1093, 2011.

[SAÏ 12] SAÏ K., TALEB L., CAILLETAUD G., "Numerical simulation of the anisotropic behavior of 2017 aluminum alloy", *Computational Materials Science*, vol. 65, pp. 48–57, 2012.

[SAÏ 14] SAÏ K., TALEB L., GUESMI F. *et al.*, "Multi-mechanism modeling of proportional and non-proportional ratchetting of stainless steel 304", *Acta Mechanica*, vol. 225, pp. 3265–3283, 2014.

[SAN 54] SANDERS J., "Plastic stress-strain relations based on linear loading functions", *2nd U.S. National Congress on Theoretical and Applied Mechanics*, American Society of Mechanical Engineers New York, pp. 455–460, 1954.

[SIM 97] SIMO J., HUGHES T., *Computational Inelasticity*, Springer-Verlag, 1997.

[SRI 10] SRIVASTAVA V., CHESTER S.A., AMES N.M. *et al.*, "A thermo-mechanically-coupled large-deformation theory for amorphous polymers in a temperature range which spans their glass transition", *International Journal of Plasticity*, vol. 26, pp. 1138–1182, 2010.

[STO 85] STOER J., *Principles of Sequential Quadratic Programming Methods for Solving Nonlinear Programs*, in SCHITTKOWSKI K. (ed.), NATO ASI Series, vol. 15, Computational mathematical programing, pp. 165–207, Springer-Verlag, Berlin, 1985.

[TAL 06] TALEB L., CAILLETAUD G., BLAJ L., "Numerical simulation of complex ratcheting tests with a multi-mechanism model type", *International Journal of Plasticity*, vol. 22, pp. 724–753, 2006.

[TAL 10] TALEB L., CAILLETAUD G., "An updated version of the multimechanism model for cyclic plasticity", *International Journal of Plasticity*, vol. 26, pp. 859–874, 2010.

[TAL 11] TALEB L., CAILLETAUD G., "Cyclic accumulation of the inelastic strain in the 304L SS under stress control at room temperature: ratcheting or creep?", *International Journal of Plasticity*, vol. 27, pp. 1936–1958, 2011.

[TAL 13a] TALEB L., "About the cyclic accumulation of the inelastic strain observed in metals subjected to cyclic stress control", *International Journal of Plasticity*, vol. 43, pp. 1–19, 2013.

[TAL 13b] TALEB L., "On the cyclic behavior of 304L and 316L stainless steels", *Key Engineering Materials*, vol. 535–536, pp. 201–204, 2013.

[TAL 14] TALEB L., CAILLETAUD G., SAÏ K., "Experimental and numerical analyses about the cyclic behavior of the 304L and 316L stainless steels at 350°C", *International Journal of Plasticity*, vol. 61, pp. 32–48, 2014.

[TAL 15] TALEB L., SAÏ K., CAILLETAUD G.,"Capabilities of the Multi-mechanism Model in the Prediction of the Cyclic Behavior of Various Classes of Metals" *From Creep Damage Mechanics to Homogenization Methods*, Springer, 2015.

[TAY 38] TAYLOR G., "Plastic strain in metals", *Journal of the Institute of Metals*, vol. 62, pp. 307–324, 1938.

[THA 83] THAKKER A., COWLES B., Low strain, long life creep fatigue of AF2IDA and INCO718, N.A.S.A, Report FR.15652, 1983.

[TOR 11] TORABINEJAD V., ZAREI-HANZAKI A., SABET M. *et al.*, "The effect of low temperature annealing on the mechanical behavior of cold rolled dual-phase twinning-induced plasticity steel", *Materials and Design*, vol. 32, pp. 2345–2349, 2011.

[TRA 17] TRANSVALOR, Z-mat user manual, MINES ParisTech – ONERA, available at: http://zset-software.com/wordpress/wp content/uploads/manuals/, 2017.

[TRE 64] TRESCA G., "Mémoire sur L'écoulement des Corps Solides Soumis à de Fortes Pressions", *Comptes Rendus de l'Académie des Sciences – Paris*, vol. 59, pp. 754–758, 1864.

[VEL 06] VELAY V., BERNHART G., PENAZZI L., "Cyclic behavior modeling of a tempered martensitic hot work tool steel", *International Journal of Plasticity*, vol. 22, pp. 459–496, 2006.

[VID 94] VIDEAU J., CAILLETAUD G., PINEAU A., "Modélisation des effets mécaniques des transformations de phases", *Journal de Physique IV*, vol. 4, pp. 227–232, 1994.

[WOL 08] WOLFF M., TALEB L., "Consistency for two multi-mechanism models in isothermal plasticity", *International Journal of Plasticity*, vol. 24, pp. 2059–2083, 2008.

[YOO 06] YOON J., BARLAT F., DICK R. *et al.*, "Prediction of six or eight ears in a drawn cup based on a new anisotropic yield function", *International Journal of Plasticity*, vol. 22, pp. 174–193, 2006.

[ZAÏ 05] ZAÏRI F., WOZNICA K., NAÏT-ABDELAZIZ M., "Phenomenological nonlinear modelling of glassy polymers", *Comptes Rendus de Mécanique*, vol. 333, pp. 359–364, 2005.

[ZAO 70] ZAOUI A., Etude de l'influence propre de la désorientation des grains sur le comportement viscoplastique de métaux polycristallins (système FC), PhD Thesis, Faculté des sciences de Paris, 1970.

[ZAR 72] ZARKA J., "Généralisation de la théorie du potentiel plastique multiple en viscoplasticité", *Journal of the Mechanics and Physics of Solids*, vol. 20, pp. 179–195, 1972.

[ZAR 78] ZARKA J., CASIER J., "Practical remarks about cyclic loadings on an elastic-plastic structure", *Nuclear Engineering and Design*, vol. 51, pp. 69–80, 1978.

[ZAR 79] ZARKA J., CASIER J., "Elastic plastic response of structure to cyclic loading: practical rules", in SIA NEMAT-NASSER (ed.), *Mechanics Today*, SIA NEMAT-NASSER, vol. 6, Pergamon Press, New York, pp. 93–198, 1979.

[ZAR 80] ZARKA J., ENGEL J., INGLEBERT G., "On a simplified inelastic analysis of structures", *Nuclear Engineering and Design*, vol. 57, pp. 333–368, 1980.

[ZHA 01] ZHANG S., MCCORMICK P., ESTRIN Y., "The morphology of Portevin–Le Chatelier bands: finite element simulation for Al-Mg-Si", *Acta Materialia*, vol. 49, pp. 1087–1094, 2001.

[ZIE 59] ZIEGLER H., "A modification of Prager's hardening rule", *Quaterly of Applied Mathematics*, vol. 17, pp. 55–65, 1959.

Index

R, S, T, V

Other titles from

iSTE

in

Materials Science

2017

CARVELLI Valter, JAIN Atul, LOMOV Stepan
Fatigue of Textile and Short Fiber Reinforced Composites
(Durability and Ageing of Organic Composite Materials Set – Volume 1)

DELHAES Pierre
Materials and Thermodynamics: Living and Economic Systems

FRESSENGEAS Claude
Mechanics of Dislocation Fields

2016

TORRENTI Jean-Michel
Materials and Infrastructures 1
Materials and Infrastructures 2

RÉMOND Yves, AHZI Saïd, BANIASSADI Majid, GARMESTANI Hamid
Applied RVE Reconstruction and Homogenization of Heterogeneous Materials

2015

JOUD Jean-Charles, BARTHÉS-LABROUSSE Marie-Geneviève
Physical Chemistry and Acid-Base Properties of Surfaces

LALEVÉE Jacques, FOUASSIER Jean-Pierre
Dyes and Chromophores in Polymer Science

2014

HAMAIDE Thierry, DETERRE Rémi, FELLER Jean-François
Environmental Impact of Polymers

LAFLEUR Pierre, VERGNES Bruno
Polymer Extrusion

2013

BATHIAS Claude
Fatigue Limit in Metals

LEXCELLENT Christian
Shape-memory Alloys Handbook

2012

DAVIM J. Paulo
Wear of Advanced Materials

DELHAES Pierre
Carbon Science and Technology: From Energy to Materials

GOUDEAU Philippe, GUINEBRETIÈRE René
X-Rays and Materials

SAANOUNI Khemais
Damage Mechanics in Metal Forming: Advanced Modeling and Numerical Simulation

VANBÉSIEN Olivier
Artificial Materials

VERDU Jacques
Oxidative Ageing of Polymers

2011

BATHIAS Claude, PINEAU André
Fatigue of Materials and Structures: Application to Damage

BATHIAS Claude, PINEAU André
Fatigue of Materials and Structures: Application to Design

PRIESTER Louisette
Grain Boundaries and Crystalline Plasticity

VIGNES Alain
Extractive Metallurgy 1: Basic Thermodynamics and Kinetics
Extractive Metallurgy 2: Metallurgical Reaction Processes
Extractive Metallurgy 3: Processing Operations and Routes

2010

BATHIAS Claude, PINEAU André
Fatigue of Materials and Structures: Fundamentals

CHATEIGNER Daniel
Combined Analysis

CHEVALIER Yvon, VINH TUONG Jean
Mechanical Characterization of Materials and Wave Dispersion V1
Mechanical Characterization of Materials and Wave Dispersion V2

DELHAES Pierre
Carbon-based Solids and Materials

2009

ALVAREZ-ARMAS Iris, DEGALLAIX-MOREUIL Suzanne
Duplex Stainless Steels

DAVIM J. Paulo
Machining Composite Materials

GALERIE Alain
Vapor Surface Treatments

2008

BATTAGLIA Jean-Luc
Heat Transfer in Materials Forming Processes

BLONDEAU Régis
Metallurgy and Mechanics of Welding

FRANÇOIS Dominique
Structural Components

REYNE Maurice
Plastic Forming Processes

TAKADOUM Jamal
Materials and Surface Engineering in Tribology

2007

BOCH Philippe, NIEPCE Jean-Claude
Ceramic Materials

CRISTESCU Constantin
Materials with Rheological Properties

2006

LASSEN Tom, RÉCHO Naman
Fatigue Life Analyses of Welded Structures

Printed and bound by CPI Group (UK) Ltd, Croydon, CR0 4YY

13/09/2026

14961076-0001